Goat nutrition

EAAP conveys its most sincere thanks to the FAO Regional Office for Europe (Rome), the International Centre for Advanced Mediterranean Studies (Paris) and the Technical Centre for Agricultural and Rural Cooperation (Wageningen) for their participation in the realization of this publication.

FAO -the Food and Agriculture Organization- is an autonomous agency within the United Nations System and has a membership of 158 member countries.
FAO carries out a major programme of technical advice and assistance for the agricultural community on behalf of governments and development funding agencies, collects, analyses and disseminates information; advises governments on policy and planning; provides opportunities for governments to meet and discuss food and agricultural problems.

CIHEAM -the International Center for Advanced Mediterranean Agronomic Studies- is an inter-governmental organization, established in 1962, which presently includes many of the Mediterranean countries, such as Marocco, Algeria, Tunisia, Egypt, Lebanon, Turkey, Greece, Yugoslavia, Italy, France, Spain, Portugal and Malta. The Center's aim is to improve the capacities of Mediterranean agriculture, through training, research and the study of the region's major problems.

CTA -the Technical Centre for Agricultural and Rural Co-operation- operates under the Lomé Convention between member states of the European Community and the African, Carribean and Pacific (ACP) states.
The aim of CTA is to collect, disseminate and facilitate the exchange of information on research, training and innovations in the spheres of agricultural and rural development and extension for the benefit of the ACP states.

Goat nutrition

Prepared under the auspices of the Food and Agriculture Organization of the United Nations (FAO), the European Association for Animal Production (EAAP), the International Centre for Studies in Mediterranean Agriculture (CIHEAM) and the Technical Centre for Agricultural and Rural Cooperation of the European Community (CTA)
(EAAP Publication No. 46, 1991)

P. Morand-Fehr (Editor)

Pudoc Wageningen 1991

This series is coordinated by Dr S. Korver and Prof. dr J. Boyazoglu

CIP-data Koninklijke Bibliotheek, Den Haag

Goat

Goat nutrition / P. Morand-Fehr (ed.). - Wageningen: Pudoc. - III -.
(EAAP publication, ISSN 0071-2477 ; no. 46)
Prepared under the auspices of the Food and Agriculture Organization of the United Nations (FAO),
the European Association for Animal Production (EAAP), the International Centre for Studies in
Mediterranean Agriculture (CIHEAM) and the Technical Centre for Agricultural and Rural Coopera-
tion of the European Community (CTA).
ISBN 90-220-1009-0 bound
SISO 633.3 UDC 636.39.084 NUGI 835
Subject headings: goats ; nutrition.

ISBN 90-220-1009-0
NUGI 835

Printed in the Netherlands.

Contents

Part 4. Feeding of young goats   257

FOREWORD

A. BOZZINI

The book you hold is a sort of summary of research work done between 1982 and 1990 by a group of experts under the Programme of the FAO Cooperative Research Sub-Network on Goat Production. Their objective was to push back the frontiers of scientific and technical expertise, particularly where this might improve the husbandry techniques and profits of goat owners.

In 1979, during a consultation in Athens, the FAO Regional Office for Europe set up a research network on sheep. It had two sub-networks, one for northern Europe and the other for the Mediterranean. They were coordinated by Dr. T.T. TREACHER. It was understood, however, that a third sub-network on goats would be established. After many preparatory meetings and contacts, the third network was launched at the Network Consultation held in Santarem, Portugal, in 1981, headed by Dr. P. MORAND-FEHR.

The new sub-network soon identified the two themes on which it was to focus its efforts: evaluation of goat breeds or populations in the context of the production system used, and the nutrition and feeding of goats. The work done on the first topic, concerning both methodology and knowledge of the different breeds, was reported in the course of a Symposium on "The Evaluation of Mediterranean Sheep and Goats" held in Santarem, Portugal, from 23 to 25 September 1987, which was covered in a 578-page report (EEC Report CD-NA-11893 ZAC, Luxembourg).

The research work done on Goat Nutrition and Feeding was presented at four seminars: Reading (Great Britain, 1982), Grangeneuve (Switzerland, 1984), Nancy (France, 1986) and Potenza (Italy, 1988). An average 25 to 35 experts from 12 countries participated: Cyprus, Egypt, France, Great Britain, Israel, Italy, Morocco, Norway, Portugal, Spain, Sweden and Switzerland. This book is a summary and compendium of the scientific and technical data presented at these four seminars.

FAO, pleased with the Working Group's concern with the dissemination of this information, decided to publish the book, especially since it combines rigorous scientific presentation of the latest data with a highly concrete presentation of practical applications. The book thus addresses a wide audience concerned with goat breeding and its industry.

The work of the sub-network and this book, which is a concrete expression of the research undertaken, are perfectly in line with FAO's objectives and policy of establishing and supporting networks.

FAO's Regional Office for Europe has indeed created a series of research networks on subjects of common interest during the past 15 years. The opportunity to meet around a table and to discuss research problems is extremely valuable: it helps to avoid the duplication of efforts and to make use of the experience and findings of other countries and go forward together. In the domain of animal husbandry, the research network on sheep and goats will be followed in future by the creation of a new network on buffalo and very probably another on wild game.

With respect to the research network on goats, this work has been made possible by the abnegation and perspicacity of all members of the Working Group, as well as the cordial atmosphere prevailing in it, and, above all, the collaboration, support and involvement the book has aroused. A list of the personalities to whom the book is indebted is found below.

For my part, I wish to particularly thank all of the international or national organizations which have supported this Working Group and either facilitated or hosted the four seminars on Goat Nutrition and Feeding: CIHEAM, EEC, EAAP, Animal and Grassland Research Institute, the Swiss Federal Research Station for Animal Production, the Ecole Nationale d'Agronomie et d'Industries Agricoles in Nancy, the Istituto sperimentale per la Zootecnia in Potenza, and the INRA, particularly its review, Annales de Zootechnie, which agreed to publish the abstracts of the communications presented at the different seminars, the organizations which have given special support to the publication of this book: EAAP, CTA and, finally, the referees who are not members of the Working Group and who meticulously analysed the various chapters and offered generous advice to the authors.

In conclusion, I can only hope that this book will, in an area where information is still quite scant, prove useful to scientists, teachers, and students of animal husbandry as well as to development experts, livestock production policy-makers, goat experts and goat breeders.

The members of the Working Group on Goat Feeding and Nutrition wish to thank the following persons who participated in many different ways in the preparation and publication of this book.

FAO :	CIHEAM :	EAAP	EEC
O. FENESAN	R. FEVRIER	J. BOYAZOGLU	M. TROISGROS
A. BOZZINI	J.L. TISSERAND		G. ROSSETTO
M. ZJALIC			
H. OLEZ			

CTA :	ECNSGP:	INRA:
M'BA	T.T. TREACHER	A. RERAT
M.T. NAIRAIN		F. WEHRLEN

ORGANIZERS OF THE SEMINARS :

T.T. TREACHER (Reading 1982)   G. BLANCHART (Nancy 1986)
J.D. OLDHAM (Reading 1982)   J. BRUN-BELLUT (Nancy 1986)
A. MOWLEM (Reading 1982)   R. RUBINO (Potenza 1988)
R. DACCORD (Grangeneuve 1984)   V. FEDELE (Potenza 1988)
J. KESSLER (Grangeneuve 1984)

REFEREES OF CHAPTERS DRAFTS

J.P. BARLET	Y. GEAY	C. LU	R. RUBINO
P. BAS	L. GUEGUEN	J.C. MALECHEK	M.R. SANZ SAMPELAYO
J. BRUN-BELLUT	T.K. GOSH	P. MORAND-FEHR	P. SCHMIDELY
Y. CHILLIARD	M. HADJIPANAYIOTOU	H. NARJISSE	P. SUSMEL
P. COLOMER-ROCHER	Ø. HAVREVOLL	E.R. ORSKOV	M. TOULLEC
R. DACCORD	B. HUBERT	B. PARAGON	T.T. TREACHER
C. DEMARQUILLY	R. JARRIGE	F. PROVENZA	M. VANBELLE
A. FALAGAN	M. LAMAND	A. PURROY	M. VERMOREL
V. FEDELE	S. LANDAU	B. REMOND	
F. GALLOUIN	J.E. LINDBERG	J. ROBELIN	

TAPE WRITING	**ENGLISH REVIEW**	**TAPE WRITING &**
D. CERCAS	K. RERAT	**PAGE SETTING**
	C. MORGAN	D. SOULILLOU
	H. COLEMAN	

INTRODUCTION

P. MORAND-FEHR

There will be very soon be 500 million goats on our planet; goat numbers have risen faster than any other livestock species in the world's herds over the past ten years. This means that in the late 20th century, a significant proportion of farmers and herders are turning to goat keeping, even though it is often regarded as a traditional practice with little promise for the future. This contradiction between the image of goat keeping and its reality calls for some attention.

There are several factors that may help to explain the present revival of goat keeping in different parts of the world.

In the first place, a number of countries have repealed laws designed to preserve forests or vegetation, and which had tended to reduce goat keeping or even eradicate it.

Secondly, the first scientific work on the goat began to make an impact in the 1970s. The value of the goat for milk and meat production has at last been recognized - especially for arid zones where these products may be the only, or almost the only, high quality protein sources available.

Furthermore, while it has been argued that goats have historically been a factor in deforestation or desertification, this reproach has been shown to be exaggerated, since man is generally the deciding factor in degradation of this kind. The important thing, in practice, is to choose management methods appropriate to the goat as a species, especially with respect to the characteristics and degree of fragility of the environment. Far from banning the goat, its behavioural features are now being harnessed to control overgrowth of brushwood and reduce the risk of bush fire.

In developing countries, it has been realized that imported cows of high genetic value are not the only possible solution to the milk deficit and that small ruminants, goats particularly, can be especially useful in making up this deficit owing to their ability to make good use of locally available fodder. Furthermore, a variety of taboos concerning goats, and the relatively low social status of the goat herd, are beginning to die out.

In the industrialized countries goat keeping has a good image owing to the gastronomic or dietetic quality of goat products and also its ecological aspects; this has helped to stabilize herd numbers after an impressive drop during the first half of the twentieth century.

A growing tourist industry may encourage small-scale, non-industrial production of such products as goat cheese, suckling kids and mohair or cashmere clothing.

So the value of the goat is not to be measured merely in goat numbers; account must also be taken of its socio-cultural role and its importance in the diet of certain populations.

Unfortunately, despite the undeniable efforts made over the last ten years, our knowledge of goat husbandry and goat products is still limited, lagging far behind cow and sheep science. This is all the more of a handicap in that there are very many, widely differing goat farming systems and breeds or populations in the world. Furthermore, the information is generally less readily available than for the other main livestock species.

For all these reasons, the FAO goat sub-network and its working group on Goat Nutrition and Feeding, after eight years' research, which has advanced the state of the art on various aspects of goat nutrition, have decided to publish this book on the subject.

The book has several aims :

- To provide an update on knowledge acquired since the last symposium on goat nutrition (held in Tours in 1981), and to supplement the information presented at the 3rd and 4th International Goat Conferences in Tucson (1982) and Brasilia (1987).

- To present scientifically rigorous information and at the same time try to draw all possible practical conclusions applicable to the different farming systems, pointing them out in a way that is practical for the reader.

- To attempt to present the problems of goat nutrition in an objective way, i.e. free of the preconceived ideas and commonplaces that still circulate on the subject of goats, pin-pointing similarities and differences in nutrition confusing the dairy goat with the dairy cow and the nursing goat with the ewe or, on the other hand, treating the goat as a case apart.

- To try to highlight what is new in the work of the FAO working group by comparing this with goat nutrition work done elsewhere.

This book, then, has a wide range of purposes. It is not the proceedings of a symposium, publishing papers in their raw state, nor a treatise on goat nutrition, nor a handbook on goat feeding. Neither is it a complete, all-round source book on goat feeding and nutrition; some subjects (eg. the feeding of goats raised for hair) are not covered, as there did not seem to be sufficient material to do so. It may be felt that some subjects such as milk production have been given fuller treatment than is warranted in the present situation, but this was a response to the great demand for information on dairy goats and the large amount of information available.

It was decided not to present the material by dealing with goat nutrition in each type of feeding system or each country in turn. That would have made a tiresome and repetitive catalogue, besides which, it was felt that the basic topics have a number of universal features that every farmer must take into account and apply to his own environment and farming system. In the end it was decided to take two livestock farming systems as examples : the intensive system based on fodder production and the extensive system using rangelands of the Mediterranean or semi-arid types. The identification of these two examples stemmed from the knowledge working group members had gleaned.

Our book is aimed at a wide-ranging readership of researchers, teachers, students and extension officers at all levels, and from decision makers through veterinarians and goat liverstock specialists to livestock farmers.

We hope the reader will feel this book is original in that, to write its twenty-four chapters, it has harnessed the energies of specialists with widely differing concerns, some looking into the nutritional mechanisms of the ruminant and taking the goat as their animal material, others who needed to carry out research to advance goat production and products in their countries. The essential point is that all wanted to cooperate and work together. Basic and applied research are complimentary, not mutually antagonistic as we are often led to believe. In fact basic research can have very valuable applications for livestock farmers; what is needed is the will to bring these to light and make them available to potential users.

Some key points stand out in all the chapters, giving the book some coherence despite the handicap of 41 different authors. For example : the goat is a highly adaptable ruminant which, because of its feeding behaviour, its digestion of high-fibre fodder in the rumen, its water metabolism and its capacity to store and later mobilize body reserves, is capable of adjusting the widely varied feeding situations. This explains why goats are so well able to sustain themselves in arid environments - generally more so than cows or sheep. But this does not make the goat a case apart among ruminants. Under certain conditions the sheep can demonstrate similar adaption mechanisms, though to a lesser extent, while the Camelidae (ruminants like the lama) can demonstrate even more marked adaptation, particularly in their water metabolism and digestion of high-fibre feedstuffs. The goat lies between the sheep and the Camelidae in its adaptability to arid environments. Rather than trying to restrict some of the goat's specific capabilities, such as its ability to store body reserves or its behaviour in selecting the plant fractions it ingests, man should make use of them to establish suitable, profitable feeding systems.

Research in the field of feeding, nutrition and year-round feeding strategies are essential for defining a production system, maintaining satisfactory herd health, obtaining satisfactory performances, and optimizing the profitability of goat products. Furthermore, nutrition is generally the production factor most easily manipulated by the livestock farmer for rapid results. Therefore, for progress in goat production, it is - along with pathology - one of the two priority areas for improvement. Research in this field has the highest chance of reaching concrete, positive results. This book is one illustration of this.

It remains for me only to beg the reader's forgiveness if the book is not perfect (our means were limited), and to thank all those who contributed to its realization. Particular thanks to all colleagues who took part in writing these twenty-four chapters, and for having put up with my constant demands.

Part I

General goat nutrition

Chapter 1

FEEDING BEHAVIOUR OF GOATS AT THE TROUGH

P. MORAND-FEHR, E. OWEN and S. GIGER-REVERDIN

INTRODUCTION

In the past, many authors have highly contributed to spread the idea that the feeding behaviour of goats is very special so that it is difficult to apply rational feeding methods. This is an understandable reaction when information is lacking on a difficult subject. It is nevertheless necessary to understand the feeding behaviour of goats since some peculiarities distinguish them from other ruminants and in some cases may explain their ability to adapt to various environments, especially arid areas (MORAND-FEHR, 1988). Besides, the feeding behaviour of goats is sometimes responsible for the degradation of these areas. This important subject is dealt with in Chapter 2.

It is also necessary to obtain fundamental data on this behaviour by studying it when the goat is fed at the trough, where the main parameters affecting feeding behaviour can be controlled. This information is important not only for intensive production systems where goats frequently consume forage in the goat-house to obtain a better yield from the grassland areas, but also in less intensive systems which can be improved by the distribution of agricultural by-products at the trough (OWEN et al., 1987).

As feeding behaviour at the trough has been analysed elsewhere (MORAND-FEHR, 1981), this chapter will update results obtained since that time.

DESCRIPTION OF THE GOAT'S MEAL

When forage is supplied at the trough, the goat's meal is divided into three phases :
- a phase of exploration, where the goat takes stock of the feed offered;
- a phase of intense feed intake where the animal satisfies most of its hunger;
- a phase of selection where the goat selects the plant fractions to be ingested, then stops often to drink, to lick the saltblock or else to eat some straw from the litter if it is fresh. It is as if the goat attempts to balance its consumption of forage by ingesting water, salts and cell-wall carbohydrates. The animal goes back and forth between the distributed forage, on the one hand, and the watering trough, saltblock or litter straw, on the other hand (MORAND-FEHR et al., 1980).

More recent observations have shown that the first phase is shorter when there is more competition among goats at the trough, when the forage is homogeneous and when the animals are used to it. On the contrary, this first phase lasts longer at individual troughs and with poor-quality forage composed of a large variety of plants, and especially if the interval from this meal and the last forage meal is rather short.

The last phase can be observed very early in kids even before weaning, but definitely after weaning. Even if the level of hay intake is low, the time spent eating hay largely exceeds that spent ingesting milk before weaning and concentrate, water and salt after weaning. Thus, this back and forth behaviour between the hay, on the one hand, and the milk, water, salt, concentrate or litter straw, on the other hand, appears very early (MORAND-FEHR et al., 1982, 1986).

Successive meals are generally separated by 1,5 to 2 hours. From our observations it was considered that the goat ate two meals when the interval between the end of the first and the beginning of the second was at least 10 min. Generally the duration of a meal, even a short meal, exceeded 15 min.

BEHAVIORAL PARAMETERS RELATED TO INGESTION AND RUMINATION

The total time spent eating is highly variable, depending on feed quality, type and mode of feeding. With usual forage and concentrate rations offered twice a day, it generally lasts 4 to 7 h per day. Similar durations have been observed in goats grazing artificial pastures (DE SIMIANE et al., 1983). The unitary eating time (UET), i.e. the time spent eating a given amount of dry matter, is generally higher in goats than in sheep (GEOFFROY, 1974); this was confirmed more recently by FOCANT (1984) and FOCANT et al. (1986a). The time spent selecting the feed explains the higher UET in this species.

In fact, if forage is offered once a day in a limited time, sheep appear to have a higher intake capacity than goats during the meal that follows the forage distribution (FOCANT et al., 1986b); indeed, in goats the time spent eating a same amount of hay is longer and the levels of feed intake lower (Table 1). However, if goats are fed forage twice a day and have it freely available all day long, the differences between the two species in the levels of intake are variable and insufficient for a distinction to be made between goats and sheep. In this case, the number of meals eaten by goats exceeds that recorded in sheep (GEOFFROY, 1974). The main meals correspond to forage feeding, but the goats eating most forage are those eating the largest number of secondary meals between the main meals (MORAND-FEHR et al., 1980). The secondary meals and their importance relative to the main meals appear to determine the goats with the highest levels of feed intake. As in sheep (DEMARQUILLY C., personal communication), the length of the main meal decreases and the number of secondary meals increases as the quality of the forage decreases.

The type of feed offered leads to substantial variations in the UET, in a ratio of 2 to 1 between a poor quality coarse hay and a fine hay with good ingestibility, and in a ratio of 1 to 10 between a hay and cereal grains (FOCANT et al., 1986a, b), explaining the important role played by forage intake in the feeding behaviour of the animals.

All factors reducing the third phase of the meal, in particular if water is not freely available, decrease markedly the forage quantity and the eating rate.

For equal levels of feed intake, goats spend less time ruminating than sheep (FOCANT et al., 1986a, b). The number of rumination periods is higher in sheep, but each one is shorter, while the mean length of a merystic cycle tends to be longer in goats. It was previously assumed that goats ruminate more during the night than sheep. However FOCANT et al. (1986a, b) showed that there was no significant difference between the two species when the animals consumed the same quantities of feed at the same times. In fact, the less forage in the diet, the less diurnal rumination in the two species.

In contrast to behavioral activities which are very early similar to those of the adult, HOOPER and WELCH (1983) observed that rumination in growing goats reached an efficiency level close to that of the adult only at the age of about 9 months.

As confirmed by FOCANT et al. (1986b), goats are satiated more rapidly than sheep and they tend to ruminate less. When the animal can select the plant fractions to be ingested, these ingestion parameters may be changed.

Table 1

COMPARISON OF DIETARY ACTIVITIES IN SHEEP AND GOATS
(From FOCANT et al., 1986b)

	Sheep (n=8)	Goats (n=8)	Level of significance
Intake of barley DM(g/kg $W^{.75}$/day)	22.0	17.1	+++
Intake of hay DM(g/kg $W^{.75}$/day)	36.7	36.3	NS
Duration of a hay meal (min)	152.6	108.3	+++
Unitary hay-eating time (min/g DM/kg $W^{.75}$)	5.4	5.8	+++
Daily time spent ruminating (min/day)	583.0	472.0	+++
Duration of a merystic cycle (s)	53.5	62.9	+++
Unitary time spent ruminating (min/g NDF/ kg $W^{.75}$)	26.9	23.4	+++
Unitary time spent masticating (min/g NDF/kg $W^{.75}$)	36.8	34.5	+

Diet : Average of feeding with hay 40 g/kg $W^{.75}$/day and barley 10 to 35 g/kg $W^{.75}$/day.

+ : $P<0.05$; +++ : $P < 0.001$; NS : non significant.

SELECTING BEHAVIOUR OF GOATS

It is now admitted that small ruminants and specially goats, select their feed (Mac CAMMON-FELMAN et al., 1981; MORAND-FEHR, 1981). For goats receiving a much larger amount of dry forage than that ingested, a supply of an even larger amount can improve the level of intake and even more so if the forage is of poor quality (MORAND-FEHR et al., 1980; OWEN et al., 1986, 1987; Figure 1). Refusals can be observed even when the amounts of forage offered are limited especially if the forage is of poor quality. But it is also with poor forages that amounts really ingested are most enhanced when the amounts offered increase. With good or excellent green forage, HUGUET et al. (1977) observed that a refusal rate exceeding 25 % in 2-year old goats and more and 35 % in 1-year old goats did not increase the level of intake. In spite of very few data available in this field, this threshold is assumed to exceed 60 % for poor-quality hay or straw.

Goats select the most nutritive fractions of the forage, the leaves more than the stems, the thin stems more than the thick ones, i.e. the fractions richest in proteins and poorest in cell-wall carbohydrates (MORAND-FEHR et al., 1980). These results have been confirmed with oat straw (OWEN et al., 1986, 1987; WAHED and OWEN, 1986) or hay of brome grass and lucerne (QUICK and DEHORITY, 1986). Sheep are just as capable as goats of selecting forage distributed at the trough (Table 2), but this is perhaps not true for other feeds such as shrubs or trees on rangeland (see Chapter 14).

The forage species modifies the level of refusals in the case of green forages. Fescue is less selected than Italian Ray Grass, violet clover less than alfalfa (DE SIMIANE et al., 1983). In addition, HUGUET et al. (1977) showed that if green forage was first chopped before feeding at the trough, its ingestibility was reduced because the goat had less chance of exerting its selecting behavior. WAHED and OWEN (1986) confirmed this finding with barley straw (Table 3). The selecting behavior is reduced with increasing proportions of concentrate in the ration.

Table 2

INTAKE AND SELECTION OF NH₃-TREATED BARLEY STRAW
BY SHEEP AND GOATS
(From WAHED and OWEN, 1986)

	Suffolk crossbred wethers	Saanen castrated goats
Straw DM offered (g/day)	1299	1477
Level of straw DM intake (g/kg $W^{.75}$/day)	45.5	58.8
Refusal rate (% of offered straw)	26	24
Chemical composition of refused straw DM :		
- nitrogen (g/kg)	11.6	12.2
- acid-detergent fibre (g/kg)	612	600

The consequences of this goat behaviour are important for the strategy of forage feeding, whatever the production system is. The main goal is to obtain the best forage utilization and hence the highest level of dry matter and, above all, energy intake. Thus, when the forage is of high quality, the goat farmer can limit refusals to between 10 and 20 % by limiting the amounts given. When provided with large amounts of poor forage, the goat will exert its selecting behaviour and increase the level of dry matter intake and the nutritive value of ingesta. Thus, this behaviour compensates the poor quality of forage.

OWEN et al. (1987) reported that the consumption of poor forage such as straw could be increased by redistributing the refusals or by treating them with ammonia. In the latter case, the ingestion level of the straw and treated refusals was identical to that obtained by the direct feeding of treated straw.

Thus, it is usually unrealistic to force goats not to refuse. It is preferable to utilize this behavioral characteristic to define a feeding strategy.

PALATABILITY OF FEEDS FOR GOATS

Goats often have a marked preference for some feeds over others and may even refuse the latter partially or totally. It is therefore of interest to predict the response of goats. BALPH and MALECHEK (1981) attempted to develop preference tests for forage using the cafeteria method, enabling a preference index for the forages tested to be established.These authors noted that green wheat was somewhat less preferred than oats.

Due to the risk of refusals of experimental concentrate feeds during nutritional trials in goats, MORAND-FEHR and HERVIEU (1988) developed a method for testing the palatability of concentrate feeds for goats. They observed that some goats of the same flock very clearly selected concentrates according to their palatability, while others accepted all of them regardless of palatability. One year-old goats discriminated better than old goats. Old goats which had been suffering from various diseases refused feeds more frequently. Concentrate refusals were most frequent during late gestation and early lactation (Figure 2).

Figure 1

EFFECT OF REFUSAL ALLOWANCE ON FORAGE INTAKE IN GOATS
(1 : from WAHED and OWEN, 1986; 2 and 3 : from MORAND-FEHR, 1981)

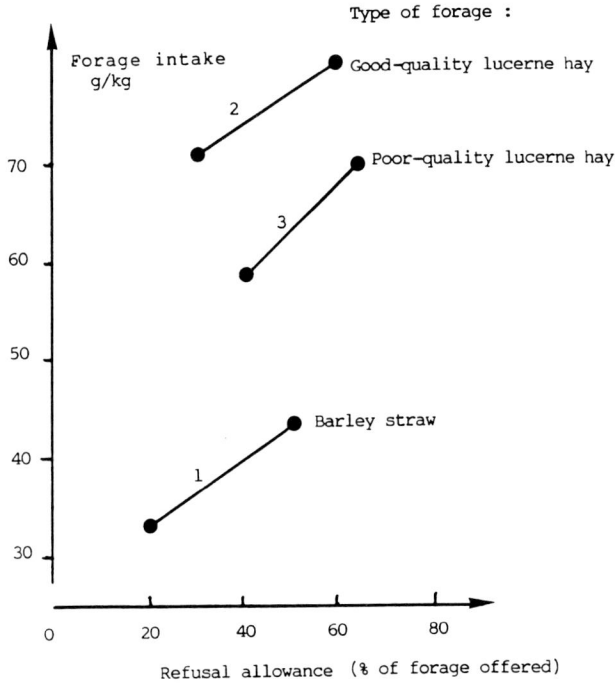

Table 3

EFFECT OF CHOPPED BARLEY STRAW ON THE LEVEL OF INTAKE IN GOATS
(From WAHED, 1987)

Characteristics	Long Straw	Chopped Straw
Number of goats	16	16
Level of straw intake DM (g/kg $W^{.75}$/day)	16.5	14.7
Proportion of offered straw refused (%)	39.3	40.8
Composition of refused straw DM :		
- nitrogen (g/kg)	4.7	5.2
- acid-detergent fibre (g/kg)	608.0	557.0

Goats with the best selecting ability have been used in an attempt to estimate the palatability of concentrate feeds for ruminants. According to the first results, the selecting ability of goats is better than that of cows, but their feed ranking is similar; nevertheless it should be confirmed by further results.

Some tests to be applied in highly controlled conditions have been developed (MORAND-FEHR et al., 1987). They are based on reproductible comparisons of paired feeds closely related to the responses of goats when they receive the concentrate alone as a forage supplement. It was thus possible to distinguish between palatabilities of less palatable families of feeds such as fats, meat meals or rapeseed oilmeals (MORAND-FEHR et al., 1985, 1987; Figure 3). In order to obtain an accurate response, feeds should be tested at different levels of incorporation in a mixture of concentrate feeds of constant composition.

Thus, the goat is a useful animal tool for determining the palatability of concentrate feeds or by-products. This is an important problem in goat nutrition to be taken into account before using new feedstuffs in this species.

Figure 2

VARIATION IN AMOUNT OF CONCENTRATE REFUSALS AROUND PARTURITION

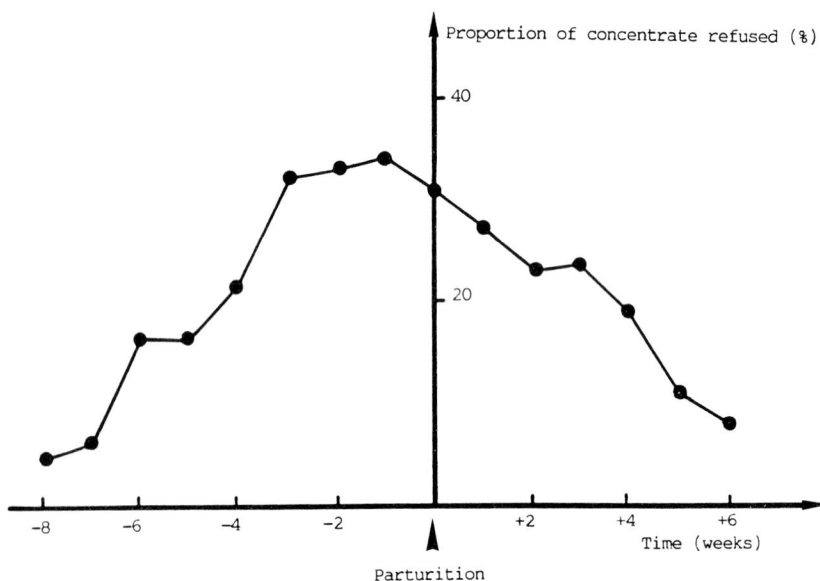

Figure 3

PALATABILITY INDEX OF CONCENTRATES INCLUDING RAPESEED OILMEALS
AT VARIOUS LEVELS OF INCORPORATION

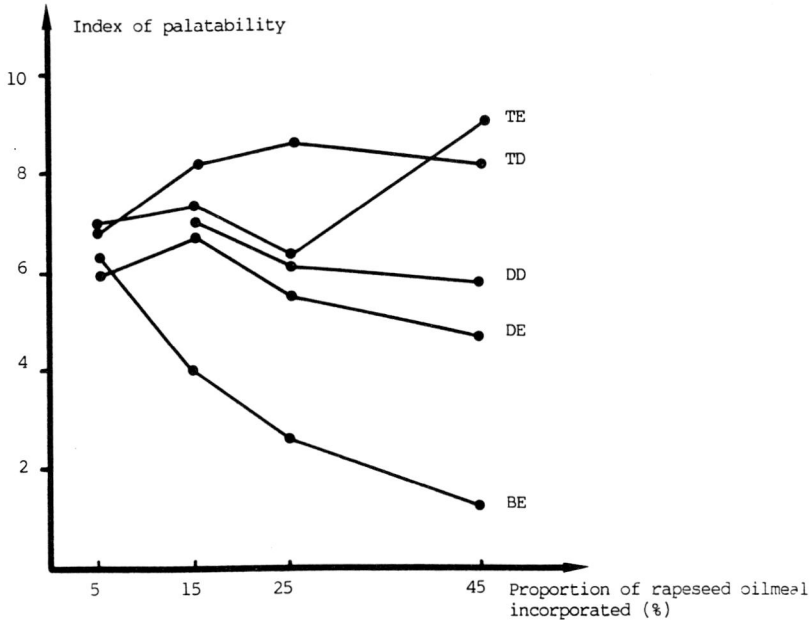

BE : Rapeseed 0, undehulled
DE : Rapeseed 00, undehulled
DD : Rapeseed 00, dehulled
TE : Rapeseed 00,very low in glucosinolates; undehulled
TD : Rapeseed 00,very low in glucosinolates; dehulled.

CONCLUSION

The feeding behaviour of goats is characterized by a very marked selecting of plant fractions and by the ranking of feeds according to their palatability. This should be taken into account for establishing adequate feeding methods. In addition, goats should usually be allowed to refuse feeds. It is advisable to use this behavioral trait to define a strategy of well accepted feed allowances. OWEN et al. (1987) even considered that this was a chance for this species to adapt to highly variable feeding conditions.

Feeding behaviour is sometimes the cause of confusing and even erroneous experimental results, as noted by VAN SOEST (1987). Comparisons between goats and other ruminant species are generally done in conditions adapted to other ruminants with for instance only 5 to 10 % refusals. It is thus not surprising that there are discrepancies in the results concerning the levels of feed intake in goats as compared to those of other ruminants (see Chapter 5).

SUMMARY

If a forage is given *ad libitum*, the level of intake depends on refusals rate that is frequently higher in goats than in other ruminants. Generally goats eat more slowly than sheep because of their very marked selecting behaviour. In goats meals are more numerous, but they do not last so long.

A meal of the goat fed on forages alone at the trough is divided into three phases: a phase of exploration where the goat takes stock of the feed offered, a phase of intense feed intake where the animal satisfies most of its hunger, and a phase of slower intake where the goat selects the plant fractions to be ingested.

Goats select the most nutritive fractions of forages, the leaves more than the stems, the thin stems more than the thick ones, the fractions richest in proteins and poorest in cell-wall carbohydrates. This selecting behaviour in goats can compensate for a poor forage quality if the forage is given *ad libitum*.

Goats may partially or totally refuse some concentrate feeds low in palatability. Their acceptability is relative to their proportion in the concentrate feed mixture.

The goat is a good experimental animal for discrimating between feeds according to their palatability.

Keywords : Adult goat, Growing goat, Feeding behaviour, Palatability, Feeding pattern, Refusals.

RÉSUMÉ

Lorsque le fourrage est distribué à volonté, le niveau de consommation de la chèvre dépend du taux de refus toléré du fourrage. En général, la chèvre consomme plus lentement que le mouton en raison de son comportement de choix très marqué vis-à-vis des aliments distribués. Le nombre de prises alimentaires est plus élevé chez la chèvre mais leur durée est plus faible.

Le repas d'une chèvre alimentée à l'auge lorsque du fourrage seul lui est distribué peut se décomposer en trois parties: une première phase d'exploration où la chèvre fait l'inventaire de ce qui lui est distribué, une seconde phase d'ingestion qui lui permet de satisfaire sa faim et une troisième d'ingestion ralentie marquée par une sélection très discriminante des fractions ingérées.

La chèvre a toujours tendance à sélectionner les parties les plus nutritives du fourrage, les feuilles par rapport aux tiges, les tiges fines par rapport aux tiges grossières, les fractions les plus riches en matières azotées et les plus pauvres en glucides pariétaux. Ainsi ce comportement de choix souvent exacerbé chez la chèvre peut permettre de compenser la médiocre qualité d'un fourrage si elle en dispose suffisamment.

Il n'est pas rare que la chèvre refuse partiellement ou totalement certains aliments concentrés peu appétibles. Ils devront être plus ou moins dilués dans des mélanges d'aliments concentrés selon leur niveau de palatabilité.

La chèvre semble être un bon matériel animal pour discriminer la palatabilité d'aliments destinés aux ruminants, notamment si leur valeur de palatabilité est proche.

REFERENCES

BALPH D.F. and MALECHEK J.C., 1981. Experimental determination of food preference in Spanish goats. J. Anim. Sci., 53, suppl. 1: 131 (Abst.).

DE SIMIANE M., HUGUET L. and MASSON C., 1983. Comportements alimentaires des chèvres à l'auge et au pâturage. Aspects liés au fourrage et à l'animal: conséquences sur les performances zootechniques (Feeding behavior of goats at the trough and on pasture: effects of kind of forages and animals on goat performance), p. 71-100. 8e Journée de la Recherche Ovine et Caprine, INRA-ITOVIC, Paris (FRANCE).

FOCANT M., 1984. Comportement alimentaire, rumination, fermentation réticulo-ruminale et acides gras volatils plasmatiques comparés chez la chèvre et le mouton; influence du régime (Feeding behavior, rumination, rumen fermentation and blood volatile fatty acids in goats and sheep: effect of diets). Reprod. Nutr. Dévelop., 24: 239-250.

FOCANT M., VANBELLE M. and GODFROID S., 1986a. Comparative feeding behaviour and rumen physiology in sheep and goats. World Rev. Anim. Prod., 33: 89-95.

FOCANT M., VANBELLE M. and GODFROID S., 1986b. Activité alimentaire et motricité du rumen chez la chèvre et le mouton pour deux régimes mixtes: foin, orge (Dietary activity and rumen motricity in hay and barley fed goats and sheep). Reprod. Nutr. Dévelop., 26: 277-278.

GEOFFROY F., 1974. Etude comparée du comportement alimentaire et mérique de deux petits ruminants: la chèvre et le mouton (Feeding behavior and rumination in goats and sheep). Ann. Zootech., 23: 63-73.

HOOPER A.P. and WELCH J.G., 1983. Chewing efficiency and body size of kid goats. J. Dairy Sci., 66: 2551-2556.

HUGUET L., BROQUA B. and DE SIMIANE M., 1977. Factors affecting green forage intake by milking goat, p. 1549-1552. 13th Intern. Grassland Congr., Section 10, Leipzig (GERMANY).

MAC CAMMON-FELDMAN B., VAN SOEST P.J., HARVATLY P. and MAC DOWELL R.E., 1981. Feeding strategies of the goat. Cornell Intern. Agric. Mimeograph 88. Cornell Univ., Ithaca (USA).

MORAND-FEHR P., 1981. Caractéristiques du comportement alimentaire et de la digestion des caprins (Characteristics of goat feeding behavior and digestion). Vol. 1, p. 21-45. In: MORAND-FEHR P., BOURBOUZE A. and DE SIMIANE M. (Eds.): Nutrition and systems of goat feeding. Symposium International, Tours (FRANCE), May 12-15, 1981, INRA-ITOVIC, Paris (FRANCE).

MORAND-FEHR P., 1988. Capacité d'adaptation des chèvres en milieu difficile (Capacity of goat adaptation in arid environment). In: La Chèvre. Ethnozootechnie, (41), p. 63-85. Publ. by "Société d'Ethnozootechnie", Paris (FRANCE).

MORAND-FEHR P. and HERVIEU J., 1988. Acceptabilité des aliments composés contenant des tourteaux de colza par des tests de préférence sur chèvres (Acceptability of concentrate feeds containing rapeseed oilmeals by cafeteria tests on goats). Reprod. Nutr. Dévelop., 28: 101-102.

MORAND-FEHR P., HERVIEU J. and SAUVANT D., 1980. Contribution à la description de la prise alimentaire de la chèvre (Contribution to the description of a meal in goats). Reprod. Nutr. Dévelop., 20: 1641-1644.

MORAND-FEHR P., HERVIEU J., BAS P. and SAUVANT D., 1982. Feeding of young goats, p. 90-104. 3rd Intern. Conf. on Goat Production and Disease, Jan. 10-15, 1982, Tucson, Arizona (USA).

MORAND-FEHR P., HERVIEU J., LEGENDRE D. and GUTTER A., 1985. Use of goats as a way for discriminating the palatability of concentrate feeds. Ann. Zootech., 34: 472 (Abst.).

MORAND-FEHR P., HERVIEU J., FAYE A. and ROUASSI M., 1986. Adaptation comportementale à l'alimentation sèche des chevrettes pendant le sevrage (Behavioral adaptation to dry feeds in young female kids during weaning). Reprod. Nutr. Dévelop., 26: 281-282.

MORAND-FEHR P., HERVIEU J., LEGENDRE D., GUTTER A. and DEL TEDESCO L., 1987. Rapid tests to assess concentrate feed acceptability by goats. Ann. Zootech., 36: 324 (Abst.).

OWEN E., WAHED R.A. and ALIMON R., 1986. Effect of amount offered on selection and intake of long untreated barley straw by goats. Ann. Zootech., 36: 324-325.

OWEN E., WAHED R.A., ALIMON R. and EL-NAIEM W., 1987. Strategies for feeding straw to small ruminants, up grading or generous feeding to allow selective feeding. African Research Network for agricultural by-products. Workshop, Oct., 1987, Bamenda (CAMEROUN).

QUICK T.C. and DEHORITY B.A., 1986. A comparative study of feeding behaviour and digestive function in dairy goats, wool sheep and hair sheep. J. Animal. Sci., 63: 1516-1526.

VAN SOEST P.J., 1987. Interactions of feeding behaviour and forage composition, Vol. 2, p. 971-987. 4th Intern. Conf. on Goats, March 8-13, 1987, Brasilia (BRAZIL).

WAHED R.A., 1987. Stall-feeding barley straw to goats: the effect of refusal rate allowance on voluntary intake and selection. PhD Thesis, Univ. of Reading (UK).

WAHED R.A. and OWEN E., 1986. Comparison of sheep and goats under stall-feeding conditions : roughage intake and selection. Anim. Prod., 42: 89-95.

Chapter 2

FEEDING BEHAVIOUR OF GOATS ON RANGELANDS

H. NARJISSE

INTRODUCTION

One of the best known attributes of goats is their ability to thrive in harsh environments which would not support other grazing livestock such as cattle or sheep. The uniqueness of goats relative to other ruminant animals as a producer of meat and milk in areas characterized by their low potential results from their assumed ability to utilize a broad range of forage species and to select from among them the material with the highest nutrient concentration.

Unfortunately, relatively few investigators have examined diets selected by goats grazing alone or in common with other livestock species. The data base specific to goats is so fragmentary that knowledge of most fundamental foraging relationships of this livestock species to its environment remains empirical and in most cases controversial.

This chapter discusses the information on hand toward understanding of the goat's feeding strategy in an attempt to illuminate the ecological conditions and adaptive mechanisms which provide goats with a nutritional advantage by comparison to other domestic ruminants.

GOAT DIET SELECTION PROCESSES

Dietary selection by grazing herbivores depends on numerous interacting factors including breed, previous dietary experience, forage availability and quality and other plant and environment-related factors. The following section is devoted to discussions of features and mechanisms underlying the grazing behavior strategy of grazing goats.

Environmental factors

Few scientists have examined the role of environmental factors in determining foraging behavior of grazing goats. Of these factors, temperature and rainfall are the most commonly investigated. SCHACHT and MALECHEK (1989) reported that under excessively wet, tropical conditions, foraging behavior and intake of goats were adversely affected. Under these conditions, aggravated by the occurrence of biting insects, grazing time was significantly reduced and foraging search was limited to sites with no standing water. The poor foraging performance of goats under cool, wet conditions is probably due to chilling resulting from their open hair coat (FELDMAN et al., 1981). Effect of thermal stress on foraging behavior of goats has never been quantified, to my knowledge. Field observations in the Moroccan high Atlas region indicate however that goats are sensitive to cold temperatures during which they reduce their grazing activities. By contrast, goats appear to adapt to arid environments better than most other ruminants. SHKOLNIK and CHOSHNIAK (1987) observed that Bedouin goats continued to graze even after four days of water deprivation and compensated for the range food scarcity and low value by expanding their grazing area. This observation was confirmed by HANJRA (1986) who reported that the distance covered by goats in search of forage in the arid Thal of Pakistan was 1.5 times more than of sheep.

Forage availability

Forage availability influences diet selection by grazing goats. NARJISSE et al. (1988) observed that in a typical mediterranean oak woodland, goats preferred grasses under low stocking intensity. This preference however did not last after the maturity of the grasses. During June and August sampling periods, the contribution of herbaceous species to the diet selected by goats, although quite high, was not influenced by the level of stocking intensity.

Another feature of this study was that even when forage scarcity was extreme, contribution of the browse species, *Quercus ilex*, to goats' diets did not exceed 51 %. Goats responded to the very low forage availability by reducing dramatically their intake to only 28 g DM/ kg $W^{.75}$ while it attained 64 g DM/ kg $W^{.75}$ when forage availability was at the highest level. This suggests that although goats select more browse than do other domestic ruminants, they tend to regulate the amount of browse eaten. On the other hand, preliminary results of on-going research on the palatability of a highly lignified perennial grass (*Stipa tenacissima*) indicate that goats eat only scanty amounts of this grass current growth, even under extremely high grazing pressure (Table 1).

Table 1.

SEASONAL CHANGE IN *STIPA TENACISSIMA* PERCENT CONTRIBUTION TO DIETS
SELECTED BY SHEEP AND GOATS UNDER THREE STOCKING INTENSITIES
(NARJISSE, MALECHEK and ELABD, 1989, unpubl.)

Stocking intensity	March		May		July	
	Goat	Sheep	Goat	Sheep	Goat	Sheep
High	12.2	21.0	13.2	14.5	*	*
Moderate	1.3	7.9	*	*	*	*
Low	1.4	1.6	*	*	1.4	*

(*) Trace.

It appears from these studies that goats are not obligatory browsers or fibrous material eaters as suggested by many authors. Instead, they tend to be flexible in dietary habit and react by lowering their level of intake when the availability of palatable forage becomes too scarce.

Forage palatability

The existing body of theory on food selection views palatability both as a tactical and strategic problem. Tactics of diet selection suggest that animals are basically hedyphagic and select food plants primarily for their flavor (ARNOLD and HILL, 1972). Another school of thought considers animals as euphagic, basing their diet selection on the optimization of total nutrient balance and avoidance of intoxication (ROZIN and KALAT, 1971, FREELAND and JANZEN, 1974). From an evolutionary point of view, it is very likely that hedyphagic properties of food are coupled with adequate nutrition and absence of toxicity. GARCIA and HANKINS (1977) conclude that food flavor is only a mediation (tactic) and that the final decision to feed or not is dictated by the internal nutritional consequences of food (strategy). Validation of these theories has however been limited to insects and monogastric species such as rats. Their applicability to goats or other free-grazing ruminant livestock generally remains uncertain.

Role of food sensory properties

In order to select a preferred food and recognize the food they should avoid, animals must find cues by which they can recognize their food. The most obvious cues detectable by an animal before ingestion are the food sensory properties. Unfortunately, very little research has been devoted to investigation of the role of these properties in diet selection by grazing livestock.

In a cafeteria trial, sheep and goats exposed during 90 minute feeding periods to alfalfa pellets contaminated by monoterpenoids odor exhibited a totally different preference (NARJISSE, 1981). Data from this experiment indicate that sheep discriminated against feed contaminated by monoterpenoids odor, while goats seemed to randomly select from both types of pellets without being significantly influenced by monoterpenoids smell (NARJISSE, 1981). Thus, goats consumed similar amounts of contaminated and control pellets (64 and 78 g respectively). In contrast, the intake of control pellets by sheep was almost double that of contaminated pellets (486 versus 253 g).

On the other hand, comparisons of anosmic sheep and goats, in which anosmia was achieved by surgically destroying the olfactory nerve fibers and the anterior portion of the olfactory bulb, suggested different responses of sheep and goats to monoterpenoids taste (NARJISSE, 1981). Indeed, anosmic goats displayed a significant preference for control alfalfa pellets over monoterpenoid treated pellets. Their intake levels were respectively 160 g and 90 g. To the contrary, no significant difference was found between consumption of monoterpenoids treated and control pellets by anosmic sheep (343 g versus 330 g).

Generalization of these findings is, however, difficult given the complexity of factors interfering in the animal's response to food sensory properties. Among these factors are 1) sensory thresholds, 2) animal age and its effect on the functionality of taste buds and 3) the phenomenon of adaptation.

Role of food nutritional properties

Several workers suggest that diet selection is based on the integration of messages contained in food sensory properties and feedbacks telling on the current nutrition state of the animal or the occurrence of some metabolic disturbances. An attempt was made to test this hypothesis in sheep and goats through assessment of the change in their intake level following the infusion of a tannin solution into their rumen (NARJISSE and EL HONSALI, 1985). The results obtained in this study suggested that infusion of tannins, compounds reputed to depress the digestion process (Mc LEOD, 1974), led to a significant reduction of intake in sheep but had no effect in goats. PROVENZA and MALECHEK (1984), however, found that condensed tannins are highly deterrent to goats. The latters, apparently learn to avoid twigs of blackbrush (*Coleogyne ramossima*) loaded with tannins by associating the flavor of these twigs to their post-ingestive consequences (PROVENZA et al., 1989). Discrepancies between these findings may be explained by recent evidence reported by CLAUSEN et al. (1989) demonstrating that not all tannins, nor tannins extracted from all plants are equally detrimental to herbivores.

Past dietary experience

Assuming that animals select for adequate nutrition through different tactics, another controversy centers around the acquisition of the ability to do so. Some suggest that it is innate and genetically programmed through the process of natural selection (COOPER et al., 1988). Others claim that it is partly learned through experience (PROVENZA et al.,

1988). The appropriateness of the first approach to domestic livestock is questionable as these animals are no longer living in their natural environment (PROVENZA and BALPH, 1988). In addition, supporters of the learning theory argue that plastic organisms can do better in variable environments and that programming of specific responses in changing and variable environment is behaviorally maladaptive and genetically costly (ROZIN and KALAT, 1971).

Some data support this latter contention for sheep, cattle and occasionally goats. Thus, ARNOLD and MALLER (1977) observed that sheep having experience with several contrasting nutritional environments displayed marked differences in their grazing preferences. In another study, NARJISSE (1981) evaluating sagebrush (*Artemisia tridentata tridentata*) contribution to the diets of sheep and goats with different nutritional background found that sagebrush was virtually absent from diets of inexperienced sheep and goats, despite the very scanty amount of alternative forage available. In contrast, experienced sheep ate substantial quantities (Table 2). Differences were also noted by MEURET and ROSENBERGER (unpubl.) in the dietary botanical composition of Alpine goats that had prior exposure to oak woodland forage compared to those with no experience with this forage. The latter consumed less oak leaves during summer and fall sampling periods. Comparison of the feeding behavior of adult and juvenile goats by PROVENZA and MALECHEK (1986) indicated, however, that the diets consumed by these two groups of animals were generally similar in their botanical composition, leaf stem ratio, protein concentration, and in vitro organic matter digestibility. In this case, the lack of forage experience in juvenile goats seemed to be compensated by longer feeding time.

These results suggest that within small ruminant species, prior dietary experience is more critical than the kind of animal species as a determinant of food selection. They also lead to the questioning of the generally held contention that goats are more tolerant of unpalatable forage species than other livestock species (DEVENDRA, 1978). The higher tolerance of unpalatable food generally assumed in caprine species is probably not an innate character but results probably from the animal's previous dietary conditioning in addition to adaptive anatomical and physiological mechanisms reviewed later in this chapter. In this connection, PROVENZA and BALPH (1988) suggested recently that diet training in livestock may be useful in reducing losses caused by poisonous plants or improving range condition by promoting consumption of less preferred species and therefore increasing the abundance of desirable ones.

Table 2

PERCENTAGES OF SAGEBRUSH IN DIETS OF EXPERIENCED SHEEP AND INEXPERIENCED SHEEP AND GOATS AT TWO LEVELS OF AVAILABLE FORAGE DURING TWO SEASONS (NARJISSE, 1981)

Period	Alternative forage availibility	Experienced sheep		Inexperienced sheep		Inexperienced goats	
		Mean	S.E.	Mean	S.E.	Mean	S.E.
August	High	7	3.7	1	0.3	0	0.1
	Low	30	6.1	5	0.9	0	0.1
November	High	0	0.2	1	0.4	0	0.1
	Low	19	6.3	6	6.3	1	0.5

BOTANICAL COMPOSITION OF THE DIET SELECTED BY GOATS

Methodological considerations

Several methods are available to assess the botanical composition of grazing goats. Among these, three are commonly used by researchers and include (I) bite counts to quantify the relative amounts of different classes of plant species selected (II) use of esophageally fistulated animals to collect samples of the grazed forage followed by microscopic determination of the botanical composition of the ingesta and (III) histological analysis of faeces collected from animals fitted with collection bags.

Each of these methods has advantages and limitations. Final choice is dictated by the objectives of the study, the topography and vegetation dominating the experimental site, and cost. In general, the first method is useful in steep terrain and when the purpose is limited to the determination of broad plant species classes selected by the grazing animals from pasture. BOURBOUZE (1982), NARJISSE and EL BARE (1986) were satisfied by the direct observation method. This method is however no longer valuable when the investigator is interested to know the particular plant species consumed, especially herbaceous ones. The coprological method allows such determination, however it tends to lead to higher estimates of evergreen species and lower estimates of forbs (VAVRA et al., 1978), and highly digestible species (VAVRA et al., 1970). There is no doubt, that the esophageal fistulæ method despite being expensive and potentially causing stress to the experimental animals is the most widely acceptable tool for determining the dietary botanical composition of grazing livestock.

General trend

The traditional classification of goats as a browse preferring species is not applicable everywhere. A number of studies can be cited that classify goats as preferring herbaceous plants over browses. LECLERC (1984) reported that goats grazing in the Corsican "maquis" exhibited a highly variable diet selection throughout the year. In this study the browse contribution to the diet reached 75 % during winter but did not exceed 15 % during spring. Also NARJISSE and EL BARE (1986) found that herbaceous species contributed significantly to the diet selected by goats grazing in the oak woodland. In this trial, the lowest contribution of herbaceous species was 32 % and was recorded in October when the forage production on the forest floor was exceedingly low (33 kg DM/ha). It appears from these studies and others (MALECHEK and LEINWEBER, 1972) that goats rely heavily on herbaceous species during the growing season. They also suggest that although goats are well known for their often high selectivity for browse, they cannot be considered obligatory browsers as generally assumed. Instead, goats appear to be flexible and opportunistic in their feeding habit. They apparently direct their food selection toward maximizing nutritional uptake.

Another feature of goats' dietary behavior is their believed reliance under conditions of low forage availability upon species of poor palatability and their ability to utilize efficiently course feedingstuffs (DEVENDRA, 1978). In reality, there are instances in which even conditions of extreme forage scarcity will not be sufficient to force goats to use unpalatable species. NARJISSE (1981) and NARJISSE et al. (unpubl.) observed such foraging behavior in plant communities dominated respectively by a woody shrub (*Artemisia tridentata tridentata*) and a perennial grass (*Stipa tenacissima*).

In summary it appears that goats have developed anatomical and physiological adaptations that allow them to feed on a wide variety of plant materials. This feeding strategy is however based primarily on food nutritive quality, as supported by the reported seasonal shift in their consumption favoring the most nutritious plant species and parts (MALECHEK and PROVENZA, 1983). This contention is also supported by the generally observed reduction on goats' dry matter intake on heavily stocked ranges, contrasting with sheep that tend to compensate for low nutritive value by increasing their ingestion of fibrous mature forage (FELDMAN et al., 1981).

Role of adaptive mechanisms

More than other grazing ruminants, goats exercise considerable selection and are adaptable to a wide range of conditions. This versatility in goat's feeding behavior seems to be the consequence of several anatomical and physiological adaptations developed by this species. Understanding of these adaptations may be useful in explaining some features of goat foraging behavior, particularly with regard to their ability to thrive in harsh environments and exhibit a high feed conversion efficiency.

Anatomical adaptations

The ability of goats to feed on a large variety of plant species and environment is likely due to their agility, mobile upper lips and tendency to assume a bipedal stance. The feature of the goat's buccal anatomy allows it to graze herbage as short as can the sheep, and feed on plant species provided with thorns and spines (VAN DYNE et al., 1980; FELDMAN et al., 1981), while its ability to stretch upward on its hind legs provides the goat with an advantage over other livestock species in feeding from a higher vegetation layer. ETIENNE et al. cited by DE SIMIANE (1987) reported that goats compensated for a decline in forage availability by increasing the height at which they browsed. Thus, the utilization rate of previously marked *Quercus ilex* branches was 22, 50 and 85 % during the first, second and third grazing passage, respectively. PFISTER and MALECHEK (1986) studied the vertical stratification of foraging by goats and concluded that this stratification may have relevance to their better survivability in plant communities dominated by trees.

Finally, goats are well known for their agility in climbing. Although, I am not aware of this property being experimentally investigated, field observations by the author in the Moroccan high Atlas demonstrate clearly that goats were the only livestock species capable of surviving in steep terrain.

Another noteworthy anatomical consideration in goats relates to their digestive tract. FELDMAN et al. (1981) reported that goats have a faster rate of passage than sheep or cattle and suggested that this property combined with rapid fermentation of browse explains why browsing is the predominant feeding strategy of goats (see Chapter 5).

Physiological adaptations

Some plants have apparently evolved chemical means of protecting themselves from grazing herbivores. These chemicals referred to as secondary metabolites include tannins, alkaloids, terpenes, cyanogenic glycosides and many others (LEVIN, 1976). Their role in chemical defense is usually related to their irritant, toxic, or unpalatable characteristics. All domestic livestock are not equally deterred by these compounds. For example, goats and sheep are not affected by Senecio alkaloids while cattle are (FELDMAN et al., 1981). Similarly, goats tolerate oak tannins better than sheep do. They ate more oak leaves throughout the year than did sheep (NARJISSE and EL BARE, 1986). Moreover, tannins extracted from oak leaves stimulated rumen microbial activity and nitrogen balance in goats but inhibited these functions in sheep (NARJISSE and EL HONSALI, 1985).

These differences in susceptibility to plant's defense, although not fully understood, may be attributed to physiological adaptations developed by resistant livestock species. Hence, it is hypothesized that goats' ability to tolerate tannins is related to their higher salivary secretion and urea recycling capacity. SETH et al. (1976), comparing sheep and goats, demonstrated that parotid salivary secretion was higher for goats. The larger salivary supply of specialized mucus may contribute to the complexing of tannins, so that protein utilization is less affected. On the other hand, COCIMANO and LONG (1967), cited by SILANIKOVE et al. (1980), reported that the amount of urea recycled by Bedouin goats was much higher than that measured in sheep under comparable conditions.

Another potential physiological adaptation relates to the presence of requisite bacteria in the rumen of goats that could increase cell protein synthesis as a sacrifice to bind tannins (VAN SOEST, 1987) and that can degrade tannins as suggested by GRANT (1976) cited by FELDMAN et al. (1981).

Dietary overlap with other animal species

As mentioned earlier, nutritional benefit appears to be the overriding factor controlling goats' feeding strategy. Under these conditions, the likelihood for interspecific competition for some plant species is real. LECLERC (1985), NARJISSE and EL BARE (1986) working respectively in the Corsican Mountain rangelands and Moroccan oak woodland confirmed the existence of dietary overlap between sheep and goats and observed that both species exhibited high preference for grasses during spring. The extent of dietary overlap declined however after the growing season. Thus, during the dry season, while the dietary contribution of grasses to sheep's diets was maintained around 70 %, this contribution did not exceed 32 % for goats during the same period (NARJISSE and EL BARE, 1986). Nevertheless, dietary overlap may continue over the dormant season under conditions where the forage alternative to the herbaceous layer consists of species of low palatability. NARJISSE et al. (unpubl.) observed that goats strictly refused to feed on unpalatable mature *Stipa tenacissima* despite the extreme scarcity of alternative forage.

Management implications

The findings reported on goats' feeding strategy suggest that although, it is recognized that goats select more browse than other domestic livestock do, dietary overlap occurs necessarily during periods of inadequate forage supply. Under such conditions, the grazing pressure on the herbaceous layer, especially its grass component, increases. The impact of this overlap on the nutritional status of the grazing animal and range condition depends primarily on the stocking rate. Competition may, therefore, well be of major concern in mixed grazing herds and grazing intensity should be carefully monitored. Interspecific interactions may however be beneficial and lead to an improvement in food resources for one or both species. For instance, bush encroachment in the bushveld region of South Africa is successfully counteracted by the use of goats which reduce the woody component of the plant community allowing grasses to regain vigor and consequently improve the feed resources for other grazing ruminants (OWEN-SMITH and COOPER, 1987). Similarly, heavy stocking by goats in the underutilized woodland of southern France may be useful in clearing the shrub-invaded forest, thus contributing indirectly to reduce fire hazard in this highly threatened Mediterranean forest (GILLET, 1986).

DIET QUALITY AND INTAKE OF GRAZING GOATS

Research on the nutritional quality of the diet selected by free ranging ruminants has generally dealt with determination of dietary crude protein and cell wall contents. Measurements of in vitro digestibility combined with faecal collection are also commonly performed to estimate dietary digestible energy and intake. Use of esphageally fistulated animals in generally recognized as the most accurate method to evaluate dietary quality. The other alternative using chemical analysis of hand harvested samples supposedly representative of the diet consumed is only indicative, and may not be an adequate predictor of diet quality, especially in plant community of high diversity.

Nutritional quality of the diet consumed by goats

The data available on dietary chemical composition analyses show that nutritive value of selected forage material is generally higher than average nutritional value of available forage for grazing goats. They also suggest a seasonal variation in the dietary content of different nutrients. BOUTTIER et al. (1983) noted a downward trend in the crude protein content and digestibility of diet ingested by lactating goats from spring through winter (Table 3).

This trend was confirmed in the Moroccan oak woodland by NARJISSE et al. (1988) who observed that dietary crude protein levels for goats varied from about 17.5 % to about 11.5 % and from 13.5 % to 11.5 % for low and high stocking intensity, respectively. In this study, where grazing sheep were also included, only minor differences were found in the crude protein content and in vitro digestibility of diet selected by sheep and goats.

Table 3

SEASONAL CHANGE IN DIETARY DIGESTIBLE PROTEIN CONTENT AND DIGESTIBILITY OF LACTATING GOATS GRAZING IN THE DROME FOREST
(BOUTTIER et al., 1983).

	April	May	June	July	August	October
Digestible protein (g/kg DM)	107	55	54	80	64	39
Digestible OM (g/kg DM)	628	531	539	493	414	443

Similar results were also reported by PFISTER and MALECHEK (1986) working in the Caatinga plant community of northeastern Brazil. These findings are not, however, sufficient to refute the hypothesis of VAN SOEST (1982) and others, that goats select diets of higher nutritional quality than sheep do. Such superiority would occur ultimately during the dry season and in sites where differentiation in nutritional value between herbaceous and browse species exists and provides an opportunity for selective feeding by animals which possess this ability. Indeed, SCHWARZ et al. (1986) reported that on Kenyan rangelands dominated by *Acacia tortilis* and low shrub *Duosperma eremohilium*, both goats and sheep selected diets of similar energy content of 10 to 11 Mcal/kg DM during the growing season. At the height of the dry season, goats were still able to select diets with 6.5 Mcal/kg DM compared with 3.5 to 4.0 Mcal/kg DM only for sheep.

Finally, no valid data are available on mineral composition of the diet being grazed. Assessment of dietary mineral levels is difficult, because of salivary contamination of samples collected by esophageally fistulated animals, and the extreme difficulties encountered to sample the herbage being grazed to the same degree as does the grazing animal. This is particularly critical, given that most forage resources available on rangelands suffer from phosphorus deficiency during the dry season (NARJISSE and EL BARE, 1986; DAMIANI and DE SIMIANE cited by DE SIMIANE, 1987).

Intake

Intake is an essential parameter for the assessment of forage quality and hence the prediction of animal performance in grazing situations. It is also a necessary tool for the planning and management of rangeland utilization by grazing animals.

Despite the critical importance of intake to forage plant breeders and to range managers, very little information is available on the amount of forage consumed by grazing herbivores, particularly goats. For the latter, most references on this topic describe amounts eaten under controlled experimental conditions with no means of making inferences from these trials to the range situation.

It is indeed recognized that measurements of intake by free ranging livestock are difficult to establish because of the complexity of selective grazing and the usually harsh topographic conditions of rangelands. These considerations make the measurement of intake by grazing animals laborious, expensive and often inaccurate. Further details about level of intake and methodology will be provided in Chapter 3.

CONCLUSION

This review reveals that present knowledge of the goat's foraging behavior remains in general too fragmentary, too site specific and in many cases controversial. It tends to indicate though, that goats have developed adaptive anatomical and physiological mechanisms that allow them a large degree of flexibility in their food selection, controlled apparently primarily by maximization of nutritional benefit.

Understanding of these adaptive mechanisms and the feeding strategy of goats is critical to land use management and efficient productivity of grazing goats. Such understanding is particularly useful in deciding what environment this species is best suited to exploit and how a plant community should be manipulated and managed to make it suitable for goats grazing alone or mixed with other domestic species. Moreover, further information is needed to illuminate the physiological mechanisms that allow goats to cope successfully with browse secondary compounds. Results of such investigations may have applications beyond the scope of goat management and provide eventually the technological basis for reducing the deterrence of these compounds to other livestock species.

SUMMARY

Information pertaining to the goat's grazing behavior strategy as well as the factors controlling it are discussed. In particular, the role of environmental factors, forage availability and palatability, and past dietary experience are reviewed. Available studies on these topics indicate that goats are not obligatory browsers or fibrous material eaters. Instead, they tend to be flexible in their dietary habit, as consequences of various adaptive anatomical and physiological mechanisms developed by this species. Among these, are features that relate to the goat's bipedal stance and buccal anatomy and assumptions attributing to goats,higher salivary secretion and urea recycling capacity. Such flexibility results in the maximization of nutritional intake, which appears to be the overriding factor controlling food selection by grazing goats. Management implications of the findings on the goat's grazing behavior are also discussed, particularly with regard to the use of goats in mixed herds and as a tool for brush control.

Keywords : Goat, Feeding behavior, Palatability, Secondary compounds, Fiber, Browser, Biological control.

RÉSUMÉ

Les informations disponibles aussi bien sur la stratégie du comportement alimentaire de la chèvre que sur les facteurs qui déterminent ce comportement sont discutées. En particulier, le rôle des facteurs de milieu, du niveau des disponibilités des ressources fourragères et de leur appetabilité ainsi que du passé alimentaire de l'animal a été passé en revue. Les études menées sur ces thèmes semblent indiquer que les caprins ne sont pas des consommateurs systématiques d'arbustes ou des espèces fortement lignifiées. Cette souplesse du comportement alimentaire des caprins est la conséquence d'adaptations anatomiques et physiologiques qui caractérisent cette espèce. Parmi celles-ci, quelques-unes ont trait aux particularités anatomiques de la cavité buccale de la chèvre et à son aptitude à se mettre en position debout. D'autres découlent d'hypothèses attribuant à la chèvre une capacité supérieure de secrétion salivaire et de recyclage de l'urée. Une telle souplesse conduit à la maximisation de la valeur nutritionnelle de la prise alimentaire qui semble être le facteur essentiel déterminant la sélectivité alimentaire des caprins sur parcours. Par ailleurs, les applications pratiques des résultats obtenus sur le comportement alimentaire du caprin sur parcours sont discutées, notamment par rapport à l'utilisation de troupeaux mixtes et à l'intérêt de la chèvre pour le débroussaillement.

REFERENCES

ARNOLD G.W. and HILL J.L., 1972. Chemical factors affecting selection of food plants by ruminants, p. 71-101. In: HARBORNE J.B. (Ed.): Phytochemical Ecology. Academic Press, New York (USA).

ARNOLD G.W. and MALLER R.A., 1977. Effects of nutritional experience in early and adult life on the performance and dietary habits of sheep. Appl. Anim. Ethol. 3: 5-26.

BOURBOUZE A., 1982. Utilisation de la végétation de type méditerranéen par des caprins (Utilization by goats of Mediterranean vegetation). Fourrages, (92), 91-106.

BOUTTIER B.A., BOURBOUZE A. and DE SIMIANE M., 1982. Composition botanique et valeur alimentaire de la ration ingérée par des chèvres laitières sur parcours dans la Drôme (Botanical composition and nutritive value of the diet ingested by grazing lactating goats in the Drome). Document ITOVIC, Paris (FRANCE).

CLAUSEN T.P., PROVENZA F.D., BURRITT E.A., BRYANT J.P. and REICHARDT P.B., 1989. Ecological implications of condensed tannin structure: a case study. J. Chem. Ecol. (in press).

COOPER S.M., OWEN-SMITH N. and BRYANT J.P., 1988. Foliage acceptability to browsing ruminants in relation to seasonal changes in the leaf chemistry of woody plants in a South African savanna. Oecologia 75: 336-342.

DE SIMIANE M., 1987. Composition botanique et valeur alimentaire de la ration ingérée par les petits ruminants. In: La forêt et l'élevage dans la région méditerranéenne française (Botanical composition and nutritive value of diet selected by small ruminants). Fourrages, numéro hors série, 167-184.

DEVENDRA C., 1978. The digestive efficiency of goats. World Rev. Anim. Prod., 14: 9-22.

FELDMAN B.M., VAN SOEST P.J., HORVATH P. and McDOWELL R.D., 1981. Feeding strategy of the goat. Cornell Intern. Agric., Mimeo, Cornell Univ., Ithaca, New York (USA).

FREELAND W.J. and JANZEN J.H., 1974. Strategies in herbivory by mammals: the role of plant secondary compounds. Am. Nat., 108: 269-289.

GARCIA J. and HANKINS W.G., 1977. On the origin of food aversion pradigms, p. 3-22. In: BARKER L.M., BEST M.R. and DOMJAN M. (Eds.): Learning Mechanisms in Food Selection, Baylor Univ. Press, WACO.

GILLET T., 1986. La remise en valeur des terres embroussaillées (Restoration of shrub invaded land). Le GIE Alpages et Forêt. In: L'animal au pâturage dans des friches et des landes. Fourrages, numéro hors série, nov. 1986, p. 41-62.

HANJRA S.H., 1986. Efficiency of grazing by sheep and Barbery goats compared with grazing by sheep and goats alone, p. 384-385. In: JOSS R.J., LYNCH P.W. and WILLIAMS O.B. (Eds.): Rangelands: A Resource Under Siege. Aust. Acad. Sci., Canberra (AUSTRALIA).

LECLERC B., 1984. Utilisation du maquis corse par des caprins et des ovins. I-Régime alimentaire des caprins (Utilization of Corsican chaparral by goats and sheep. I-Goat's diet). Acta Oecologia, Ecol. Applic., 5: 383-406.

LECLERC B., 1985. Utilisation du maquis corse par des caprins et des ovins. II-Comparaison du régime des ovins et des caprins (Utilization of Corsican chaparral by goats and sheep. II-Comparison of diet selected by sheep and goats). Acta Oecologia, Ecol. Applic., 6: 303-314.

LEVIN D.A., 1976. The chemical defense of plants to pathogens and herbivores. Ann. Rev. Ecol. Syst., 7: 121-159.

MALECHEK J.C. and LEINWEBER C.L., 1972. Forage selectivity of goats on lightly and heavily grazed ranges. J. Range Manag., 25: 105-111.

MALECHEK J.C. and PROVENZA F.D., 1983. Feeding behavior and nutrition of goats on rangelands. World Anim. Rev., 47: 38-48.

MAC LEOD M.N., 1974. Plant tannins- their role in forage quality. Nutr. Abst. Rev., 44: 803-815.

NARJISSE H., 1981. Acceptability of big sagebrush to sheep and goats: Role of monoterpenes. Ph. D. Dissertation. Utah State University, Logan (USA).

NARJISSE H. and EL HONSALI M., 1985. Effect of tannin on nitrogen balance and microbial activity of rumen fluid in sheep and goats. Ann. Zootech., 34: 482.

NARJISSE H. and EL BARE B., 1986. Seasonal changes in the dietary botanical composition of sheep and goats grazing in an oak forest. In: JOSS P.J., LYNCH P.W. and WILLIAMS O.B. (Eds.): Rangelands: A Resource Under Siege, p. 369-371. Austr. Acad. Sci., Canberra (AUSTRALIA).

NARJISSE H., ANNASSER Z. and HEMISSI A., 1988. Comportement alimentaire des ovins et des caprins sur un parcours forestier à base de chêne vert (Grazing behavior of sheep and goats in oak woodland). Seminar of FAO subnetwork on Goat Nutrition and Feeding, Oct. 3-5, 1988, Potenza (ITALY).

OWEN-SMITH N. and COOPER S.M., 1987. Foraging strategies of browsing ungulates: Comparison between goats and African wild ruminants, Vol. 2, p. 957-968. 4th Intern. Conf. on Goats, March 8-13, 1987, Brazilia (BRAZIL).

PFISTER J.A. and MALECHEK J.C., 1986. Nutrition and feeding behavior of goats and sheep in a deciduous shrub woodland of Northeastern Brazil, p. 411-412. In: JOSS P.J., LYNCH P.W. and WILLIAMS O.B. (Eds.): Rangelands: A Resource Under Siege, p. 411-412. Austr. Acad. Sci., Canberra (AUSTRALIA).

PROVENZA F.D. and MALECHEK J.C., 1984. Diet selection by domestic goats in relation to blackbrush twig chemistry. J. Appl. Ecol., 21: 831-841.

PROVENZA F.D. and MALECHEK J.C., 1986. A comparison of food selection and foraging behavior in juvenile and adult goats. Appl. Anim. Behav. Sci., 16: 49-61.

PROVENZA F.D. and BALPH D.F., 1988. The development of dietary choice in livestock on rangeland and its implications for management. J. Anim. Sci., 66: 2356-2368.

PROVENZA F.D., BALPH D.F., OLSEN J.D., DWYER D.D., RALPHS M.H. and PFISTER J.A., 1988. Toward understanding the behavioral responses of livestock to poisonous plants, p. 407-424. In: JAMES L.F., RALPHS M.H. and NIELSEN D.B. (Eds.): The Ecology and Economic Impact of Poisonous Plants on Livestock Production. Westview Press, Boulder (USA).

PROVENZA F.D., BURRITT E.A., CLAUSEN T.P., BRYANT J.P., REICHARDT P.B. and DISTEL R.A., 1989. Conditioned flavor aversion: a mechanism for goats to avoid condensed tannins in blackbrush. Am. Nat. (in press).

ROZIN P. and KALAT J.W., 1971. Specific hungers and poison avoidance as adaptive specializations of learning. Physiol. Rev., 78: 459-486.

SCHACHT W.H. and MALECHEK J.C., 1989. Nutrition of goats as influenced by thinning and clearing of deciduous woodland in Northeastern Brazil. J. Range Manag., (in press).

SCHWARZ H.J., HERLOCKER D.J. and SIAD A.N., 1986. Forage intake on semi-arid to arid pastures in northern Kenya, p. 385-386. In: JOSS P.J., LYNCH P.W. and WILLIAMS O.B. (Eds.): Rangelands: A Resource Under Siege. Austr. Acad. Sci., Canberra (AUSTRALIA).

SETH O.N., RAI G.S., YADAV P.C. and PANDEY M.C., 1976. Saliva secretion. A note on the rate of secretion and chemical composition of parotid saliva in sheep and goats. Indian J. Anim. Sci., 46: 660-663.

SHKOLNIK A. and CHOSHNIAK I., 1987. Goats and the desert ecosystem, Vol. 1, p. 115-129. 4th Intern. Conf. on goats. March 8-13, 1987, Brazilia (BRAZIL).

SILANIKOVE N., TAGARI H. and SHKOLNIK A., 1980. Gross energy digestion and urea recycling in the desert black Bedouin goat. Comp. Biochem. Physiol., 67: 215-218.

VAVRA M., RICE R.W. and HANSEN R.M., 1970. Esophageal vs. fecal sampling for the botanical determination of steer diets. Amer. Soc. of Anim. Sci. Proc. Western Sec., 21: 291.

VAVRA M., RICE R.W. and HANSEN, 1978. A comparison of esophageal fistula and fecal material to determine steer diets. J. Range Manag., 31: 11-13.

VAN SOEST P.J., 1982. Nutritional Ecology of the Ruminant, p. 374. O & B Books, Corvalis, Oregon (USA).

VAN SOEST P.J., 1987. Interaction of feeding behavior and forage composition, Vol. 2, p. 971-987. In: 4th Intern. Conf. on Goats, March 8-13, 1987, Brasilia (BRAZIL).

VAN DYNE G.M., BROCKINGTON N.R., SZROCS Z., DUEK J. and RUBIC C.A., 1980. Large herbivore subsystems, p. 269-537. In: BREYMEYER A.J. and VAN DYNE G.M. (Eds.): Grasslands, Systems Analysis, and Man. International Biological Programme, Cambridge Univ. Press, London (UK).

Chapter 3

DRY MATTER INTAKE OF ADULT GOATS

D. SAUVANT, P. MORAND-FEHR and S. GIGER-REVERDIN

INTRODUCTION

In adult female goats as in other ruminants, the level of voluntary dry matter intake (VDMI) is the main determinant of nutrient supplies. The VDMI varies according to numerous factors, the influences of which have to be identified and quantified to obtain a precise feeding management. Available information concerning VDMI in goats is insufficient. Most of the published allowances are given within too large a range of variation (Table 1) to be applied in practice.

The present paper was designed to provide more accurate information on VDMI by focusing attention on experimental data in relation to animal and dietary factors of variation. Moreover utilization of the French Fill Unit System for calculation of goat diets is suggested (DULPHY et al., 1987).

VDMI VARIATIONS RELATED TO ANIMAL FACTORS

Dynamic aspects : the physiological stage.

Pregnancy

The VDMI level remains relatively constant at the beginning of pregnancy when expressed on a live weight (LW) basis, but it decreases regularly for about 20 weeks before parturition at a mean rate of 300 g / 100 kg LW / month.

In goats of the same size, the differences in prolificacy do not significantly modify this variation in the level of (DMI), until two weeks before parturition.

Lactation

The range of VDMI variation is maximum during lactation, and the highest weekly mean VDMI measured in the Alpine breed is 3.6 kg/d or 6.8 % of LW or 181 g/kg $W^{.75}$ (SAUVANT, 1978). The VDMI rises just after parturition and reaches a maximum between 6 and 10 weeks of lactation. However, this increase is not linear and according to our experimental data base, the DMI was equal to 72, 83, 90, 95 %, respectively of the maximum value during the first 4 weeks of lactation (Figure 1). During this period variation in the level of VDMI was almost parallel to the lactose and raw milk yield (RMY) and did not follow the variation in the yield of fat or protein (Figure 1). After reaching its maximum, the DMI decreased rather linearly with a mean rate of about 25 g/animal/week.

This overall variation in the level of VDMI throughout lactation was described in goats by SAUVANT and MORAND-FEHR (1981), and RANDY and SNIFFEN (1982).

Table 1

REFERENCE VALUES OF DRY MATTER INTAKE OF GOATS UNDER VARIOUS CLIMATES

	Dry matter intake	
	% live weight	g/kg $W^{.75}$
Tropical conditions (DEVENDRA and BURNS, 1983)		
Meat goat	1.8 - 3.8	40 - 128
Local dairy goats (Jamnapari)	2.0 - 4.7	41 - 131
Exotic dairy goats (Alpine, Saanen...)	2.8 - 4.9	62 - 142
Mean value for maintenance	1.4 - 1.7	43 - 50
Temperate conditions :		
Lactating goats (Mac KENZIE, 1970)	5	
	4 (max 6.8)	
Lactating goats (SAUVANT, 1978)	1.6 - 6.8	47.1 - 181.2

Static aspects : the within-stage VDMI variations

Pregnancy

In a population of 393 Alpine goats, the mean VDMI value for the last 6 weeks before parturition was 1.52 kg/d (SD = 0.26). During this period, the parity, LW and prolificacy were positively related and influenced simultaneously and similarly the DMI to the same extent. This effect was particularly marked during the first two years of life (Figure 2).

Lactation

Milk yield (MY)

Individual MY variations accounted for more than half of the VDMI variations. At the onset of lactation, the VDMI variations were more related to the lactose or raw milk yield (RMY) than to the fat corrected milk yield (FCMY), or the fat or protein yields. These relationships together with the above mentioned dynamic variations in the same parameters (Figure 1) may explain why the DMI level prediction from a weekly pooled data base was much more accurate when using RMY rather than the other production parameters. From our experimental data concerning 2067 weekly means of daily measurements during the first 8 weeks of lactation, the equation of VDMI prediction was :

$$\text{VDMI} = 164.7 + 368.6\ \text{RMY} + 34.8\ W^{.75}\ (1) \quad (n = 2067;\ R = 0.79;\ \text{RSD} = 299\ \text{g})$$
$$\text{(g/d)} \qquad\qquad \text{(kg/d)} \qquad \text{(kg)}$$

the statistical values of these three criteria being :

	VDMI (g/d)	RMY (kg/d)	$W^{.75}$ (kg)
- mean :	2116	3.30	20.75
- standard deviation :	485	0.96	2.09
- minimum value :	912	0.13	11.88
- maximum value :	3 460	6.69	38.50

The accuracy of this equation increased from week 1 to 8. This trend emphasizes the fact that the VDMI regulation during the first days of lactation was less dependent of the energy expenses for maintenance and production. The regression coefficient of RMY was similar to values observed by LU et al. (1987), i.e. from 350 to 370 g/kg RMY according to the period under consideration.

During the decreasing phase of lactation the DMI was explained with a similar level of accuracy from RMY or from 3.5 % FCMY. The mean DMI of 204 goats was explained by the following model (SAUVANT et al., unpubl.) from weekly means of daily measurements during 8 to 19 weeks of the decreasing phase of lactation :

$$\text{VDMI} = 533 + 305.2 \text{ RMY} + 13.3 \text{ LW} + a \text{ (2)} \qquad (n = 204; R = 0,93; RSD = 141 \text{ g})$$
(g/d)   (kg/d)   (kg)

Figure 1

EVOLUTION AT THE ONSET OF LACTATION OF THE LEVELS OF DRY MATTER INTAKE,
RAW MILK YIELD (RMY), FAT (FY) AND PROTEIN YIELDS (PY)
(SAUVANT and MORAND-FEHR, unpubl.)

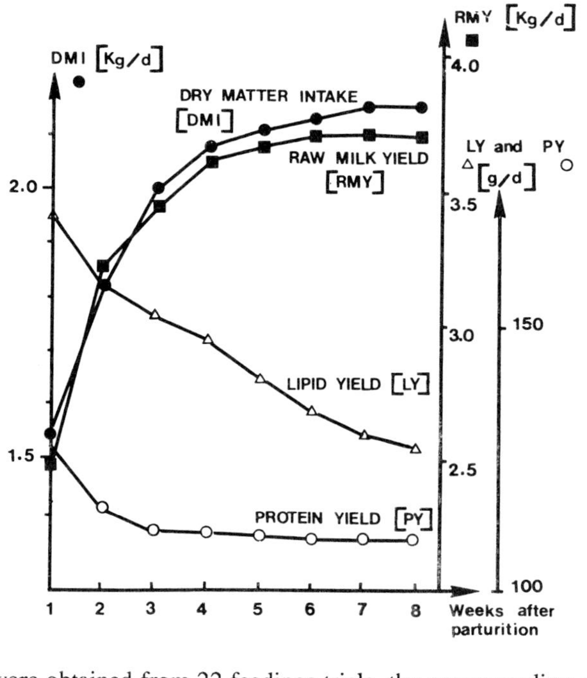

These data were obtained from 22 feedings trials, the corresponding effect (a, i = 1 to 22) was taken into account in the calculation of equation (2). A slight but significant improvement was obtained by including the LW change (LWC, in g/week) :

$$\text{VDMI} = 507.4 + 303.8 \text{ RMY} + 12.8 \text{ LW} + 0.171 \text{ LWC} + a \text{ (3)}$$

(n = 204; R = 0.94; RSD = 138 g).

Expressed on the basis of 3.5 % FCMY, the VDMI variation was larger, i.e. 340.2 instead of 303.8 g/kg.

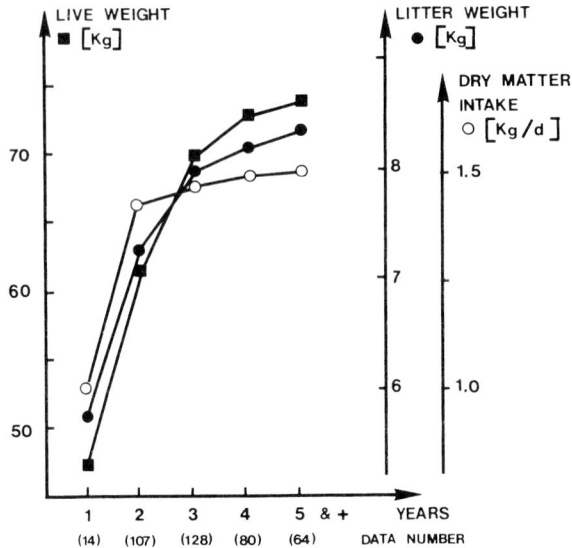

Figure 2

INFLUENCE OF AGE ON LIVE WEIGHT, LITTER WEIGHT AND INTAKE CAPACITY
IN GOATS AT THE END OF PREGNANCY
(SAUVANT et al., unpubl.)

Live weight (LW)

Differences in LW accounted for 10 to 30 % of the VDMI individual variations. A difference of intake capacity of 10 g of DMI/kg LW was observed (SAUVANT and MORAND-FEHR, 1977). Equation (1) confirms this value, but equations (2) and (3) and GIGER and al. (1988), indicate a corresponding value of about 13 g, while LU et al. (1987) obtained a much higher value (20 g) on the basis of one feeding trial. These differences cannot be related to dietary effects.

Energy balance

The energy status of dairy goats can also influence the intake capacity (SAUVANT, 1981; see Chapter 6). Thus, SAUVANT et al. (1979) observed that for the same LW and MY a higher lipomobilization level at the onset of lactation was associated with a significantly lower level of VDMI. However, from recent data on 155 goats (unpubl.), it was not possible to confirm this effect on the mean values of the first 6 weeks of lactation. This is probably because high yielding animals exhibit both higher DMI and lipomobilization levels, when estimated by the calculated energy balance and blood levels of NEFA and glucose.

The level of body lipid storage also affects the VDMI level; thus, CHILLIARD (1985) observed that pregnant and lactating goats with more body fat have a significantly lower intake capacity.

DMI VARIATIONS RELATED TO DIETARY FACTORS

Forage quality

VDMI of goats reared in similar conditions can vary largely according to the forage species and quality (see Chapter 13). Several studies (DE SIMIANE et al., 1981; BLANCHART et al., 1980) have shown that bucks and wethers exhibit a rather similar hierarchy of forage VDMI. Moreover the data obtained at the Lusignan experimental station (ITOVIC, INRA) and summarized in Figure 3 indicate that the forage vegetative stage influences similarly the VDMI of lactating goats, kept on a straw litter and supplemented daily with 35 g concentrate DM/kg $W^{.75}$, and the "standard sheep" used for 25 years in France to evaluate forage digestibility and ingestibility (DEMARQUILLY et al., 1978) :

GF DMI = - 12.8 + 1.07 SF DMI + b_1 (4)   (n = 61; R = 0.90; RSD = 6.9 g/kg $W^{.75}$)

GF DMI : goat forage VDMI (g/kg $W^{.75}$)

SF DMI : standard sheep forage VDMI (g/kg $W^{.75}$)

b_1 : forage effect (22 levels).

Figure 3

RELATIONSHIPS BETWEEN THE VOLUNTARY DRY MATTER INTAKE OF FORAGES
BY STANDARD SHEEP AND LACTATING GOATS
(from results of LUSIGNAN STATION - INRA - ITOVIC)

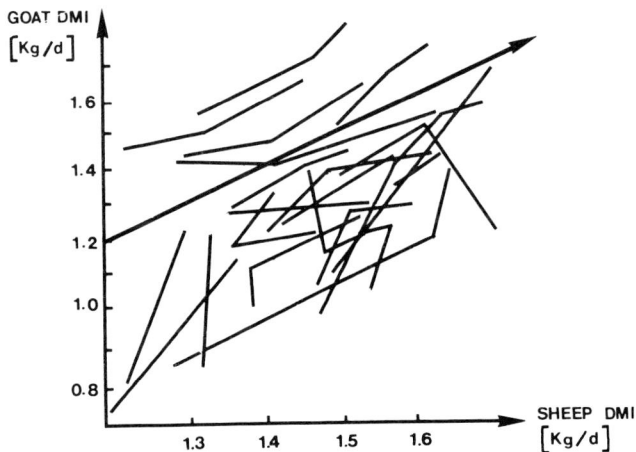

This relationship in VDMI between goats and sheep justifies the utilization of the Fill Unit system in goats (see below). The level of intake of standard herbage (77 % organic matter digestibility, VDMI level for standard sheep 75 g/kg $W^{.75}$) offered to lactating goats is 67.5 g VDMI/kg $W^{.75}$. This value corresponds to a total VDMI/kg $W^{.75}$ of 102.5 g which takes into account a mean value of 35 g concentrate DMI and 14 g straw VDMI/kg $W^{.75}$. If the probable substitution rates between these feeds are used, the estimation of standard VDMI of grass fed to goats is about 123 g/kg $W^{.75}$. The forage effect b_1 is highly significant and indicates that the goat and standard sheep do not have similar levels of forage intake when influenced by plant species or variety.

MORAND-FEHR et al. (1987) observed an increase in hay VDMI of 11 g/d with increasing crude protein content. However, knowledge on goats is not sufficient yet to obtain reliable models for prediction of forage VDMI from analytical criteria as in sheep.

The leaves of tree or bush biomass may represent a large proportion of the goat diet. The VDMI estimation of such forages presents practical difficulties, and hence only a few studies have been conducted until now to simulate rangeland conditions with fresh products (see Chapter 14). According to several authors, the VDMI of fresh products is highly variable, from 51 to 123 g/kg $W^{.75}$, and no data are available to depict the relationship between the digestibility and ingestibility values of this biomass (MEURET, 1989).

Level of refusals and forage choice

The refusal level of *ad libitum* fed forage influences significantly its ingestibility and digestibility (see Chapter 1), this simultaneous influence was reported for hay (MORAND-FEHR et al., 1978). GIGER et al. (1987a) showed that a refusal level of 15 % for lucerne hay improved by 0.6 point its measured organic matter digestibility. A similar trend has been observed with poor forages, thus OWEN et al. (1987) indicated that a 10 % increase in the refusal level of barley straw improved its ingestibility by about 3.8 g/kg LW and its digestibility by 1.5 points. This phenomenon is probably enhanced in range feeding conditions.

A close relationship between the offered and ingested quantities per time unit was demonstrated for the intake of tree leaves. Moreover, MEURET (1989) showed that the rate of intake (g DM/mn) varied inversely with the leaf biomass availability on oaks. This means that the length of time for grazing on the range has to be adapted to the quantity of potentially available edible biomass (see Chapter 14).

Concentrate allowances

Forage VDMI (F DMI) is also related to the level of concentrate intake (SAUVANT and MORAND-FEHR, 1977, 1978; MOWLEM et al., 1987). Figure 4 illustrates the general substitution phenomenon between concentrates and forage when the concentrate supplies vary. From these data, obtained from 22 experiments with Alpine or Saanen goats producing a mean of 2.60 kg 35 % FCM, it appears that a 100 g increase in concentrate DMI (C DMI) induced a decrease of 111 g F DMI, the subsequent substitution coefficient of 1.11 appearing in the prediction equation of forage DMI derived from the same data set.

F DMI = 395.3 + 328 RMY + 16.7 LW - 1.11 C DMI (5)

(n = 204; R = 0.86; RSD = 222 g).

This equation confirms the previous one proposed by SAUVANT and MORAND-FEHR (1981) with a smaller number of data (98 vs 204). In fact, when more than two levels of concentrate supply were compared within the same trial the substitution coefficient increased with the level of concentrate (MOWLEM et al., 1985), as clearly demonstrated with dairy cows (DULPHY et al., 1987).

It should be emphasized that when concentrate allowances were based on individual supply according to the milk energy output, the between-animal substitution coefficient was equal to 0. If this was not the case, the substitution coefficient was positive or negative when the supply of energy derived from concentrates was higher or lower than the energy requirement for milk production. The type of concentrate is also able to modify the substitution coefficient. Thus, GIGER et al. (1987b) showed in 6 trials that the same quantity of a fibrous concentrate led to a higher DMI (+ 6.26 g/kg $W^{.75}$, 135 g/goat/day) as compared with a starchy one (see Chapter 15).

Figure 4

SUBSTITUTION RATE BETWEEN FORAGE AND CONCENTRATE INTAKE (DMI) IN GOATS
(SAUVANT and GIGER, 1988)

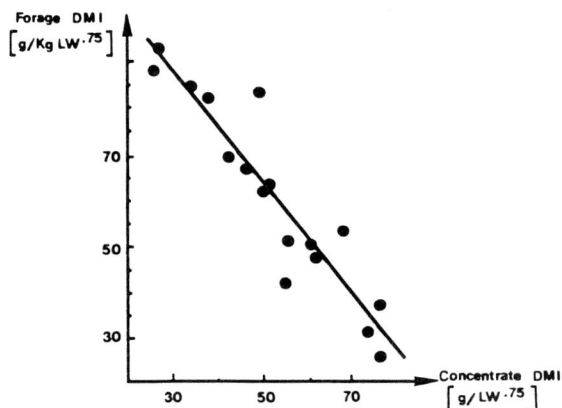

Diet energy density (ED)

When complete diets are given offered, which has become a current management procedure, it is recommended to consider the OM digestibility (OMD) or energy density as a predictive parameter of goat VDMI.

During early lactation, SAUVANT et al. (1987) observed that an enhanced diet OMD favoured an increase in VDMI from about 100 g/d per unit of OMD. In the same trial it was also demonstrated that the OMD increase simultaneously improved the nutritional energy status, most likely by the favorable cumulated effects of OMD and DMI increases.

When the diet energy density is large enough to satisfy energy requirements, the relationship between DMI and ED becomes negative. Thus, we observed a mean decrease of 66 g VDMI/d when the proportion of hay DM in the diet decreased by 10 % (SAUVANT and MORAND-FEHR, 1978). EL BEDDAWY (1985) observed the same trend under mediterranean conditions with Zaraibi goats, which consumed 3.9 to 5.3 kg DM/100 kg LW with a 40/60 roughage / concentrate ratio and only 3.6 to 4.2 kg for a 20/80 ratio. More recently using mixed diets with a mean OMD value of 69.1 % (SD = 5.5 %), corresponding to 2.42 Mcal ME/kg DM (SD = 0.23), GIGER (1987) and GIGER et al. (1988) observed a DMI decrease of 22 g/d per point of OMD increase. These data are comparable to those of the NRC (1986) which indicate 2 mean values of goat VDMI according to diet ED, i.e. : 1.09 and 0.91 kg/d for diets with an ED of 2.0 and 2.4 Mcal/kg DM, 59 and 69 % OMD or 1.15 and 1.45 Mcal milk NE. LU et al. (1987) reported several feeding trials which confirm these trends (Table 2).

Influence of the previous diet

This effect is particularly marked around parturition. At the onset of lactation, total VDMI is largely influenced by the proportion of hay intake in the diet during late pregnancy (SAUVANT et al., 1979). This mean effect is equal to 45 g total VDMI/100 g hay DMI and remains significant, at least during the first two months of lactation. This delayed influence suggests a lower level of rumen involution due to a more fibrous diet before parturition.

Table 2

INFLUENCE OF THE DIET ENERGY DENSITY
ON THE DAIRY GOAT INTAKE AND PERFORMANCE
(LU et al., 1987)

Parity	Primiparous		Multiparous	
	Energy density (Mcal ME/kg MS)			
	2.31	1.98	2.46	2.01
DM intake (kg/day)	2.02	2.63	2.76	3.61
ME intake (Mcal/day)	4.74	5.21	6.79	7.26
Theorical LW change (g/day)	64	53	77	85
4 % FC Milk yield (kg/day)	2.30	2.27	3.20	2.94

THE FILL UNIT SYSTEM FOR GOATS

To integrate the numerous sources of variation in VDMI through additive units, INRA proposed to use a Fill Unit System (FU). The first version (DEMARQUILLY et al., 1978, JARRIGE et al., 1979) was recently corrected and specially adapted to dairy cows through the milk FU (MFU, DULPHY et al., 1987). A main advantage of this FU system is to profit from the data bank of several hundreds of intake measurements of temperate and mediterranean forages obtained by DEMARQUILLY (INRA) with standard sheep.

Although the data are scarce on goat DMI with various diets, we proposed (MORAND-FEHR et al., 1987) to use the MFU for dairy goats on account of the similarity of production level (PL), based on energy (E maintenance + E production / E maintenance), between goats and cows. Thus, a 60 kg LW goat, producing 4.0 kg of 3.5 % FCMY has the same PL as a cow of 600 kg LW producing 25 kg of 4.0 % FCM chosen as a standard cow (DULPHY et al., 1987). This goat, also chosen as a standard, has a DMI level of 123 g/kg $W^{.75}$ of the standard grass which has a fill value of 1 MFU.

For other milk production levels, it was assumed that the marginal variations in feed intake capacity (IC) per kg milk yield (DIC/DMY) in goats (g) and cows (c) (DULPHY et al., 1987) were similar :

(DIC/DMY)c = 0.165 exp (- 0.02 MY 4)

(DIC/DMY)g = 0.165 exp (0.125 MY 3.5)

Variations in IC according to the physiological stage, live weight and level of production were established from our data (Table 3). Goats consumed rations composed of pasture hay or legume crops, maize silage or beet pulps, and concentrates based on cereals, oilmeals and by-products. The proportion of forages varied from 100 to 30 %.

Using the experimental values of the forage to concentrate substitution rate obtained with cows, Table 4 shows the fill values of concentrates according to milk yield, forage energy and fill values. With this system, the intake capacity (IC) of a goat (Table 3) is equal to the sum of the product terms between DMI and FU of the different ingredients (i) of the diet :

$$IC = \sum DMI_i \times FU_i$$

Such an equation can be associated with other equations related to constraints of energy, protein and minerals to calculate the VDMI values of a balanced diet. Nowadays with the INRATION programme (INRATION, 1988) it is possible to quickly perform these calculations on a PC type computer.

Table 3

VARIATIONS IN THE CAPACITY OF INTAKE OF THE ADULT GOAT

*	**Maintenance**	- Goat of 60 kg LW : 1.72 MFU
		- Variations for other LW : 0.10 MFU/10 kg LW
*	**Pregnancy**	- Months 1 to 4 : same values as maintenance
		- 5th month : maintenance value - 0,11 MFU
*	**Lactation**	- After the second month : 2.65 MFU for a 60 kg LW goat producing 4 kg of 3.5 % fat corrected milk (FCM)
		- Variation according to milk yield : 0.23 MFU per kg of 3.5 % FCM.
		- 1st week of lactation : previous value x 0.72
		- 2nd week of lactation : previous value x 0.83
		- 3rd week of lactation : previous value x 0.90
		- 4th week of lactation : previous value x 0.95

Table 4

FILL VALUE OF THE CONCENTRATE
ACCORDING TO THE MILK YIELD AND FORAGE FILL VALUE
(MORAND-FEHR, SAUVANT and BRUN BELLUT, 1987)

Forage energy value FU (1)	Forage fill value MFU/kg DM	Milk yield (kg 3.5 % FCM)				
		2	3	4	5	6
0.95	0.95	0.83	0.74	0.59	0.47	0.40
	1.15	0.78	0.61	0.49	0.41	0.37
0.90	0.95	0.79	0.67	0.51	0.44	0.37
	1.15	0.71	0.55	0.46	0.40	0.37
0.85	0.95	0.70	0.55	0.46	0.39	0.35
	1.15	0.62	0.50	0.43	0.39	0.35
0.80	1.00	0.63	0.51	0.42	0.37	0.33
	1.20	0.58	0.48	0.42	0.37	0.33
0.75	1.05 – 1.25	0.55	0.45	0.40	0.36	0.33
0.70	1.05 – 1.25	0.50	0.43	0.37	0.36	0.33
0.65	1.10 – 1.30	0.48	0.42	0.37	0.36	0.33

(1)   1 Feed Unit is equal to 7.11 Milk MJ

CONCLUSION

Experimental data provide reliable key values for assessing the influence of body size and production level of goats on their level of VDMI. Among the other animal factors, the effects of body condition and gut capacity need more information. The diet factors seem to be as important as in other ruminant females, however there is a lack of experimental data on goats in comparison with sheep and cattle. As a consequence, systematic studies on the influence of forage ingestibility, the refusal level and concentrate supply need to be made to obtain a better fit of the MFU system to goats.

Until now most data on goat feed intake have been obtained from milking animals housed under experimental conditions. As a large proportion of the goat flock worldwide is at least partly kept on rangeland, more attention should be paid to the VDMI level of these kinds of feeds and animals in relation to the natural environment of goats.

SUMMARY

The voluntary dry matter intake (VDMI) of goats decreases regularly before parturition and increases thereafter to reach a maximum between 6 and 10 weeks of lactation. In pregnancy, live weight and prolificacy positively influence goat VDMI. In lactation, more than 50 % of VDMI variations are explained by milk yield and 30 % by live-weight.

At the onset of lactation, a higher mobilization is associated with a significantly lower level of VDMI for the same liveweight and milk yield. But in pregnant and lactating goats, a high body fat content results in a lower feed intake capacity. Forage VDMI depends on forage species and quality and also on the level of concentrate intake.

The dietary energy density due to a high organic matter digestibility increases VDMI and milk yield in early lactation. But with low forage/concentration ratios, a reduction of hay percent in the diet can reduce VDMI.

This chapter suggests the application of the INRA Fill Unit system (FU) to goats and the use of the value of MFU (Milk Fill Unit) because the VDMI level in dairy goats and cows seems to be influenced by the same factors of variation under the same feeding conditions.

Keywords : Adult goat, Level of intake, Level of refusals, Fill Unit System, Forage choice, Concentrate allowances.

RÉSUMÉ

L'ingestion volontaire de matière sèche (IVMS) des chèvres décroît régulièrement avant la mise bas et croît ensuite pour atteindre un maximum entre 6 et 10 semaines de lactation. En gestation, le poids vif et la prolificité influencent positivement l'IVMS des chèvres.

Au début de la lactation, une mobilisation plus élevée s'accompagne d'un niveau d'IVMS significativement plus faible à mêmes poids vif et production laitière. Mais chez les chèvres en gestation et en lactation, une teneur lipidique élevée de l'organisme conduit à une capacité d'ingestion plus faible. L'IVMP de fourrage dépend de l'espèce végétale, de sa qualité et aussi du niveau de complémentation. La densité énergétique alimentaire liée à une digestibilité en matière organique élevée accroît l'IVMS et la production laitière en début de lactation. Mais, avec des rapports fourrage / concentré bas, une réduction du pourcentage de foin dans le régime peut réduire VDMI.

Ce chapitre propose d'appliquer le système des unités d'encombrement (UE) aux caprins en utilisant les valeurs des unités d'encombrement lait (UEL) parce que les niveaux d'IVMS des chèvres et des vaches laitières semblent être influencés par les mêmes facteurs de variation dans les mêmes conditions d'alimentation.

REFERENCES

BLANCHART G., BRUN BELLUT J. and VIGNON B., 1980. Comparaison des caprins aux ovins quant à l'ingestion, la digestibilité et la valeur alimentaire de diverses rations (Comparison of ingestion, digestibility and nutritive value of diets in sheep and goats). Reprod. Nutr. Dévelop., 20: 1731-1737.

CHILLIARD Y., 1985. Métabolisme du tissu adipeux, lipogenèse mammaire et activités lipoprotéine-lipasiques chez la chèvre au cours du cycle gestation-lactation (Adipose tissue metabolism, mammary lipogenesis and lipoprotein-lipasic activities in pregnant and lactating goats during gestation and lactation), Thèse de doctorat d'Etat, Université Paris 6 (FRANCE).

DEMARQUILLY C., ANDRIEU J. and SAUVANT D., 1978. Composition et valeur nutritive des aliments (Composition and nutritive value of feeds), p. 469-518. In: L'alimentation des ruminants. INRA Publ., Versailles (FRANCE).

DE SIMIANE M., GIGER S., BLANCHART G. and HUGUET L., 1981. Valeur nutritionnelle et utilisation des fourrages cultivés intensivement (Nutritive value and utilization of cultivated forages). Vol 1, p. 274-299. In: MORAND-FEHR P., BOURBOUZE A. and DE SIMIANE M. (Eds.): Nutrition and systems of goat feeding. Symposium International, Tours (FRANCE), May 12-15, 1981, INRA-ITOVIC, Paris (FRANCE).

DEVENDRA C. and BURNS M., 1983. Goat production in the tropics. Commonwealth Agricultural Bureaux, Slough (UK).

DULPHY J.P., FAVERDIN P., MICOL D. and BOCQUIER F., 1987. Révision du système des unités d'encombrement (UE) (Actualization of Fill Unit (FU) system). Bull. Techn. CRZV Theix, INRA, (70), 35-48.

EL BEDDAWY T.M., 1985. Nutrition and feeding system using different energy and roughage levels for milk and meat production by goats. Ph. D. Thesis, Cairo University (EGYPT).

GIGER S., 1987. Influence de la composition de l'aliment concentré sur la valeur alimentaire des rations destinées au ruminant laitier (Effect of concentrate composition on nutritive value of diets for dairy ruminants). Thèse de docteur-ingénieur, INAPG, Paris (FRANCE).

GIGER S., SAUVANT D., HERVIEU J. and DORLEANS M., 1987a. Valeur alimentaire du foin de luzerne pour la chèvre (Nutritive value of lucerne hay in goats). Ann. Zootech., 36: 129-152.

GIGER S., SAUVANT D. and HERVIEU J., 1987b. Influence de la nature de l'aliment concentré complémentaire sur l'encombrement de rations de la chèvre en lactation (Effect of kind of concentrates on diet filling in dairy goats). Reprod. Nutr. Dévelop. 27: 201-202.

GIGER S., SAUVANT D., HERVIEU J. and DORLEANS M., 1988. Influence of diet type on the level of voluntary dry matter intake by dairy goats. Seminar of FAO Subnetwork on Goat Nutrition and Feeding, Oct. 3-5, 1988, Potenza (ITALY).

INRATION 1988. Computer program of diet calculation based on the 1988 recommandations. INRA, Paris (FRANCE).

JARRIGE, 1979. Le système des unités d'encombrement pour les bovins (Fill Unit system for cattle). Bull. Tech. CRZV, Theix, INRA, (70), 213-22.

LU C.D., SAHLU T. and FERNANDEZ J.M., 1987. Assessment of energy and protein requirements for growth and lactation in goats. Vol. 2, p. 1229-1247. 4th Intern. Conf. on Goats, March 8-13, 1987, Brasilia (BRAZIL).

Mac KENZIE D., 1970. Goat husbandry. Faber & Faber, London (UK).

MEURET M., 1989. Valorisation par les caprins laitiers de rations ligneuses prélevées sur parcours (Utilization of lignous diets from rangelands for dairy goats). Thèse de Docteur en sciences agronomiques, Faculté des sciences agronomiques, Gembloux (BELGIUM).

MORAND-FEHR P., SAUVANT D. and DE SIMIANE M., 1978. La consommation alimentaire des chèvres laitières : comportement alimentaire, capacité d'ingestion et niveau d'ingestion des aliments (Ingestion of dairy goats : Feeding behaviour, capacity and level of ingestion), p. 54-72. 4e journées de la recherche ovine et caprine, INRA-ITOVIC, Paris (FRANCE).

MORAND-FEHR P., GIGER S., SAUVANT D. and BROQUA B. and DE SIMIANE M., 1987. Utilisation des fourrages secs par les caprins (Dry forage utilization for goats), p. 391-422. In: DEMARQUILLY C. (Ed.): Les fourrages secs : récolte, traitement, utilisation. INRA Publ., Paris (FRANCE).

MOWLEM A., OLDHAM J. and NASH S., 1985. Effect of concentrate allowance on *ad libitum* hay consumption by lactating British Saanen goats. Ann. Zootech. 34: 474 (Abst.).

OWEN E., WAHED R.A. and ALIMON R., 1987. Effect of amount offered on selection and intake of long untreated barley straw by goats. Ann. Zootech., 36: 324-325.

RANDY H.A. and SNIFFEN C.J., 1982. Dairy goat dry matter intake and forage preference as influenced by stage of lactation, p. 334. 3th Intern. Conf. on Goat Production and Disease, Jan. 10-15, 1982, Tucson, Arizona (USA).

SAUVANT D., 1978. La capacité d'ingestion de la chèvre (Capacity of ingestion in goats) p. 55-69. In: CAAA, L'alimentation des caprins, ADEPRINA, Paris (FRANCE).

SAUVANT D. and MORAND-FEHR P., 1981. Prediction of voluntary food intake of goat during lactation. Intern. Goat Sheep Res., 1: 274-281.

SAUVANT D. and MORAND-FEHR P., 1977. Influence du niveau d'apport d'aliments concentrés en pleine lactation sur les performances de la chèvre (Effect of concentrate supplies on performances of goats in mid-lactation), p. 174-183. Symp. on Goat Breeding in Mediterranean Countries, Oct. 3-7, 1977, Malaga, Grenada, Murcia (SPAIN).

SAUVANT D. and MORAND-FEHR P., 1978. Adaptation du niveau des apports d'aliments concentrés au stade physiologique de la chèvre (Concentrate supplies related with physiological status of goats), p. 93-115. 4e journées de la recherche ovine et caprine, INRA-ITOVIC, Paris (FRANCE).

SAUVANT D., HERVIEU J., CHILLARD Y. and MORAND-FEHR P., 1979. Facteurs influençant la quantité de matière sèche ingérée par la chèvre en début de lactation (Factors influencing the level of dry matter intake of goats in early lactation), 30th EAAP Annual Meeting, Harrogate (UK).

SAUVANT D., 1981. Alimentation énergétique des caprins (Energy nutrition of goats), Vol. 1, p. 55-79. In: MORAND-FEHR P., BOURBOUZE A. and DE SIMIANE M. (Eds.): Nutrition and systems of goat feeding, Symposium International, Tours (FRANCE), May 12-15, 1981, INRA-ITOVIC, Paris (FRANCE).

SAUVANT D., HERVIEU J., GIGER S., TERNOIS F. and MANDRAN N., MORAND-FEHR P., 1987. Influence of the diet organic matter digestibility on the goat nutrition and production at the onset of lactation. Ann. Zootech., 37: 335 (Abst.).

Chapter 4

WATER METABOLISM AND INTAKE IN GOATS

S. GIGER-REVERDIN and E. A. GIHAD

INTRODUCTION

Water is quantitatively the most important body constituent as it makes up about two thirds of the body mass of animals. As life is impossible without water, water consumption is a key-point in animal feeding, particularly in arid regions. The ready availability of water has caused many nutritionists to overlook the real value of this indispensable nutrient. However, the metabolism of water is complex and difficult to understand.

Water requirements of goats are a function of water input and output. Studies concerning water intake and metabolism are both scarce and very heterogenous as goats are farmed under a variety of conditions. Some studies investigate intensive production in temperate climates (GIGER et al., 1981), most of them are related to requirements of animals living under desert conditions in which animals survive in arid conditions and consume feeds of very low nutritive value leading to a low level of production (KHAN and GHOSH, 1981 ; SHKOLNIK and SILANIKOVE, 1981). Moreover, studies involve not only heterogenous climatic conditions, but also different breeds of goats. Publications concerning water utilization in temperate climates are far less common than publications on water utilization in arid zones.

In order to better understand the water requirements of goats, the most recent data about water metabolism will be summarized before studying factors concerning variations in water requirements and the effects of water restriction on zootechnical performances.

WATER METABOLISM

Water content of goat body

As in other ruminant species, the water content of the goat body varies considerably depending on the age and amount of fat of the animal. However, water content (%) of the fat-free adult body is relatively constant, whatever the age, averaging from 71 to 73 % of body weight (BAS, pers. comm.) as in other species (CHURCH, 1974).

Body water may be divided into three parts:
- intracellular water representing about two thirds of the total water,
- extracellular water including blood and lymphatic fluid,
- interstitial water.

The quantity of intracellular water cannot be altered without lethal consequences.

Water inputs and outputs

Water inputs have three origins:

- water drunk,
- water contained in the feed ingested,
- metabolic water resulting from oxidation of energy sources.

The input of metabolic water which is derived from digested nutrients may be calculated by the indirect method of TAYLOR (1970), which has been proposed for East African ungulates living in desert conditions :

Metabolic water (ml) = total digested carbohydrates (g) x 0.62 + total digested fat (g) x 1.1 + total digested protein (g) x 0.42.

Water outputs also have three origins:

- water loss in faeces and urine linked to the digestive and metabolic utilization of feed,
- evaporative loss of water through the lungs and body surfaces, which is very seldom measured,
- water retention and output like in the foetuses or in milk production.

The division of water losses into these three compartments is variable depending upon production (milk production and faecal dry matter), the environmental temperature (evaporative loss of water), and renal output.

Regulation of water balance

Under normal conditions, water balance is zero which means that water input is equal to water output. Water loss in faeces depends upon the level of non digestible organic matter, and therefore upon both the level of dry matter intake and digestibility of the diet. The physiological status of the animal, however, has a significant influence. Lactating goats have about 7 % less dry matter in faeces than dry goats (GIGER et al., unpubl.). This observation is in agreement with the work done by FORBES (1968) in ewes, which suggests that in lactation and especially in the early part, lactating animals have a higher metabolic rate than dry ones, and so loose more water per unit of dry matter intake for vaporization and excretion. It is also very difficult in this case to impute the influence linked to the physiological status of the animal to its level of production or intake, and to the diet composition.

Urine concentration is tightly regulated by several mechanisms including water input, sodium ion concentration in plasma and blood pressure and is under the control of hormones. Water loss is also related to the amount of metabolites generated during normal body functions.

The kidneys control water losses in two ways:

- by controlling the absolute urine concentration and
- by reduction in urine volume.

Evaporative losses of water through the lungs and the skin depend on environmental temperature and respiratory rate. As the water concentration in milk is fairly constant, water loss in milk is a direct function of the level of production.

For goats, as for other ruminants renal water balance mechanisms are tightly linked to electrolyte concentration. Restriction in water intake generally causes water resorption and retention of sodium (Na^+) with excretion of potassium (K^+), in order to maintain plasma osmolality (WITTENBERG et al., 1986) and homeostasis (KHAN et al., 1978).

Both thirst and water homeostasis are maintained by the nervous system and by hormonal regulation, particularly through the action of vasopressin, plasma renin activity and plasma aldosterone at the level of the renal tubule (DAHLBORN and KARLBERG, 1986 ; WITTENBERG et al., 1986 ; DAHLBORN et al., 1988). Lactating goats become dehydrated more rapidly than non-lactating ones during water deprivation, due to loss of water into the milk. When goats were given saline water, plasma volume increased with plasma sodium concentrations, but plasma renin activity and plasma aldosterone levels decrease (DAHLBORN and KARLBERG, 1986).

Heat does not significantly influence renal hormones as vasopressin or aldosterone (AUGUSTINSSON et al., 1986), although rectal temperature and respiratory rate increase and respiratory alkalosis may follow.

Black Bedouin goats which are used to live in desert conditions possess a unique adaptive mechanism of water regulation. They can loose a large portion of their rumen water (EL HADI, 1986) and then drink large quantities of water (up to 40 % of their live-weight) (SHKOLNIK and SILANIKOVE, 1981), which are first retained for up to several hours in the rumen. The water is then distributed in a manner which prevents over-hydration, 85 % being retained in the rumen of a fasting goat even 7 h following drinking (SHKOLNIK and SILANIKOVE, 1981). During these periods of large variations in water input, body dry matter content and blood volume remain constant. There is no increase in urinary water excretion or significant change in blood osmotic concentration during rehydration (CHOSHNIAK and SHKOLNIK, 1978 ; DAHLBORN and KARLBERG, 1986). Moreover, sudden rehydration does not alter rumen microflora digestive functions (BROSH et al., 1983).

WATER   INTAKE

Water intake (WI) is the sum of water consumed voluntarily by drinking and water contained in the feeds. In many environments, the water intake through forage may be high relative to other species because of their ability or willingness to browse. The result is that goats are less dependent on free water sources than other domestic species, but do not equal certain wild animal species in this respect (NRC, 1981). Water consumption of goats is influenced by factors related to the main losses of water including absorption of nutrients from the digestive tract, hydrolytic processes of digestion acting as a solvent, excretion of metabolic wastes and toxins, synthesis, secretion and metabolism of products including enzymes, hormones, peptides and biochemical substances, thermoregulation, cellular osmotic pressure which gives their shapes to the cells, chemical solvent for biochemical reactions.

Diet factors

Dry matter intake

Dry matter (DMI) and water intakes (WI) are highly correlated (GIGER, 1987). When goats are deprived of feed for slightly more than 24 h (28 h., see DAHLBORN and KARLBERG, 1986), water intake decreases as does the production of urine and milk.

Dry matter percentage

Water intake is strongly influenced by the water content of feeds. The dietary percentage of dry matter (% DM) is a good additional predictor of intake in lactating goats, in temperate conditions (GIGER, 1987):

$$\text{WI (g/kg W}^{.75}) = \quad 2.98 \text{ DMI (g/kg W}^{.75}) + \quad 0.854\ \% \text{ DM}$$
$$(0.13) \qquad\qquad\qquad (0.209)$$
$$(r = 0.82,\ n = 171,\ RSD = 54)$$

This equation has been obtained with lucerne hay and/or maize silage as basal diet, generally in association with a compound feed. It may not be used for green forages in which the dry matter content is lower than 20 %.

When % DM is less than 30 %, goats do not need to drink in order to satisfy their water intake needs (GHOSH, 1982), but actually they do drink when water is available.

Mineral and protein content

Many minerals influence water intake, but sodium chloride (NaCl), either added to feed or distributed as saline water, is the most significant as it will increase water intake.

Rich protein diets also increase water intake in goats as in other ruminants (SENGAR and MUGDAL, 1982). High protein diets generate increased catabolic by-products including minerals and nitrogenous end products, such as urea for which water serves as a solvent, thereby enhancing the water secretion.

Taste factors

Taste factors will also affect normal water intake. But, goats are more tolerant to high concentrations of some taste stimulants (acetic acid, quinone hydrochloride, sucrose and sodium chloride, i.e. bitter, sour, sweet and salty taste, respectively) than sheep or cattle (GOATCHER and CHURCH, 1970).

Animal factors

Live-weight

Water intake may be expressed either as kg of water per kg of dry matter or as kg of water per kg of metabolic body weight. The latter formula is most useful when comparing water requirements between species. However, there is some controversy concerning the exponent of metabolic body weight. Some investigators prefer $W^{.75}$, while others use $W^{.82}$ (GHOSH, 1987 ; GIHAD et al., 1988).

Physiological status and level of production

In temperate climates, water intake for maintenance is 107 g/kg $W^{.75}$ in dry and non-pregnant goats (GIGER, 1987). Pregnancy increases the intake to about 139 g/kg $W^{.75}$ at mid-pregnancy, indicating that pregnancy creates specific water needs observed by GHOSH and MOITRA (1984) in Beetal goats or by GIGER (1987) in Alpine and Saanen breeds. The water intake during pregnancy increases by 25 % at the third month and by 40 to 50 % at the end of pregnancy. The status of lactation also increases water needs, at a mean level of 165 g/kg $W^{.75}$ for goats producing about 148 g milk / kg $W^{.75}$ 10 weeks after parturition (GIGER, 1987).

For each kg of milk produced, goats ingest about 1.28 kg of water (GIGER, 1987). This value is higher than the water output of milk (0.89 kg H_2O/kg of milk). This indicates that metabolism increases in pregnant and lactating animals and that more catabolites are excreted through the kidneys. This is similar to other ruminants (FORBES, 1968 ; DAVIES, 1972).

From a practical point of view, it is important to stress the fact that at parturition, water intake may increase two or three fold on a single day.

Environmental factors

Temperature

Goats adapt to high temperatures more easily than other domestic ruminants (SCHOEN, 1968). They also appear to be less subject to high temperature stress than other species of domestic livestock and require less water evaporation to control body temperature (NRC, 1981). They keep the same dry matter intake level, reduce their water losses in urine and faeces and increase their pulmonary ventilation. Therefore, water intake increases markedly at temperatures above 30°C. Water intake (kg/kg DM) is approximatively of 3.15 kg/kg DM at 23°C, 3.14 kg/kg DM at 30°C, 4.71 kg/kg DM at 35°C and 4.85 kg/kg DM (GHOSH, 1982). CONSTANTINOU (1987) indicates that the upper limit of the thermoneutral zone for Damascus goats living in Cyprus is between 25 and 30°C, but may be higher for some "desert breeds" such as Nubian goats. Water consumption is considerably reduced when temperature is below 5°C, as observed in winter by Norwegian authors. Losses related to respiratory evaporation may be half the total water losses in some tropical conditions and may exceed the sum of urinary, faecal and milk losses (MALTZ et al., 1982).

Frequency of drinking

The frequency of drinking is of practical importance as watering interval in some middle-eastern desert countries may range from 3 to 7 days, depending on the season, availability of water and the distance travelled between the watering points (EL HADI, 1986). Lack of water modifies water equilibrium in the body: cell and gut water decrease considerably during water deprivation which prevent blood from thickening and allow to dissipate the internal body heat through the circulatory medium (KHAN and GHOSH, 1981). Therefore, water deprivation may have an influence on milk yield and composition, as studied by MALTZ and SHKOLNIK (1984): these parameters have been measured on Black Bedouin goats during 4 days of dehydration followed by 2 days of rehydration. Milk was maintained during the first two days of dehydration, fell to 35 % of the initial value over the third and fourth days and recovered fully during the 2 days of rehydration. As milk yield fell, milk osmolality, and milk fat and protein concentrations rose. Finally, production was about 70 % of the normal one with significant consequences on the growth of kids. A thrice weekly watering allowance is significantly better than a twice weekly (MITTAL, 1985).

Even if it is not of the same importance under temperate climates, a twice daily allowance of water seems to be necessary for animals at a mean level of production. High yielding animals need to drink more often and present a high inter-individual variety. With an ad libitum supply, the animals use to drink just after milking and at the same time as they eat.

WATER RESTRICTION

A decrease in water intake, even for a single day, gradually reduces secretion of urine and milk (DAHLBORN and KARLBERG, 1986). Breeds well-adapted to desert conditions, such as Black Bedouin goats, may loose up to 25 or 30 % of their live-weight, if they drink only once every four days. Even under these adverse conditions, goats may still produce 1 or 2 kg of milk and they continue to eat normally. Saanen goats are more sensitive than Black Bedouin goats to the effect of dehydration, as they may significantly decrease their dry matter intake and their level of production under similar conditions (SILANIKOVE, 1985). This example suggests that the different genotypes may differ in their ability to meet water requirements (NRC, 1981).

During periods of reduced water intake, goats may decrease the amount of dry matter ingested, especially with food of good quality. The ingested material is retained in the rumen for longer periods of time resulting in increased nitrogenous digestion due to enhanced bacterial activity and increased mixing of rumen contents (BROSH et al., 1986 ; BROSH et al., 1987).

WATER SUPPLY

In temperate zones water is abundant and inexpensive, while it is scarce in arid climates. Consideration must be given to not only the amount of water available, but to the frequency of water availability and water purity.

Drinking water from open ponds, reservoirs and wells must be tested for impurities and contaminations. Water from approved municipal facilities may be excluded from testing. Good water is colorless and tasteless. The presence of organic nitrogen in water indicates decay of organic matter. Excess nitrates in water may be toxic.

Goats also are quite sensitive to the tidiness of troughs. In some cases, it happens that they do not drink at all what influences their performances. Therefore, goats well-adapted to some dry tropical climates may drink some briny water containing up to 10 % of minerals. These goats use to live near salted lakes bordering the deserts.

CONCLUSION

Regarding herd management in intensive milk production systems, it is not possible to have high levels of production without offering water *ad libitum*. However, in areas where water is scarce, attempts must be made to maximize production, while utilizing water as efficiently as possible and still meet the requirements of the animal. Some breeds have a good ability of adaptation to some very hard living conditions as deserts or dry areas.

SUMMARY

Nutritional biochemistry of water which is the most important body constituent in goats as in other animals, is complex. Moreover, studies on this topic are scarce and very heterogenous, as some concern goats herded in intensive production in temperate climates, but most of them are related to animals surviving in arid conditions.

Therefore, the main factors affecting water metabolism and free water intake of goats are analogous to other ruminants: dry matter intake (water consumption is about three times the dry matter intake), composition of the feeds and especially their water, salt and mineral contents, taste factors, weight, level of production (1.28 kg/kg of milk) and physiological status of the animals (maintenance consumption expressed in g/kg $W^{.75}$ is 107 for a dry and non-pregnant goat, 140 at mid-pregnancy and 165 at mid-lactation) and environmental factors. Regulation of water balance is under the control of hormones and of the nervous system and is also linked to plasma ion concentration. Acute problems occur when goats are unable to maintain water balance or to control body temperature.

Animals which are used to live in desert conditions possess a unique adaptative mechanism of water regulation. Goats seem to be less sensitive to water deprivation than other ruminants and some breeds can adapt quite well to some desert conditions.

Water of good quality is necessary to reach high levels of production, but in areas where it is scarce, it is necessary to use it as efficiently as possible and to maximize production.

Keywords : Goat, Water metabolism, Water intake, Frequency of drinking, Water restriction, Water supply.

RÉSUMÉ

L'eau qui est, chez la chèvre comme chez les autres animaux, le constituant quantitativement le plus important, présente un métabolisme complexe. Les études concernant ce sujet sont peu nombreuses et très hétérogènes: certaines concernent des animaux élevés en système intensif et en climat tempéré, mais la plupart ont trait à des animaux vivant en zones arides.

Cependant, les principaux facteurs de variation du métabolisme hydrique et de la consommation en eau sont les mêmes pour les chèvres que pour les autres ruminants: la quantité de matière sèche ingérée (la consommation en eau est égale à environ trois fois la quantité de matière sèche ingérée), la composition chimique de la ration, et plus particulièrement sa teneur en eau, en sels et en minéraux, le poids vif, le niveau de production (1,28 kg d'eau/kg de lait) et le stade physiologique des animaux (la consommation exprimée en g/kg $P^{0.75}$, et correspondant à l'entretien, est de 107 pour un animal tari et vide, de 140 en milieu de gestation et de 165 en milieu de lactation) ainsi que les conditions climatiques. La régulation du métabolisme hydrique est sous contrôles neural et hormonal, et est liée à la concentration en ions du plasma. De sérieux problèmes peuvent apparaître, quand les chèvres ne peuvent plus contrôler leur bilan hydrique ou leur température corporelle.

Les animaux, adaptés à vivre dans les conditions désertiques, possèdent des mécanismes spéciaux d'adaptation de leur système de régulation. Les chèvres semblent moins sensibles à une restriction en eau que les autres ruminants, et certaines espèces peuvent s'adapter relativement bien à des conditions désertiques.

Une eau de bonne qualité est nécessaire à l'obtention de hauts niveaux de production, mais dans les régions où elle est rare, il est nécessaire d'optimiser son utilisation afin de maximiser les productions obtenues.

REFERENCES

AUGUSTINSSON O., HOLST H., FORSGREN M., ANDERSSON H. and ANDERSSON B., 1986. Influence of heat exposure on acid-base and fluid balance in hyperhydrated goats. Acta physiol. scand. 126: 499-503.

BROSH A., CHOSHNIAK I., TADMOR A. and SHKOLNIK A., 1986. Infrequent drinking, digestive efficiency and particle size of digesta in Black Bedouin goats. J. Agric. Sci. (Camb.) 106: 575-579.

BROSH A., SHKOLNIK A. and CHOSHNIAK I., 1987. Effects of infrequent drinking on the nitrogen metabolism of Bedouin goats maintained on different diets. J. Agric. Sci. (Camb.) 109: 165-169.

BROSH A., SNEH B. and SHKOLNIK A., 1983. Effect of severe dehydration and rapid rehydration on the activity of the rumen microbial population of Black Bedouin goats. J. Agric. Sci. (Camb.) 100: 413-421.

CHOSHNIAK I. and SHKOLNIK A., 1978. The rumen as a protective osmotic mechanism during rapid rehydration in the Black Bedouin goat. In: BENZON A. (Ed.): Osmotic and volume regulation. Symposium XI, Munksgaard, 1978.

CHURCH D.C., 1974. Water metabolism and requirements, Vol. 2, p. 401-412. In: CHURCH D.C. (Ed.): Nutrition: Digestive physiology and nutrition of ruminants. O. & B. Books, Corvallis, Oregon (USA).

CONSTANTINOU A., 1987. Goat housing for different environments and production systems, p. 241-268. 4th Intern. Conf. on goats. March 8-13, 1987, Brazilia (BRAZIL).

DAHLBORN K. and KARLBERG B.E., 1986. Fluid balance during food deprivation and after intraruminal loads of water or isotonic saline in lactating and anoestral goats. Q. J. exp. physiol. 71: 223-233.

DAHLBORN K., HOLTENIUS K. and OLSSON K., 1988. Effects of intraruminal load of saline or water followed by voluntary drinking in the dehydrated lactating goat. Acta physiol. scand. 132: 67-73.

DAVIES P.J., 1972. A note on the water intake of ewes in late pregnancy and early lactation. Anim. Prod. 15: 307-310.

EL HADI H.M., 1986. The effect of dehydration on Sudanese desert sheep and goats. J. Agric. Sci. (Camb.) 106:17-20.

FORBES J.M., 1968. The water intake of ewes. Br. J. Nutr. 22: 33-43.

GHOSH T.K., 1982. Ph. D Thesis, Fac. Vet. BCKV, India. Cited by GHOSH, 1987.

GHOSH T.K., 1987. Water requirement and water feeding strategies in goats, Vol. 2, p. 1267-1274. 4th Intern. Conf. on goats, Mar. 8-13, 1987, Brasilia (BRAZIL).

GHOSH T.K. and MOITRA D.N., 1984. Studies on nutrient utilisation and water requirements by pregnant and dry Black Bengal does. Kerala J. Vet. Sci. 15: 131-134.

GIGER S., 1987. Influence de la composition de l'aliment concentré sur la valeur alimentaire des rations destinées au ruminant laitier (Effect of concentrate composition on nutritive value of dairy ruminant diets). Thèse de Docteur Ingénieur, INAPG, Paris (FRANCE).

GIGER S., HERVIEU J., SAUVANT D. and MORAND-FEHR P., 1981. Facteurs de variation et modèles de prévision de l'ingestion d'eau par la chèvre en lactation en climat tempéré (Factors of variation and prevision models of water intake of lactating goats in temperate conditions), Vol. 1, p. 254-262. In: MORAND-FEHR P., BOURBOUZE A. and DE SIMIANE M. (Eds.): Nutrition and systems of goat feeding. Symposium International, Tours (FRANCE), May 12-15, 1981, INRA-ITOVIC, Paris (FRANCE).

GIHAD E.A., EL GALLAD T.T., SOUD A., FARID M.A. and ABOU EL NASR H., 1988. Feed and water intakes by camel, sheep and goats fed low protein agro-industrial desert by-products. In: Proc. 1st Intern. Seminar of Camel Nutrition, Mar., 1988, Ourgla (ALGERIA).

GOATCHER W.D. and CHURCH D.C., 1970. Taste reponses of ruminants. J. Anim. Sci. 31: 373-382.

KHAN M.S. and GHOSH P.K., 1981. Water use efficiency in the Indian desert goat, Vol. 1, p. 249-253. In: MORAND-FEHR P., BOURBOUZE A. and DE SIMIANE M. (Eds.): Nutrition and systems of goat feeding, Symposium International, Tours (FRANCE), May 12-15, 1981, INRA-ITOVIC, Paris (FRANCE).

KHAN M.S., GHOSH P.K. and SASIDHARAN T.O., 1978. Effect of acute water restriction on plasma proteins and on blood and urinary electrolytes in Barmer goats of the Rajasthan desert. J. Agric. Sci. (Camb.) 91: 395-398.

MALTZ E. and SHKOLNIK A., 1984. Milk composition and yield of the Black Bedouin goat during dehydration and rehydration. J. Dairy Res. 51: 23-27.

MALTZ E., SILANIKOVE N. and SHKOLNIK A., 1982. Energy cost and water requirements of Black Bedouin goats at different levels of production. J. Agric. Sci. (Camb.) 98: 499-504.

MITTAL J.P., 1985. Impact of water stress on goat production under Indian arid zone, p. 796-798, Vol. 2. Proc. 3rd AAAP Anim. Sci. Congress, May 6-10, 1985, Seoul (COREA).

NRC, 1981. Nutrient requirements, p. 2-9. In: Nutrient requirements of goats: Angora, dairy and meat goats in temperate and tropical countries. National Academy Press, Washington, DC (USA).

SCHOEN A., 1968. Studies on the water balance of the East African goats. East Afr. Agric. Forest. J. 36: 256-262.

SENGAR S.S. and MUGDAL V.D., 1982. Effect of feeding treated and untreated proteins on the growth and milk production in goats ; effect on feed utilization, and quality and quantity of milk produced. Indian J. Anim. Sci. 52: 1047-1051.

SHKOLNIK A. and SILANIKOVE N., 1981. Water economy energy metabolism and productivity in desert ruminants,Vol. 1., p. 236-248. In: MORAND-FEHR P., BOURBOUZE A. and DE SIMIANE M. (Eds.): Nutrition and systems of goat feeding, Symposium International, Tours (FRANCE), May 12-15, 1981, INRA-ITOVIC, Paris (FRANCE).

SILANIKOVE N., 1985. Effect of dehydration on feed intake and dry matter digestibility in desert (Black Bedouin) and non-desert (Swiss Saanen) goats fed on lucerne hay. Comp. Biochem. Physiol. A. 80: 449-452.

TAYLOR R., 1970. Strategies of temperature regulation: effect on evaporation in East African ungulates. Am. J. Physiol. 219: 1131-1135.

WITTENBERG C., CHOSHNIAK I., SHKOLNIK A., THURAU K. and ROSENFELD J., 1986. Effect of dehydration and rapid rehydration on renal function and on plasma renin and aldosterone levels in the Black Bedouin goat. Pfluegers Arch. 406: 405-408.

Chapter 5

DIGESTION IN GOATS

J.L. TISSERAND, M. HADJIPANAYIOTOU and E.A. GIHAD

INTRODUCTION

The development of both intensive (Chapter 17) and extensive (Chapter 19) production systems attests to the adaptability of goats to a very wide variety of climate and vegetation. The goat is considered as a unique animal for efficient production in areas where other domestic ruminants cannot produce adequately (DEVENDRA, 1980; DEVENDRA and BURNS, 1980). In temperate areas goats receive good quality forage supplemented with small quantities of concentrates (MORAND-FEHR and SAUVANT, 1978), while in semi-arid areas high-yielding goats eat concentrate rich diets (HADJIPANAYIOTOU, 1987). Although, goats are well adapted to a broad variety of environmental conditions, their nutrient requirements, voluntary feed intake and the physiology of digestion process have not yet been studied extensively (RAUN, 1982). Research efforts have recently been undertaken by teams participating in the FAO European research network on goats (BLANCHART et al., 1980; MORAND-FEHR, 1981; LOUCA et al., 1982; HADJIPANAYIOTOU and ANTONIOU, 1983; ALRAHMOUN et al., 1986; MASSON et al., 1986, 1989; TISSERAND et al., 1986) aimed at the better understanding of the digestion process in goats. The objective of this chapter is to review data on goats digestion obtained by these teams and compare them with those obtained from other places and with other ruminant species.

ANATOMY OF THE DIGESTIVE SYSTEM

The digestive tract of goats is very similar to that of other ruminants (CASTLE, 1956; CHURCH, 1976), accounting for 7-8 per cent of the weight of an adult goat. The mean size of the different compartments is as percentage of the total digestive tract: stomach 48 %, small intestine 30 %, large intestine 22 %. In comparison to the abomasum, the reticulo-rumen is larger in goats than in sheep (Table 1).

It must be underlined however, that the volume and weight of the stomachs vary with the level of feed intake, diet composition and feeding behavior (CHURCH, 1976). There are no significant differences between the rumen volume of sheep and goats receiving the same food (MALOIY, 1974; ALRAHMOUN, 1985) (Table 2). The total length of the digestive tract of goats (about 43 m) is comparable to that of sheep (SPEDDING, 1975) but is less than that of cattle (53 m) (CHURCH, 1976). The rumen of goats is longer and narrower than that of sheep. Rumen papillae are more broad like leaves in goats, and in sheep more narrow like tongues (BATTACHARYA, 1980).

Table 1

VOLUMES OF DIFFERENT STOMACH COMPARTMENTS RELATIVE TO WHOLE STOMACH AND OF ALL INTESTINES RELATIVE TO THE WHOLE DIGESTIVE TRACT

Species	Reticulum(%)	Rumen (%)	Omasum (%)	Abomasum (%)	All intestines	References
Sheep	11	62	5	22	51-56	CHURCH (1976)
Cows		64 **	25	11	37.7	CHURCH (1976)
Goats :						
- Saanen		84 **	4	12	52.6	TAMADA (1973)
- Angora*	9	76	4	11		BATTACHARYA (1980)
- Kil*	8	77	4	11		BATTACHARYA (1980)

(*)   Based on water-fill volumes ; the other values on the basis of fresh weight.
(**)   Reticulo-rumen.

Table 2

VOLUMES* OF THE RETICULO-RUMEN OF GOATS AND SHEEP RECEIVING DIFFERENT DIETS (ALRAHMOUN, 1985)

	Species	Pasture hay	NaOH treated straw	NaOH treated straw + soya oilcake	NaOH treated straw + urea
Volume (1)	Goats	10.4 ± 1.6	10.2 ± 0.6	10.5 ± 1.0	10.0 ± 2.1
	Sheep	8.4 ± 0.2	10.2 ± 0.3	10.9 ± 0.4	8.6 ± 0.9
Volume (1/kg W$^{.75}$)	Goats	0.5 ± 0.1	0.5 ± 0.1	0.5 ± 0.1	0.5 ± 0.1
	Sheep	0.4 ± 0.1	0.6 ± 0.1	0.6 ± 0.1	0.5 ± 0.1

(*)   Measured with polyethylene glycol (HYDEN, 1955).

TRANSIT

There are no differences in the mean retention time (MRT) of feed particles in the digestive tract of goats and sheep receiving the same quantity of a good quality forage. When sheep and goats eat a tropical forage (DEVENDRA, 1981) or straw (MASSON et al.,1986), the MRT in the entire digestive tract is higher in goats than in sheep (Table 3).

The transit time in the reticulo-rumen is particularly high in this case (MASSON et al., 1986). The overall retention time in the whole tract is thus higher in goats than in sheep (Table 3).

Table 3

MEAN RETENTION TIME (h.) OF DIGESTA IN THE DIGESTIVE TRACT OF SHEEP AND GOATS

Species	Method	Diet	Reticulo-rumen	Whole digestive tract	Reference
Goats	Stained particles	Straw + concentrate	22	32-45	CASTLE (1956)
Goats	} Cr_2O_3	Tropical hay	-	114	DEVENDRA (1981)
Sheep			-	81	
Goats	} Stained particles	Concentrate + urea	-	40	SREEMANNARAYANA and MAHAPATRO(1979)
Sheep			-	39	
Goats	} Cr_2O_3	Alfalfa hay	-	43-78	HARYANTO and JOHNSON (1983)
Sheep			-	50-74	
Goats	} Labelled forage	Alfalfa hay + cocksfoot	27	42	MASSON et al. (1986)
Sheep			30	41	
Goats	} Labelled forage	NaOH treated straw	58	73	MASSON et al. (1986)
Sheep			62	67	
Goats	}	Low quality hay	28	-	VAN SOEST (1982)
Sheep			35	-	
Goats	}	Savana range	38	-	VON ENGELHARDT et al. (1986)
Sheep			46		

In goats as in other ruminants, the MRT in the digestive tract and in the reticulo-rumen decreases as the level of intake increases (LINDBERG, 1988).

In the case of Bedouin goats receiving alfalfa, the MRT varies with the drinking frequency. MRT of Cr mordanted digesta in the rumen, and the entire gut of goats was 37.8 ± 73 and 26.4 ± 3.4 and 63.1 and 47.3 when water was given either one every 4 days or once daily respectively (BROSH et al., 1986b). Nevertheless, BROSH et al. (1989) showed that the drinking rhythm affected the MRT more with an alfalfa hay-based diet than with a straw-based diet.

LEVEL OF INTAKE

Although there are no substantial differences between goats and sheep in terms of voluntary forage intake when they receive good quality forage capable of covering their requirements, notably those of energy and nitrogen, the same is not true when the forage is rich in lignified cell walls and poor in nitrogen. In the latter case, and if the forage is offered *ad libitum,* goats tend to consume more than sheep (Table 4). Based on the observations of

ALRAHMOUN (1985) this may be at least partially due to a lower water consumption by goats than by sheep (Table 5). FOCANT et al. (1988) also observed higher straw intake in goats than in sheep.

Table 4

EFFECT OF FIBRE AND PROTEIN CONTENTS ON THE INGESTION OF FORAGE BY GOATS AND SHEEP

	Composition (% DM)		Intake (g DM/kg $W^{.75}$)		Reference
Forage	Fibre	Protein	Goats	Sheep	
Pasture hay	35.6	9.2	58.9	51.7	ALRAHMOUN (1985)
NaOH-treated barley straw	46.1	3.7	53.7	35.0	ALRAHMOUN (1985)
Barley straw	45.9	3.9	47.7	38.0	MASSON et al. (1989)
NaOH-treated barley straw	46.1	3.7	62.2	55.0	MASSON et al. (1989)

Table 5

WATER CONSUMPTION (ml/g DM intake) BY GOATS AND SHEEP RECEIVING DIFFERENT DIETS

	Pasture hay	NaOH-treated straw
Sheep	4.1 ± 0.6	7.2 ± 0.4
Goats	3.2 ± 0.5	5.5 ± 0.8

As a matter of fact, two cases should be considered, i. e. either forage is available in restricted amounts and there are no substancial differences between sheep and goats, or they are freely available and goats tend to select much more, especially when the forage is long (see Chapter 1).

The capacity of higher consumption of poor forages, however, is characteristic of adult goats. In young growing goats, the consumption of lignin-rich forages is lower than that of young sheep (MASSON et al., 1986).

Adult goats are very selective when grazing resulting in large refusals (MASSON and DE SIMIANE, 1980) (see Chapter 13). More generally, goats prefer leaves to stems and seeds to leaves (MORAND-FEHR, 1981; BOURBOUZE, 1982). As a result of this, the nutritive value of the forage actually ingested by goats is higher than that of the forage offered.

The unitary eating time (min/g DMI/kg $W^{.75}$) tends to be longer in sheep consuming good forage, but goats have the possibility of considerably increasing this time when they eat straw (Table 6). DULPHY and CARLE (1986) and FOCANT et al. (1986) reported comparable results with timothy hay and with different diets.

Table 6

FEEDING BEHAVIOUR OF GOATS AND SHEEP
(MASSON et al., 1986)

	Ray grass hay		NaOH-treated barley straw	
	Goats	Sheep	Goats	Sheep
Unitary eating time (min/g DMI/kg $W^{.75}$)	3.0 ± 0.3	4.0 ± 0.4	6.4 ± 0.4	4.4 ± 1.1
Unitary rumination time (min/g DMI/kg $W^{.75}$)	4.2 ± 0.5	5.3 ± 0.5	9.3 ± 1.0	10.9 ± 3.2

The unitary rumination time (min/g DMI/kg $W^{.75}$) is always shorter in goats (DULPHY and CARLE, 1986). This could at least partially favor the retention of poor forages in the rumen of goats. In addition, the number of meals tends to be higher in goats (GEOFFROY, 1974).

The chopping of forage in most experiments more or less substantially limits the selectivity by goats.

Studies carried out with animals fitted with permanent esophageal cannulas (MASSON et al., 1989) show that the esophageal contents of sheep and goats are richer in water and ash than the ingested forages. The nitrogen levels of the esophageal contents are similar to those of the ingested matter when the forage is nitrogen rich as alfalfa but are more nitrogen rich when the forage is nitrogen deficient as wheat straw (Table 7). This results from the enrichment of the ingested nitrogen-poor forage by salivary urea. This important role of the saliva in recycling blood urea in the digestive tract has been stressed by BROSH et al. (1986a), SILANIKOVE (1986), OBARA and SHIMBAYASHI (1987).

When the forage is lignified (wheat straw), there is more grinding by mastication in both species, but the esophageal contents of goats have more fine particles than those of sheep (Table 8).

SEHT et al. (1976) reported that saliva secretion (ml/kg DMI) is significantly higher in goats than in sheep receiving the same diet (847 ± 32.5 vs 502 ± 69). The mean quantity of saliva secreted (ml/g DMI) is 1.2 in goats vs 0.9 in sheep. This was not confirmed by the observations of MASSON et al. (1986), but caution is required, since restitutions by the esophageal cannula are not very precise. The urea content of saliva is not significantly different between the two species.

OVERALL DIGESTION

Research made with forages produced in temperate climates showed no difference in digestibility between sheep and goats (CHENOST, 1972; GEOFFROY, 1974; BLANCHART et al., 1980; ALAM et al., 1985; MORAND-FEHR et al., 1987). With these good quality forages, sheep sometimes have even better digestibility than goats (ADEMOSUM, 1970; WILSON, 1977).

On the contrary, poor quality forages are better digested by goats than by sheep (EL HAG, 1976; GIHAD, 1976; WILSON, 1977; DEVENDRA, 1981; ANTONIOU and HADJIPANAYIOTOU, 1985). Based on a literature review, LOUCA et al. (1982) concluded that moderate to good quality forages are digested identically by sheep and goats, while poor quality forages are better digested by goats (Table 9).

Table 7

COMPARATIVE CHEMICAL COMPOSITION OF FORAGE OFFERED AND THE ESOPHAGEAL
CONTENTS IN GOATS AND SHEEP
(MASSON et al., 1989)

	Alfalfa hay		Wheat straw	
	Goats	Sheep	Goats	Sheep
Dry matter (% Crude matter)				
- Forage offered	89.8 ± 2.3	89.8 ± 2.3	92.8 ± 1.8	92.8 ± 1.8
- Esophageal contents	20.1 ± 2.6	17.5 ± 2.3	33.7 ± 5.6	33.4 ± 4.2
Ash (% DM)				
- Forage offered	8.9 ± 1.2	8.9 ± 1.2	9.2 ± 1.2	9.2 ± 1.2
- Esophageal contents	10.7 ± 2.3	12.0 ± 4.1	13.6 ± 1.8	14.9 ± 1.7
Crude Protein (% DM)				
- Forage offered	16.2 ± 2.6	16.2 ± 2.1	3.4 ± 0.8	3.4 ± 0.8
- Esophageal contents	15.8 ± 2.1	16.8 ± 1.3	9.5 ± 4.6	12.3 ± 5.6

Table 8

COMPARATIVE PARTICLE SIZE OF ESOPHAGEAL CONTENTS IN GOATS AND SHEEP
(ARISTA, 1989)

	PARTICLES (% total particles)			
	< 0.8 mm		< 0.4 mm	
Forages	Goats	Sheep	Goats	Sheep
Alfalfa hay	26.2	25.1	9.7	9.2
NaOH-treated wheat straw	49.4	38.2	-	-
Straw + soya oilcake	33.1	32.6	21.3	17.9

In both kid and lambs an increase in feeding level did not affect the *in vivo* digestibility of DM (74 ± 2.1), NDF (79 ± 2.8) and nitrogen (66 ± 1.6) of high quality forage (ALAM et al., 1987).

As stated above, with an *ad libitum* feeding of cell-wall rich and nitrogen poor forages, goats eat and are more selective than sheep. As a result, the forage truly ingested by goats is of higher nitrogen and of lower cell wall contents. This can explain why dry matter and crude fiber digestibility coefficients are higher in goats than in sheep. These results confirm the observations of CUDDEFORD and WAARD (1982), DEVENDRA (1977), EL HAG (1976), GIHAD and EL BEDAWY (1980), GIHAD (1981) and SHARMA and RAJORA (1977).

Table 9

APPARENT DIGESTIBILITY OF FORAGES BY DAMASCUS HE-GOATS AND CHIOS RAMS

(LOUCA et al., 1982)

	Chemical composition (% DM)		Quantity ingested (g/kg W$^{.75}$)	Apparent digestibility (%)		
	CP	CF		DM	CP	CF
Alfalfa hay	23.1	24.1				
- Goats			67.8	67.4	72.1	51.3
- Sheep			67.0	68.6	79.0	56.2
Barley hay	9.8	31.4				
- Goats			48.4	56.7	55.7	52.7
- Sheep			45.1	58.2	58.8	58.6
Barley straw	5.1	40.4				
- Goats			24.9	48.9	18.9	56.7
- Sheep			21.7	45.0	19.3	52.8

DIGESTION IN THE RUMEN

In spite of limited knowledge on digestion in the different parts of the digestive tract of goats, the breakdown of foodstuffs most likely occurs primarily in the reticulo-rumen. PORTER and SINGLETON (1971) reported that digestion in the rumen involves 84 % of organic matter, 92 % of cellulose and 89 % of hemicelluloses. In addition, KAMEOKA and MORIMOTO (1959) gave values of 62 and 85 % for digestion of organic matter in the rumen, which is higher than those of CHURCH (1976) for sheep (52 to 75 %), but with different diets.

A number of authors (DEHORITY and GRUBL, 1977; EL HAG, 1976; HADJIPANAYIOTOU and ANTONIOU, 1983) have noted a higher concentration of bacterial protein in the rumen of goats in comparison to sheep.

Furthermore, GIHAD and EL-BEDAWY (1980) observed a higher concentration of cellulolytic bacteria in the rumen contents of goats than in those of sheep receiving poor tropical forage (6.6 % CP and 38 % CF). This could explain the higher digestibility of crude fiber in goats in comparison to sheep (GIHAD et al., 1981).

Work by BELLET (1984) confirmed this higher capacity of goats and above all showed that with a diet composed exclusively of NaOH treated wheat straw, the bacterial population of the rumen remained relatively constant in goats, while it decreased considerably in sheep (Table 10).

The concentration of volatile fatty acids (VFA) resulting from the breakdown of carbohydrates constitutes a good criterion for microbial activity in the rumen. According to EL HAG (1976), it was higher in goats than in sheep, regardless of the diet. The difference increased even further as the forage quality decreased. On the other hand, WATSON and NORTON (1982), CABRERA et al. (1983) and HADJIPANAYIOTOU and ANTONIOU (1983) found no significant differences between the two species in terms of the concentration of volatile fatty acids in the rumen (Table 11).

Table 10

EFFECT OF TREATING STRAW WITH SODIUM HYDROXIDE AND OF THE NITROGEN SUPPLY ON THE MICROBIAL POPULATION OF GOATS AND SHEEP
(TISSERAND et al., 1986)

Population	Straw + soy oilmeal		NaOH treated straw		NaOH treated straw + soya oilmeal	
	Goats	Sheep	Goats	Sheep	Goats	Sheep
Anaerobic (10^8)	13.4±05.6	6.0±03.0	9.4±05.9	0.5±03.1	9.9±00.8	3.5±01.9
Cellulolytic (10^7)	74.5±14.8	36.8±16.4	46.5±17.5	2.8±01.7	53.6+17.6	29.?±09.2
Amylolytic (10^6)	99.7±33.4	57.5±23.5	62.2±15.2	V.0±02.2	65.5±17.6	21.6±08.1
Proteolytic (10^7)	61.5±30.2	25.7±11.8	19.4±24.8	4.2±02.5	57.1±25.5	22.9±13.0

Table 11

CONCENTRATION OF VOLATILE FATTY ACIDS (VFA) IN THE RETICULO-RUMEN OF GOATS (G) AND SHEEP (S)
(HADJIPANAYIOTOU and ANTONIOU, 1983)

DIET	Total VFA (moles/l)		Acetic acid (moles %)		Propionic acid (moles %)		Butyric acid (moles %)	
	G	S	G	S	G	S	G	S
Barley hay alone	54	75	73	75	18	17	8	8
+ Concentrate (50/50)	83	80	67	67	19	19	14	14
Alfalfa hay	102	106	80	78	15	16	6	6
+ Concentrate (50/50)	135	115	59	68	28	20	12	15
Barley straw alone	28	30	74	76	17	18	8	7
+ Concentrate (50/50)	74	82	68	70	21	20	12	10

Results by ALRAHMOUN et al. (1986) showed no difference in VFA concentrations in the rumen of goats and sheep fed a diet composed of NaOH treated straw and soybean oilmeal; the replacement of oilmeal by urea and even its suppression, led to considerable differences between the two species. Goats fed *ad libitum* with non supplemented NaOH treated straw produced more VFA and had a statistically higher molar percentage of acetic acid than sheep (Table 12).

Cellulolytic activity assessed by the breakdown of ADF (VAN SOEST, 1982) in nylon bags placed in the rumen for 48 h was significantly higher in goats than in sheep receiving NaOH-treated wheat straw supplemented with soybean oilmeal or more so without supplementation. Furthermore, this study showed that the addition of a nitrogen source (urea or soybean oilmeal) did not significantly change the cellulolytic activity of goats.

Experiments with growing goat kids (MASSON et al., 1986) showed that the lactose enrichment of a nitrogen-poor straw-based diet increased the recycling of blood urea more in goats than in sheep (Table 13).

Table 12

BIOCHEMICAL CHARACTERISTICS OF THE RUMEN OF GOATS AND SHEEP (MEAN OF FOUR
SAMPLES BEFORE AND 1, 3 AND 7 HOURS AFTER THE MORNING MEAL)
(ALRAHMOUN et al., 1986)

| DIET | pH | Total VFA | | C_2* | C_3* | C_4* |
		(moles/l)	(M/100g DMI)	(moles %)		
NaOH treated straw :						
Goats	6.4±0.2	60.8±4.2	79.6±4.9	72.2	22.0	4.9
Sheep	7.1±0.1	27.1±1.1	51.8±7.2	67.3	28.1	4.9
NaOH treated straw + urea :						
Goats	6.6±0.2	84.6±4.6	77.8±8.8	74.9	19.1	4.4
Sheep	7.0±0.1	59.4±4.1	46.0±7.5	72.6	20.3	6.3
NaOH treated straw + soya oilmeal :						
Goats	6.4±0.2	96.8±6.6	73.4±9.1	72.9	18.4	5.2
Sheep	6.6±0.1	73.3±6.2	71.3±2.6	72.0	19.5	5.8

(*) C_2: acetic acid; C_3: propionic acid; C_4: butyric acid.

Table 13

UTILIZATION OF NITROGEN BY GOATS AND SHEEP RECEIVING DIFFERENT FORAGES
(MASSON et al., 1986)

| | Ray grass hay | | NaOH treated straw | | NaOH treated straw + lactose | |
	Goats	Sheep	Goats	Sheep	Goats	Sheep
CP ingested (g/kg $W^{.75}$)	4.6±0.4	5.11±0.03	0.77±0.05	1.0±0.1	0.73±0.8	0.96±0.2
Fecal nitrogen (g/kg $W^{.75}$)	2.2±0.2	2.4±0.2	1.0±0.1	1.4±0.1	1.4±0.3	1.4±0.2
Urinary nitrogen (g/kg $W^{.75}$)	1.2±0.1	1.4±0.4	0.5±0.1	0.4±0.6	0.4±0.0	0.5±0.1
Plasma Urea (mg/100 ml)	17.7±2.2	18.6±3.6	10.3±1.0	9.1±1.0	5.6±1.6	8.5±0.8
Urinary Urea (g/day)	4.0±0.4	5.7±2.2	2.2±0.4	2.8±0.5	0.7±0.2	2.5±0.4

DIGESTION IN THE INTESTINE

The retention time of digesta may be longer in goats than in sheep fed poor-quality non supplemented forages (MASSON et al., 1986).

Almost no data are available concerning the possible particularities of digestion in the small intestine of goats. However, it is likely that the main characteristics found in other ruminants apply to goats.

The digestion of nutrients in the small intestine, and above all of bacterial protein synthesized in the rumen, is very efficient in goats.

The only available study (ALAM et al., 1987) showed that there are no major differences between lambs and kids in site or extent of digestion of a high quality forage capable of supporting live-weight gain. Whether this situation holds for low quality feeds, however, is unknown. In both kids and lambs an increase in feeding level was associated with reduction of digestibility in the stomachs, almost no change in digestion in the small intestine and increased digestion in the large intestine.

Preliminary data by MASSON and FAURIE (1989) concerning digestion in the cecum of goats and sheep showed that the pH and concentrations of ammonia and total volatile fatty acids were slightly higher in goats than in sheep (Table 14). On the contrary, there were no significant differences in cellulolytic activity measured in nylon bags (Table 15).

Table 14

CONCENTRATION OF AMMONIA AND TOTAL VOLATILE FATTY ACIDS (VFA)
IN THE CECUM OF GOATS AND SHEEP
(MASSON and FAURIE, 1989)

	Alfalfa + cocksfoot hay			Barley straw		
	pH	NH_3 (mg/100ml)	VFA (g/l)	pH	NH_3 (mg/100ml)	VFA (g/l)
Goats	7.9±0.2	25.8±6.5	5.7±1.8	8.1±0.3	6.5±1.3	1.2±0.2
Sheep	7.4±0.2	15.4±1.8	4.4±1.2	8.0±0.3	4.8±0.9	0.9±0.2

Table 15

DISAPPEARANCE (%) OF DRY MATTER FROM NYLON BAGS
AFTER 12, 24 AND 48h IN THE CECUM OF GOATS AND SHEEP
(MASSON and FAURIE, 1989)

	Alfalfa + cocksfoot hay			Barley straw		
Length of incubation (h)	12	24	48	12	24	48
Goats	32.3±5.9	44.0±4.0	55.9±2.3	18.4±2.2	29.4±2.5	35.9±2.9
Sheep	31.1±4.3	41.5±7.6	54.8±3.0	19.9±1.8	29.5±8.5	35.6±3.0

CONCLUSION

Digestion studies in goats are relatively limited compared to sheep and cattle. Comparative digestibility of feedstuffs by various ruminants should be considered with caution because such comparisons were carried out in different environments and at different time intervals and without consideration of variables such as age, breed and condition of animals. In most cases with good quality forages or balanced rations no digestibility differences among domestic ruminants are observed. However, with forages low in nitrogen content and high in cell walls and not property supplemented, goats have better digestive efficiency than other ruminants. This has been ascribed to the longer MRT of digesta, higher concentration of cellulolytic bacteria and maintenance of balanced microbial population in the rumen and their higher efficiency for recycling urea from blood to the rumen. The lower water consumption and the higher dry matter intake of poor quality roughages by goats compared to sheep associated with longer retention of digesta give goats special advantages in utilizing the vegetation of semi-arid and arid areas.

SUMMARY

The digestive tract of goats is very similar to that of other ruminants. The volume and weight of the stomachs vary with the level of feed intake, diet composition and feeding behavior. There is no difference in the mean retention time (MRT) of feed particles in the whole digestive tract of sheep and goats eating the same quantity of good quality forages, but the MRT of goats receiving poor quality forage is longer than that of sheep fed the same diet. Factors modifying the MRT in goats are similar to those observed in other ruminants.

A very large proportion of digestion occurs in the reticulo-rumen. With moderate to high quality forages, digestion in goats, sheep and cattle is similar, but goats are more capable than sheep for using cell-wall rich and nitrogen poor forages. In the latter case, goats retain nutrients for a longer time in the digestive tract, have a higher concentration of cellulolytic bacteria in the rumen and are more efficient in recycling of blood urea. In harsh conditions, goats consume less water and more feed dry matter than sheep.

Keywords : Goat, Digestion in the rumen, Rumen microflora, Digestion in the intestine, Level of intake, Quality of forages.

RÉSUMÉ

L'appareil digestif des caprins est très semblable à celui des autres ruminants. Le volume et le poids des estomacs varie avec le niveau d'ingestion, la nature du repas et le comportement alimentaire. Il n'y a pas de différence en ce qui concerne le temps de rétention moyen (TRM) des particules alimentaires dans la totalité du tube digestif chez des moutons et des chèvres consommant la même quantité de fourrages de bonne qualité mais les chèvres recevant des fourrages de mauvaise qualité ont un TRM plus long que les moutons au même régime. Les facteurs qui modifient le TRM chez les chèvres sont semblables à ceux constatés chez les autres ruminants.

Une partie très importante de la digestion se situe dans le réticulo-rumen. Pour des fourrages de qualité moyenne à bonne, la digestion est similaire chez les chèvres, les moutons et les bovins mais les caprins sont plus aptes que les ovins à utiliser des fourrages riches en parois et de faible teneur en azote. Dans ce dernier cas, les chèvres présentent un temps de rétention dans l'appareil digestif plus élevé, une plus grande concentration de bactéries cellulolytiques dans le rumen et un recyclage plus efficace de l'urée sanguine. Les chèvres consomment moins d'eau et plus de matière sèche que les ovins dans les conditions climatiques des zones arides.

REFERENCES

ADEMOSUN A.A., 1970. Nutritive value of Nigerian forages. I - Digestibility of Pennisetum purpureum by sheep and goats. Nigerian Agric. J., 7: 19-26.

ALAM M.R., POPPI D.P. and SYKES A.R., 1985. Comparative intake of digestible organic matter and water by sheep and goats. Proc. New-Zealand Soc. Anim. Prod., 45: 107-111.

ALAM M.R., LAWSON G.D., POPPI D.P. and SYKES A.R., 1987. Comparison of the site and extent of digestion of nutrients of a forage in kids and lambs. J. Agric. Sci. (Camb.) 109: 583-589.

ALRAHMOUN W., 1985. Utilisation digestive comparée chez les caprins et les ovins: Effets de la nature du régime, du traitement des pailles par la soude, de la nature de la source azotée (Digestive utilization in goats and sheep: Effect of diet, NaOH treated straws, kind of nitrogen sources). Thèse Université de Dijon (FRANCE).

ALRAHMOUN W., MASSON C. and TISSERAND J.L., 1986. Étude comparée de l'activité microbienne dans le rumen chez les caprins et les ovins. II - Effet du niveau azoté et de la nature de la source azotée (Microbial activity in goat and sheep rumen. II - Effect of nitrogen level and source). Ann. Zootech., 35 : 109-120.

ANTONIOU T. and HADJIPANAYIOTOU M., 1985. The digestibility by sheep and goats of five roughages offered alone or with concentrates. J. Agric. Sci. (Camb.), 105: 633-671.

ARISTA P., 1989. Digestion comparée dans la cavité buccale chez les ovins et chez les caprins. Étude des bols œsophagiens (Digestion in goats and sheep rumen microflora: Effects of the diets). Mémoire DEA, ENSBANA, Université de Dijon (FRANCE).

BATTACHARYA A.N., 1980. Research on goat nutrition and management in Mediterranean Middle East and adjacent areal countries. J. Dairy Sci., 63: 1681-1700.

BELLET B., 1984. Étude des variations de la microflore ruminale des ovins et caprins en fonction de différents régimes (Variation of goats and sheep rumen microflora: Effects of the diets). Thèse de Doctorat de 3e cycle, ENSSAA, Université de Dijon (FRANCE).

BLANCHART G., BRUN-BELLUT J. and VIGNON B., 1980. Comparaison des caprins aux ovins quant à l'ingestion, la digestibilité et la valeur alimentaire de diverses rations (Comparison of intake, digestibility, dietary value of various rations in goats and sheep). Reprod. Nutr. Dévelop., 20: 1731-1737.

BOURBOUZE A., 1982. Utilisation de la végétation de type méditerranéen par des caprins (Utilization of mediterranean vegetation in goats). Fourrages, 92: 91-106.

BROSH A., SHKOLNIK A. and CHOSHNIAK I., 1986a. Metabolic effects of infrequent drinking and low quality feed on Bedouin goats. Ecology 67: 1086-1090.

BROSH A., CHOSNIAK I., TADMOR A. and SHKOLNIK A., 1986b. Infrequent drinking digestive efficiency and particle size of digesta in black Bedouin goats. J. Agric. Sci. (Camb.), 106: 575-576.

BROSH A., CHOSNIAK I. and SHKOLNIK A., 1989. VFA production in the rumen of Bedouin goats. Effect of drinking food quality and feeding time. International Symposium , Copenhagen (DENMARK), Acta Vet. Scand., suppl. 86: 140-142.

CABRERA R., VILLARROEL P. and VIALE CASTILLO A., 1983. Rumen fermentation activity in the goat and sheep. S. Afr. J. Anim. Sci., 13: 213-219.

CASTLE E.J., 1956. The rate of passage of foodstuffs through the alimentary tract of the goat. I - Studies on adult animals fed on hay and concentrates. Brit. J. Nutr., 10: 15-23.

CHENOST M., 1972. Observations préliminaires sur la comparaison de potentiel digestif et de l'appétit des caprins et des ovins en zone tropicale et en zone tempérée (Preliminary observations on digestive capacity and appetite of goats and sheep in tropical and temperate areas). Ann. Zootech., 21: 107-111.

CHURCH D.C., 1976. Digestive physiology and nutrition of ruminants. In: CHURCH D.C. (Ed.): Digestive physiology. Vol. 1, O. & Books, Corvallis, Oregon (USA).

CUDDEFORD C. and DE WAARD T., 1981. Effect of urea supplementation on intake and utilization of a diet composed of whole barley straw by immature goats and sheep. Vol. 1, p. 160-167. In: MORAND-FEHR P., BOURBOUZE A. and DE SIMIANE M. (Eds.): Nutrition and systems of goat feeding. Symposium International, Tours (FRANCE), May 12-15, 1981, INRA-ITOVIC, Paris (FRANCE).

DEHORITY B.A. and GRUGL J.A., 1977. Characterization of the predominant bacteria occuring in the rumen of goats (*Capra hircus*). Appl. Environ. Microbiol., 33: 1030-1036.

DEVENDRA C., 1977. Studies on the intake and digestibility of two varieties (Sirdong and Coloniao) of Guinea grass (Panicum maximum) by goats and sheep. I - Long grass. Malays. Agric. Res. Devel. Inst. Res. Bull., 5: 91-109.

DEVENDRA C., 1980. Feeding and nutrition of goats, Vol. 2, p. 239-256. In: CHURCH D.C. (Ed.): Digestive physiology and nutrition of ruminants. O. & Books Corvallis, Oregon (USA).

DEVENDRA C., 1981. The utilization of forages from cassava, pigeon, pea Leucaena and groundnut by goats and sheep in Malaysia, Vol. 1, p. 338-342. In: MORAND-FEHR P., BOURBOUZE A. and DE SIMIANE M. (Eds.): Nutrition and systems of goat feeding. Symposium International, Tours (FRANCE), May 12-15, 1981, INRA-ITOVIC, Paris (FRANCE).

DEVENDRA C. and BURNS M., 1980. Goat production in the tropics. Commonwealth Agricultural Bureaux, Slough (UK).

DULPHY J.P. and CARLE B., 1986. Activités alimentaires et méryciques comparées des bovins, des caprins et des ovins (Dietary and merycic activities in cattle, goats and sheep). Reprod. Nutr. Dévelop. 26: 279-280.

EL HAG G.A., 1976. A comparative study between desert goat and sheep efficiency of feed utilization. World Rev. Anim. Prod., 12: 43-48.

FOCANT M., VANBELLE M. and GODFROID S., 1986. Activités alimentaires et motricité du rumen chez la chèvre et le mouton pour deux régimes mixtes foin-orge (Dietary activities and rumen motricity in hay and barley fed goats and sheep). Reprod. Nutr. Dévelop., 26: 277-278.

FOCANT M., VANBELLE M. and JULY A., 1988. Effets de l'ingestion de paille traitée à la soude sur certaines caractéristiques fermentaires du rumen des ovins et des caprins (Effect of NaOH treated straw on goats and sheep rumen fermentations). Reprod. Nutr. Dévelop. 28: 121-122.

GEOFFROY F., 1974. Étude comparée du comportement alimentaire et mérycique de deux petits ruminants : la chèvre et le mouton (Feeding behavior and rumination in goats and sheep). Ann. Zootech., 23: 63-73.

GIHAD E.A., 1976. Intake digestibility and nitrogen utilization of tropical natural grass hay by goats and sheep. J. Anim. Sci., 43: 879-883.

GIHAD E.A., 1981. Utilization of poor forage by goats, Vol. 1, p. 263-271. In: MORAND-FEHR P., BOURBOUZE A. and DE SIMIANE M. (Eds.): Nutrition and systems of goats feeding. Symposium International, Tours (FRANCE), May 12-15, 1981, INRA-ITOVIC, Paris (FRANCE).

GIHAD E.A. and EL-BEDAWY T.M., 1980. Fiber digestibility by goats and sheep. J. Dairy Sci., 63: 1701-1709.

GIHAD E.A., EL-BEDAWY T.M. and ALLAM S.M., 1981. Comparative efficiency of utilization of untreated and NaOH treated poor quality roughages through *in situ* digestion by sheep, goats and buffaloes, Vol. 1, p. 327-337. In: MORAND-FEHR P., BOURBOUZE A. and DE SIMIANE M. (Eds.): Nutrition and systems of goat feeding. Symposium International, Tours (FRANCE), May 12-15, 1981, INRA-ITOVIC, Paris (FRANCE).

HADJIPANAYIOTOU M., 1987. Intensive feeding systems for goats in the Near East, Vol. 2, p. 1109-1141. 4th Intern. Conf. on Goats, March 8-13, 1987, Brazilia (BRAZIL).

HADJIPANAYIOTOU M. and ANTONIOU T., 1983. A comparison of rumen fermentation patterns in sheep and goats given a variety of diets. J. Sci. Food Agric., 34: 1319-1322.

HARYANTO B. and JOHNSON W.L., 1983. Digestibility and retention time of alfalfa and coastal bermudagrass hays by sheep and goats as influenced by level of intake. J. Anim. Sci., 57 (Suppl. 1): 435 (Abst.).

HYDEN S., 1955. A turbidimetric method for the determination of higher polyethylene glycol in biological material. Kungliza Landbrukshögskalarung. Annaler 22: 139-142.

KAMEOKA K. and MORIMOTO H., 1959. Extent of digestion in the rumen reticulum-omasum of goats.J. Dairy Sci., 42: 1187-1197.

LINDBERG J.E., 1988. Retention time of small particles and of water in the gut of dairy goats fed at different levels on intake. J. Anim. Physiol. Anim. Nutr., 59: 173-181.

LOUCA A., ANTONIOU T. and HADJIPANAYIOTOU M., 1982. Comparative digestibility of feedstuffs by various ruminants, specifically goats, p. 112-127. 3rd Intern. Conf. on Goat Production and Disease. Tucson, Arizona (USA).

MALOIY M.O., 1974. Digestion and renal function in East African goats and haired sheep. East Afr. Agric. Forestry J., 40: 177-188.

MASSON C. and DE SIMIANE M., 1980. Utilisation du pâturage rationné par la chèvre laitière (Utilization of rationned pasture in dairy goats). Fourrages, 84: 43-56.

MASSON C. and FAURIE F., 1989. Contribution à l'étude comparative de la digestion caecale chez les ovins et les caprins : premières observations sur pH azote ammoniacal et acides gras volatils (A comparative study of caecal digestion in sheep and goats : first observations on pH, NH_3 and volatile fatty acids), Ann. Zootech. 38: 1-4.

MASSON C., ALRAHMOUN W. and TISSERAND J.L., 1986. Étude comparée de la quantité ingérée, de la digestibilité, de l'utilisation de l'azote, du temps moyen de rétention et du comportement alimentaire chez les jeunes caprins et ovins recevant différents régimes (Comparison of level of intake, digestibility, nitrogen utilization, average retention time and feeding behavior in young goats and sheep given various diets). Ann. Zootech., 35: 49-60.

MASSON C., KIRILOV D., FAURIE F. and TISSERAND J.L., 1989. Comparaison des activités alimentaires et méryciques d'ovins et de caprins recevant de la paille d'orge traitée ou non à la soude (Feeding behaviour and rumination in sheep and goats fed with hydroxyde treated or untreated barley straw). Ann. Zootech., 38 : 73-82.

MORAND-FEHR P., 1981. Caractéristiques du comportement alimentaire et de la digestion des caprins (Feeding behavior and digestion in goats), Vol. 1, p. 21-45. In: MORAND-FEHR P., BOURBOUZE A. and DE SIMIANE M. (Eds.): Nutrition and systems of goat feeding. Symposium International, Tours (FRANCE), May 12-15, 1989, INRA-ITOVIC, Paris (FRANCE).

MORAND-FEHR P. and SAUVANT D., 1978. Nutrition and optimum performances of dairy goats. Livest. Prod. Sci., 5: 203-213.

MORAND-FEHR P., GIGER S., SAUVANT D., BROQUA R. and DE SIMIANE M., 1987. Utilisation des fourrages secs par les caprins, p. 391-422 (Utilization of dry forages by goats). In: DEMARQUILLY C. (Ed.): Les fourrages secs : récolte, traitement, utilisation. INRA, Paris (FRANCE).

OBARA Y. and SHIMBAYASHI K., 1987. The appearance of recycled urea in the digestive tract of goats fed a high protein ration. Spain J. Zootech. Sci., 58: 611-617.

PORTER P. and SINGLETON A.G., 1971. Digestion of hay carbohydrates in small ruminants. Brit. J. Nutr., 26: 77-88.

RAUN N.S., 1982. The emerging role of goats in world food production, p. 133-141. 3rd Conf. on Goat Production and Disease, Tucson, Arizona (USA).

SETH O.N., RAI G.S., YADAV P.C. and RANDEY M.D., 1976. A note on the rate of secretion and chemical composition of parotid saliva in sheep and goats. Indian J. Anim. Sci., 46: 660-663.

SHARMA V.V. and RAJORA N.K., 1977. Voluntary intake and nutrient digestibility of low grade roughage by ruminants. J. Agric. Sci. (Camb.), 88: 79-88.

SILANIKOVE N., 1986. Feed utilization energy and nitrogen balance in goats. World Rev. Anim. Prod., 22: 93-96.

SPEDDING C.R.W., 1975. The biology of agricultural systems. Academic Press. New York (USA).

SREEMANNARAYANA O. and MAHAPATRO B.B., 1979. Rate of passage of digesta in sheep and goats fed on nitrogen from different sources. Indian Vet.J., 50: 772-777.

TAMADA T., 1973. Feed utilization and anatomical adaptation of kids after removal of compartments of forestomach or omasal laminae. J. Dairy Sci., 56: 473-483.

TISSERAND J.L., BELLET B. and MASSON C., 1986. Effet du traitement des fourrages par la soude sur la composition de l'écosystème microbien du rumen des ovins et des caprins (Effect of NaOH treated roughages on composition and microbial ecosystem of goats and sheep rumen). Reprod. Nutr. Dévelop. 26: 313-314.

VAN SOEST P.J., 1982. Nutritional ecology of the ruminant. O & Books Corvallis, Oregon (USA).

VON ENGELHARDT W., WEYRETER H., HELLER R., LECHNER M., SCHWARTZ H.J., RUTAGWENDAR D. and SEHULTKA W., 1986. Adaptation of indigenous sheep, goats and camels in harsh grazing conditions, p. 105-115. Nuclear and related techniques for improving productivity of indigenous animals in harsh environments. Intern. atomic energy Agency, Vienna (AUSTRIA).

WATSON C. and NORTON B.W., 1982. The utilization of pangola grass hay by sheep and angora goats. Proc. Aust. Soc. Anim. Prod., 14: 467-473.

WILSON A.D., 1977. The digestibility and voluntary intake of the leaves of tree and shrubs by sheep and goats. Aust. J. Agric. Res., 28: 501-508.

Chapter 6

ENERGY REQUIREMENTS AND ALLOWANCES OF ADULT GOATS

D. SAUVANT and P. MORAND-FEHR

INTRODUCTION

Energy supplies represent the most expensive dietary component and the main limiting factor for goat production as with other ruminant species. It is therefore necessary to accurately determine the requirements for energy and its allowances. This topic was already treated in Tours (SAUVANT, 1981), so the purpose of the chapter is to discuss data from recent publications and to focuse attention on the requirements for maintenance, lactation and pregnancy, on the significance of liveweight changes for energy requirements and on energy density allowances during early lactation.

REQUIREMENTS

Maintenance

Table 1 summarizes the main experimental data published on maintenance requirements of the goat. They are classified according to the two main methods used, involving 14 feeding trials and 9 calorimetric measurements. The mean values obtained with both methods did not differ significantly. The two sets of data can therefore be pooled and result in an average net energy requirement for maintenance (NEm) of 77 kcal (0.322 MJ)/kg $W^{.75}$. Converted into metabolisable energy (ME), by assuming an efficiency of ME conversion into maintenance NE of 0.72, the value is 106 kcal ME/kg $W^{.75}$ (0.445 MJ ME/kg $W^{.75}$). These values are similar to those adopted in France by MORAND-FEHR and SAUVANT (1978; 78.5 kcal for maintenance or 65.4 milk kcal of NE/kg $W^{.75}$) or in 1981 by the NRC (101.4 kcal ME/kg $W^{.75}$). Recently AGUILERA et al. (1989) obtained a similar value (0.421 MJ ME/kg. $W^{.75}$) by calorimetric measurements.

In the future, these mean values will probably not be greatly modified. More information is needed, however, on the causes of the large variation from one reference to another which appear in Table 1. In addition, it is urgent to have precise data on the effects of factors such as climate, breed and physical activity on energy requirements. For example WILKINSON and STARK (1987) proposed to increase the maintenance energy requirement by 25 % and 50 % in goats grazing flat and mountain areas, respectively, without, however, giving any explicit scientific information on the basis of the calculation for these proposals.

Lactation

The mean goat milk energy values reported in published literature are listed in Table 2. These values are derived from direct calorimetric measurements on milk samples or from feeding trials. In the context of a calorimetric approach which provides more reliable data, models were proposed to predict the milk energy value from its fat content. For a 4 % fat corrected milk (FCM) the mean energy content was 710 kcal/kg (0.297 MJ/kg), which is lower than the French proposal of 1978 based on a set of calorimetric measurements (760 kcal), or that of the NRC (1246 kcal ME = 748 kcal milk NE). Consequently MORAND-FEHR and SAUVANT (1988 and 1989) recently proposed to take into account a decrease in the energy value of goat FCM milk in France.

Table 1

ENERGY REQUIREMENT FOR MAINTENANCE IN GOATS

References	Requirement kcal/kg $W^{.75}$	(1)
MORGEN, BERGER and FINDERLING (1906)	96.4	C
ORR (1923)	73.5	C
RITZMAN et al. (1936)	50.7	C
BRODY (1945)	89.1	C
MAJUMBAR (1960)	96.7	F
MITCHELL (1962)	75.0	C
ARMSTRONG and BLAXTER (1964)	52.2	C
DEVENDRA (1967)	63.2	F
OPSTVEDT (1967)	91.3	F
STOHMANN and al. (1968)	79.2	F
FUJIHARA et al. (1973)	85.2	F
AKINSOYINU (1974)	67.6	F
SKJEVDAL (1974)	84.2	F
SAUVANT and MORAND-FEHR (1977)	78.4	F
ITOH et al. (1978)	61.9	C
RAJPOOT et al. (1978)	73.8	F
SINGH and SENGAR (1978)	82.0	F
MOHAMMED and OWEN (1980)	75.6	F
MOHAMMED and OWEN (1981)	72.5	F
RAJPOOT et al. (1981)	61.5	F
GIGER et al. (1987)	83.9	F

(1) C : Calorimetric study; F : Feeding trial.

It should be noticed that feeding trials generally provide higher values: 751 kcal (0.314 MJ)/kg 4 % FC milk but this difference might be due to digestive or metabolic interactions, which were not taken into account.

Variations in the milk energy content according to the fat content (FC) are similar from one study to another (Table 2), with a mean value of 11.13 (0.047 MJ) kcal NE/g fat/l. Expressed on a ME basis and assuming an efficiency of 0.6 of ME into milk NE, the value is 18.55 kcal (0.078 MJ/g fat/l). The recent proposal of WILKINSON and STARK (1987) is slightly lower (0.060 MJ ME/g fat/l), however the basis of calculation is not indicated. Otherwise it is very surprising that the corresponding NRC proposal is only 3.26 kcal ME/g fat/l.

We propose to use the following relationship to calculate goat milk NE.

Milk NE = 710 + 11.13   (FC - 40)
(kcal)   (g/l)

or

Milk NE = 0.297 + 0.047 (FC - 40)
(MJ)   (g/l)

Table 2

ESTIMATION OF THE ENERGY VALUE OF GOAT MILK

Références	Energy value (E) (kcal/kg 4 % FC milk)	Observations	Data number
PETERSON and TURNER (1939)	778.8	E = 396.2 + 9.56 FC	18
HAENLEIN (1950)	625.2	Literature synthesis	
KALAISSAKIS (1959)	641.5	E = 83.1 + 13.96 FC	70
OPSTVEDT (1967)	696.0	E = 273,2 + 10.57 FC	480
MACKENZIE (1967)	809.4	Feeding trial	
GRAF and al. (1970)	694.8	E = 242.8 + 11.30 FC	295
SAUVANT and al. (1974, INRA (1978)	759.7	E = 312.9 + 11.17 FC	128
SKJEVDAL (1974)	687.5	E = 277.5 + 10.25 FC	120
SENGAR (1980)	748.0	Feeding trial	
NRC (1981)	748.0	3 feeding trials	
LU (1987)	695.4	Feeding trial	
WILKINSON and STARK (1987)	781.0	Literature synthesis (?)	
GIGER (1987)	756.0	Feeding trial	

(1) For each calorimetric study relationships were proposed between the milk calorific value (E, kcal/kg) and fat content (FC, ‰).

Pregnancy

The main references concerning the energy requirements of goats during pregnancy are given in Table 3. Attempts are made to obtain a greater precision in relation to litter weight and stage of pregnancy. In France, it is now suggested to use the general procedure applied to ewes, for which more experimental data are available. Data concerning the fœtus or litter weight during the last 6 weeks of pregnancy have been provided by ROBELIN et al. (1978):

$$FW = PFW \exp 0.265 (1 - \exp (- 0.0103t))$$

with   FW = Fœtus Weight (kg)
          PFW = Parturial Fœtus Weight (kg)
          t = time in weeks before parturition (t = - 1, - 2 etc.).

Table 3

ENERGY REQUIREMENT OF PREGNANT GOATS

HUSTON, SHELTON and ELLIS (1971)	127 à 324 kcal ME	Angora goats (22.6 to 45.3 kg LW)
OYENUGA and AKINSOYINU (1976)	173.6 kcal ME	Indian goats
RAJPOOT (1978)	182.0 kcal ME	Indian goats
SAUVANT (1981)	166.0 to 182.0 kcal ME	Alpine goats (50 to 70 kg LW)

This Gompertz growth model was derived from lamb data, but it seems reasonable to assume that the goat foetuses present the same kind of growth profile at the end of gestation. The gravid uterus weight (GUW) is assumed to be equal to 1.6 x litter weight and the most probable litter or fœtus daily growth (FG) can be expressed as a fonction of the parturition weight:

FG = 30.2 exp (0.1609t)
(g/d)   (wk)

As specific data on the composition of goat fœtus are not yet available, a lamb relationship is used (ROBELIN et al., 1978) to estimate the energy fixed into the gravid uterus (EFGU) according to the litter weight gain:

EFGU = 0.278 ((0.0302 exp (1.609 t)) PFW)$^{1.336}$
(kcal/d)

These models enable us to calculate the specific energy requirement for maintenance + pregnancy.

in kcal NE: 77(LW - GUW)$^{.75}$ + EFGU

in kcal ME: $\dfrac{77}{0.72}$(LW - GUW)$^{.75}$ + $\dfrac{EFGU}{0.13}$

This approach was applied to our experimental data base from 394 animals during the last 6 weeks of pregnancy. Figure 1 shows the results and their mean variations according to parity and week number before kidding. Net energy requirements remained fairly constant for both parameters, while ME varied much with the time. This is explained by the large difference in the efficiency of ME transformation into maintenance and uterus energy.

Figure 1

ENERGY REQUIREMENTS OF PREGNANT GOATS : VARIATIONS ACCORDING
TO THE PARITY NUMBER AND WEEKS BEFORE KIDDING

PARITY NUMBER   WEEKS BEFORE KIDDING

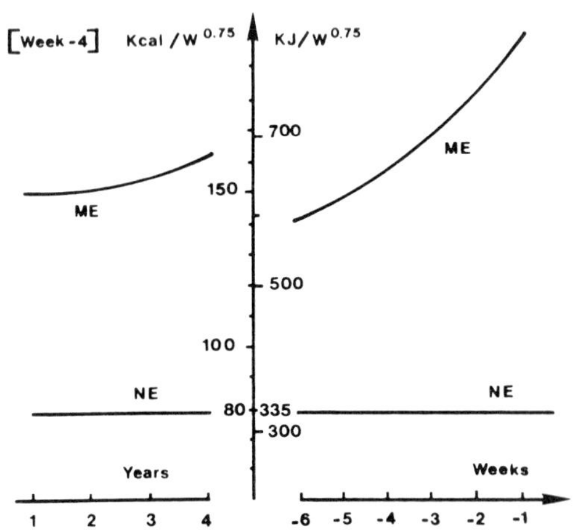

Values obtained from 394 Alpine, Saanen goats studied during the 6 last weeks of pregnancy.

ENERGY CONTENT OF LIVEWEIGHT CHANGES

Liveweight loss

At the onset of lactation, the goat mobilizes a part of its energy reserves, particularly the adipose tissues. The intensity of this phenomenon increases with the milk yield (SAUVANT et al., 1984) and can be routinely evaluated with blood content of non esterified fatty acids (NEFA) and the milk content of long chain fatty acids. However, at present there is no data concerning variations of the body composition which would allow us to quantify the lipomobilization. Thus, our experimental data from feeding and nutritional trials at the onset of lactation were used to estimate indirectly the energy value of the liveweight loss. These data are derived from daily measurements of intake and production of 88 goats. In weeks 1, 2, 4 and 6 of lactation, the milk fatty acid productions, blood contents of NEFA and beta-hydroxybutyrate were measured simultaneously. Using the dietary ether extract content and organic matter digestibility, it was possible to calculate the levels of intake of digestible organic matter and ether-extract. From the obtained data, the production of each of the 10 main milk fatty acids was explained with a stepwise regression according to the representative parameters of the two origins of milk fatty acids (SAUVANT et al., 1987a, 1988):

- either endogenous (lipomobilization), from the blood NEFA and beta-hydroxybutyrate contents.

- or dietary, from the level of intake of digestible organic matter and ether-extract long chain precursors.

The interpretation of these equations led to an individual estimation of milk fat from lipomobilization. Thus, during the first 6 weeks, at the extremes cumulated values ranged were from less than 0.5 up to 6 kg; (Figure 2). About half the milk fat produced appeared to originate from the lipomobilization process. In the same goats we measured the mean liveweight change (LWC) during the same period as a function of the milk fat derived from the lipomobilization (MFL).

LWC = 2.92 - 0.52 MFL (n = 88, R = 0.89)
(kg)   (kg)

When 0.52 kg of MFL are secreted, the goat losses one kg of LW. Assuming a lipomobilization energy efficiency of 0.82, the mobilized fat is of 0.63 kg/kg LW loss. This value can be compared to the data of BOCQUIER et al. (1987) who observed from trials using the D_2O technique, that 1 kg of LW loss in 83 ewes at the onset of lactation was equal to 0.4 to 0.5 kg of lipids.

Assuming that the simultaneous protein mobilization represents 10 % of the weight of the lipid mobilization, the accepted value for cows and ewes (CHILLIARD et al., 1987), it appears that the energy value of 1 kg of LW loss of adult goats is 6.26 Mcal (= 26.18 MJ) milk NE. This value is higher than that suggested by DEVENDRA (1967): 2 042 kg DOM = 8.84 Mcal ME = 5.30 Mcal Milk NE/kg LW loss, but it is close to the 28 MJ/kg LW loss of the NRC (1981) cow proposal adopted for goats by WILKINSON and STARK (1987).

From a literature review CHILLIARD et al. (1987) estimated that the energy value of corrected LW loss in the dairy cow ranged from 6.8 to 10.2 Mcal/kg (28.4 and 42.7 MJ). In ewes, the estimation of BOCQUIER et al. (1987) ranged from 2.8 to 8.8 Mcal (11.7 and 36.8 MJ)/kg LW loss. These authors observed a wide range of literature references from 3.4 to 21.5 Mcal (14.2 to 89.9 MJ)/kg LW loss. This wide range of published data is mainly caused by differences in the weight of digesta between times and diets.

Figure 2

HISTOGRAM OF THE NUMBER OF GOATS WITH VARIOUS LEVELS
OF ESTIMATED MILK FAT PRODUCTION FROM LIPOMOBILIZATION
BETWEEN PARTURITION AND THE 6th WEEK OF LACTATION
(SAUVANT et al., 1987a)

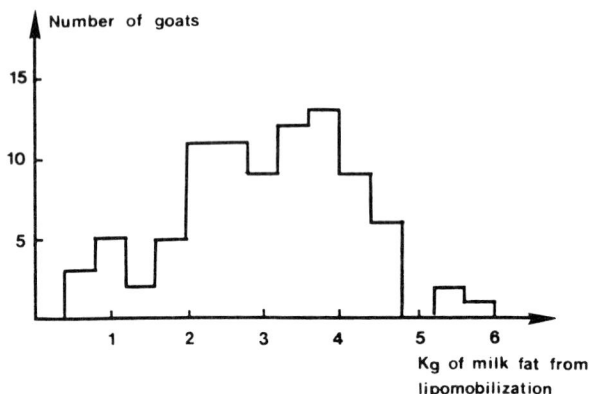

Liveweight gain

The recovery of body reserves mobilized at the onset of lactation takes place in the period after the milk yield has peaked and when the energy balance becomes positive. To estimate the energy value of the LW gain we used the data of our feeding trials for the declining phase of production. From daily measurements of intake and production, and weekly measurements of liveweight in 94 goats for 15 to 19 weeks (SAUVANT, 1981), a mean value of 8.25 Mcal (= 34.51 MJ) milk NE or 13.75 Mcal (= 52.25 MJ) of ME per kg corrected LW gain was estimated. This correction corresponds to the conventionnal subtraction of 3.7 x (dry matter intake) from the LW.

WILKINSON and STARK (1987) suggested the adoption of the ARC cow proposal (34 MJ ME/kg LW gain), but this value is noticeably lower than ours. Values from the literature for 3 experiments on goats are also lower:

DEVENDRA (1967):   (9.9 Mcal ME/kg LW gain),
OYENUGA and AKINSOYINU (1978): (5.14 Mcal ME/kg LW gain),
RAJPOOT (1978):   (5.12 Mcal ME/kg LW gain).

However, our proposal is within the range of Mcal milk NE values (7.0 to 8.5) proposed for dairy cows by CHILLIARD et al. (1987) on the basis of numerous experimental data.

ENERGY DENSITY ALLOWANCE AT THE ONSET OF LACTATION

Early lactation is an important period for energy nutrition and feeding of dairy goats. As the dietary energy supply has a determining role at that moment, we compared three diets of different OM digestibility or energy density (Ed in kcal ME/kg DM) (SAUVANT et al., 1987b):

- Diet L: based on *ad libitum* lucerne hay and restricted amounts of maize silage and concentrate (Ed = 2.53 Mcal ME/kg DM).

- Diet M: based on *ad libitum* maize silage and restricted amounts of lucerne hay and concentrate (Ed = 2.73 Mcal ME/kg DM).

- Diet H: based on *ad libitum* silage beet pulp and restricted amounts of lucerne hay and concentrate (Ed = 2.81 Mcal ME/kg DM).

The number of goats were 30, 38 and 59 respectively for groups L, M and H.

An increase in the Ed of the diet significantly increased the levels of dry matter intake (Figure 3) and milk production (Figure 4). Moreover, a rise in the dietary Ed increased the blood glucose content. This aspect is important considering that one of the main energy nutrition problems in dairy goats is gluconeogenesis during this period of lactation. The level of lipomobilization also depends on the dietary Ed because the blood NEFA and milk C18 fatty acid percentage or secretion decreases from diet L to H. This is confirmed by applying the indirect method of estimation of lipomobilization described above. The improvement in Ed between L and H diets led to a saving of about 1 kg body lipids in goats producing a mean level of 4 kg of 3.5 % FC milk during the first 6 weeks of lactation.

Another way of increasing the diet energy density is to use fat supplementation. Several trials on this topic during early lactation have shown that fat supplementation increases the milk fat yield and content (SAUVANT et al., 1983). The actual rate depended on the diet OM digestibility or Ed, and was much lower with a high dietary OM digestibility. Supplementation with fat did not seem however to improve the goat energy status evaluated through the blood levels of NEFA and glucose.

Figure 3

EVOLUTION OF THE DRY MATTER INTAKE WITH DIETS OF DIFFERING ENERGY DENSITY
(SAUVANT et al., 1987b)

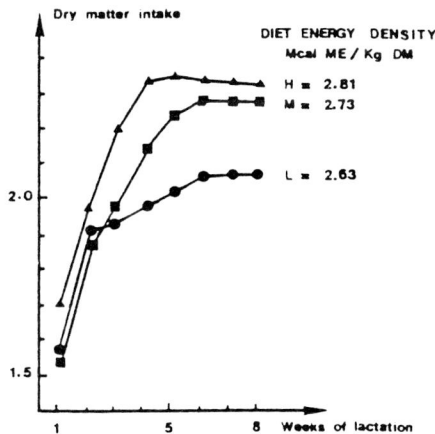

CONCLUSION

A large number of data are available on the energy requirements of goats for maintenance and milk production, but there are large differences from one reference in the literature to another. Further research is needed to elucidate the causes of these variations. In contrast, as data are available on pregnancy requirements, the effects of prolificacy and stage of pregnancy should be taken into account, using data from ewes to calculate energy requirements in pregnant goats. However, specific goat data are obviously needed for a more accurate calculation of requirements.

The energy content of liveweight change was estimated in goats using indirect values close to values for cows and ewes. Indirect approaches for body composition estimation have been developed by using methods such as the D_2O or urea (BROWN and TAYLOR, 1986; BAS et al., 1988, 1989) dilution. We hope that these methods will lead to a more reliable estimation of the energy content of liveweight change.

At the start of lactation the OM digestibility of the diet should be maximized to increase the milk output with a simultaneous improvement of the nutritional status. According to our data it is recommended that a diet with an energy density of 2 800 Kcal (11.7 MJ) ME/kg DM, or an organic digestibility of 77 % should be fed in early lactation. However it should be kept in mind that a high ED diet can cause problems involving inappetence and latent in the rumen acidosis (see Chapter 12).

Figure 4

EVOLUTION OF THE MILK YIELD WITH DIETS OF DIFFERING ENERGY DENSITY
(SAUVANT et al., 1987b)

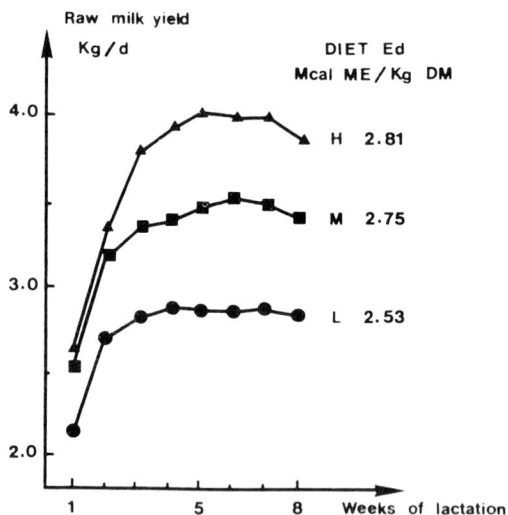

No effect on milk protein and fat contents.

SUMMARY

This chapter deals mainly with recent data on the energy requirements of adult goats for maintenance, lactation and pregnancy and on the significance of liveweight changes and energy density allowances during early lactation.

The mean value of maintenance requirements from 14 feeding trials and 9 calorimetric measurements indicated in the literature is 77 kcal NEm/kg $W^{.75}$. In the future, this value will probably not be greatly modified. From data in the literature, the mean value of energy content in 4 % fat corrected milk is 710 kcal/kg. Taking into account the variation of milk energy value according to fat content, goat milk net energy (mcal) is: 710 + 11.13 x (fat content (g/l) - 40).

For pregnancy requirements, it was proposed to use the procedure applied to ewes which takes into account litter weight increase during the last 6 weeks of pregnancy. From our experimental data on 394 pregnant goats, the total energy requirements of pregnant goats are almost constant during the last 6 weeks of pregnancy (about 335 KJ/W$^{.75}$).

From our observations, 1 kg of live weight loss in early lactation corresponds to 6.26 Mcal milk NE and 1 kg corrected weight gain in mid lactation corresponds to a mean energy value of 8.25 kcal milk NE.

The increase in the digestible energy density of 0.3 Mcal ME/kg diet DM improves goat milk performance in early lactation and allows to a saving of 1 kg of body lipids in goats producing about 4 kg 3.5 % FCM. From our data, it is recommended that the energy density and the organic mater digestibility of diets for high yielding goats should be 2 800 kcal ME/kg DM and 77 %, respectively.

Keywords : Adult goat, Energy requirements, Live weight changes, Energy density of diets, Maintenance, Gestation, Lactation.

RÉSUMÉ

Ce chapitre traite principalement des données récentes sur les besoins énergétiques des caprins adultes pour l'entretien, la lactation et la gestation, sur la signification des variations du poids vif et de la densité énergétique de la ration en début de lactation.

La valeur moyenne du besoin d'entretien venant de 14 expériences d'alimentation et de 9 mesures calorimétriques publiées est de 77 kcal EN entr./kg P$^{0.75}$. A l'avenir, cette valeur ne serait pas susceptible d'être modifiée sensiblement. A partir des données bibliographiques, la moyenne de la valeur énergétique d'un kg de lait de chèvre contenant 4 % de matières grasses est de 710 kcal. En tenant compte des variations de cette valeur selon le taux butyreux (TB), l'énergie nette (kcal) du lait de chèvre est égale à: 710 + 11,13 x TB (g/l) - 40).

Pour les besoins de gestation, il a été proposé d'utiliser une méthode appliquée aux brebis qui tient compte de l'augmentation du poids à la naissance au cours des 6 dernières semaines de gestation. D'après nos données expérimentales sur 394 chèvres en gestation, les besoins énergétiques totaux de la chèvre gestante sont presque constants durant les 6 dernières semaines de gestation (environ 335 KJ/P$^{0.75}$).

D'après nos observations, une perte d'un kg de poids vif en début de lactation correspond à 6.26 Mcal EN lait, et à 1 kg de gain de poids au milieu de la lactation à une valeur moyenne de 8.25 kcal EN lait.

L'accroissement de la densité d'énergie digestible de 0.3 Mcal EM/kg de matière sèche dans la ration améliore les performances en début de lactation et permet d'économiser 1 kg de lipides de réserve chez des chèvres produisant 4 kg de lait à 4 % de TB. En conséquence, d'après nos observations, il est recommandé que la densité énergétique et la digestibilité de la matière organique de la ration pour des chèvres fortes productrices soit de 2 800 kcal EM/kg MS et 77 % respectivement.

REFERENCES

AGUILERA J.F., PRIETO C., FONOLIA J., 1989. Energy metabolism of lactating goats of the Granadina breed. 11th Symp. on Energy metabolism. EAAP publ. n° 42. Pudoc, Wageningen (THE NETHERLANDS).

AKINSOYINU A.O., 1974. Studies on protein and energy utilization by West African dwarf goats. Ph. D. Thesis, Univ. of Ibadan (NIGERIA).

ARMSTRONG D.G. and BLAXTER K.L., 1964. Effects of acetic and propionic acids on energy retention and milk secretion in goats. Proc. 3rd Symp. on Energy metabolism, EAAP. Acad. Press, London (UK).

BAS P., MORAND-FEHR P., SAUVANT D. and HERVIEU J., 1988. Prévision de l'eau corporelle de la chèvre laitière par injection d'urée (Prediction of body water in dairy goats by urea injections). Reprod. Nutr. Dévelop., 28: 185-186.

BAS P., CHILLIARD Y., MORAND-FEHR P., SCHMIDELY P. and SAUVANT D., 1989. Estimation *in vivo* de l'état d'engraissement des chèvres à partir des méthodes de l'urée ou de l'eau lourde (*In vivo* estimation of fattening conditions in goats with urea and D_2O methods). Reprod. Nutr. Dévelop., suppl 2, 253s - 254s.

BOCQUIER F., THERIEZ M. and BRELURUT A., 1987. Recommandations alimentaires pour les brebis en lactation (Feeding allowances for lactating ewes). Bull. Tech. CRZV Theix, (70), 199-211.

BRODY S., 1945. Bioenergetic and growth, 1974 reprint, Hafner Press, Publ. (USA).

BROWN D.L. and TAYLOR S.J., 1986. Deuterium Oxide Dilution Kinetics to predict body composition in dairy goats. J. Dairy Sci., 69: 1151-1155.

CHILLIARD Y., REMOND B., AGABRIEL J., ROBELIN J. and VERITE R., 1987. Variations du contenu digestif et des réserves corporelles au cours du cycle gestation-lactation (Variations of digesta and body reserves during gestation-lactation cycle). Bull. Tech., CRZV Theix, (70), 117-131.

DEVENDRA C., 1967. Studies in the nutrition of the indigenous goat of Malaysia. II - The maintenance requirements of pen-fed goats. Malays. Agric. J., 46: 80-97.

FUJIHARA T., TASAKI I. and FURUHASI T., 1973. Energetic utilization of starch introduced into abomasum of goats, p. 67-70. 6th Symp. on Energy metabolism, EAAP, Univ. Hohenheim (GERMANY).

GIGER S., 1987. Influence de la composition de l'aliment concentré sur la valeur alimentaire des rations destinées au ruminant laitier (Effect of concentrate composition on dietary value of the diets for dairy ruminants). Thèse de docteur-ingénieur, INAPG-Paris (FRANCE).

GRAF P., OSTERKORN K., PAUTZ J., FRAHM K. and GALL C., 1970. Der Energiegehalt der Ziegenmilch. I - Die Beziehung zwichen Fett une Energie, Heft 8, für Tierz. Univ., Munchen (GERMANY).

HAENLEIN G.F.W., 1950. Cited by HAENLEIN, 1980.

HAENLEIN G.F.W., 1980. Status of world literature on dairy goats. Introductory remarks. J. Dairy Sci., 63: 1591-1599.

INRA, 1978. Alimentation des ruminants (Nutrition of Ruminants). INRA Publ., Versailles (FRANCE).

HUSTON J.E., SHELTON H. and ELLIS W.C., 1971. Nutritional requirements of the Angora goat. Texas agricultural experimental stations Bull., 1105.

KALAISSAKIS P., 1959. Langfristige Untersuchungen zum äusseren und inneren Stoffwechsel von graviden und laktierenden Ziegen - 3 - Mitt. Z. Tierphys. Tierenähr und Futtermittelkd, 14: 303-311.

LU C.D., TILAHUN S. and FERNANDEZ J.M., 1987. Assessment of energy and protein requirements for growth and lactation in goats, Vol. 2, p. 1229-1247. 4th Intern. Conf. on Goats, March 8-13, 1987, Brasilia (BRAZIL).

MAJUNBAR B.N., 1960. Studies on goat nutrition. II - Digestible protein requirements for maintenance from balance studies. J. Agric. Sci. (Camb.), 54: 335-340.

MITCHELL, 1962. Cited by DEVENDRA, 1967.

MORGEN A., BEGER C. and FIGERLING G., 1906. Weitere Untersduchungen über die Wiking der einzelnen Nährstoffe and die Milchproduktion. Landwirtschaftliche Versuchs Stationen, 64: 93-242.

MOHAMMED H.H. and OWEN E., 1980. Comparison of the maintenance energy requirements of sheep and goats. Anim. Prod., 30: 479 (Abst.).

MOHAMMED H.H. and OWEN E., 1981. Comparison of maintenance energy requirements of sheep and goats fed dried lucerne of dried grass. Vol. 1, p. 18-27. In: MORAND-FEHR P., BOURBOUZE A. and DE SIMIANE M. (Eds.): Nutrition and systems of goat feeding. Symposium International, Tours (FRANCE), May 12-15, 1981, INRA-ITOVIC, Paris (FRANCE).

MORAND-FEHR P. and SAUVANT D., 1978. Caprins (Goats), Chapter 15, p. 449-467. In: JARRIGE R. (Ed.): L'Alimentation des ruminants. INRA Publications, Versailles (FRANCE).

MORAND-FEHR P. and SAUVANT D., 1988. Caprins (Goats), Chapter 14, p. 282-304. In: JARRIGE R. (Ed.): L'Alimentation des bovins, ovins et caprins. INRA, Paris (FRANCE).

MORAND-FEHR P. and SAUVANT D., 1989. Goats, Chapter 11, p. 169-180. In: JARRIGE R. (Ed.): Ruminant Nutrition, recommanded allowances and Feed tables. INRA, John Libbey, London (UK).

NRC., 1981. Nutrient requirements of domestic animals, n° 15 : Nutrient requirements of goats. National Academic Press, Washington, DC (USA).

OPSTVEDT J., 1967. Feeding experiments with goats. I - Studies on the effect of energy level and feed combination on feed intake and milk yield, composition and organoleptic properties. Bull. Tech., n° 134. Agric. Coll. of Norway, As (NORWAY)

ORR J.B., 1923. The application of the indirect method of calorimetry to ruminants. J. Agric. Sci. (Camb.), 13: 447-461.

OYENUGA V.A. and AKINSOYINU A.O., 1976. Nutrient requirements of sheep and goats of tropical breeds, p. 505-511. 1st International Symposium: Feed composition, Animal Nutrient requirements and computerisation for diets. Utah State Univ., Logan, Utah (USA).

PETERSON V.E. and TURNER C.W., 1939. The energy of goat milk. J. Nutr., 17: 293.

RAJPOOT R.L., 1978. Energy and Protein in Goat Nutrition. Ph. D. Thesis Dept. Anim. Husb. Dairy RBS Coll., Bichpuri (INDIA).

RAJPOOT R.L., SENGAR D.P.S. and SINGH S.N., 1981. Energy and Protein in Goat Nutrition, Vol. 1, p. 101-124. In: MORAND-FEHR P., BOURBOUZE A. and DE SIMIANE M. : Nutrition and systems of goat feeding. Symposium International, Tours (FRANCE), May 12-15, 1981, INRA-ITOVIC, Paris (FRANCE).

RITZMAN E.G., WASHUBURN L.E. and BENEDICT F.G., 1936. The basal metabolism of the goat. New Hampshire Agricultural Experiment Station Technical Bulletin, 66.

ROBELIN J., PETIT M. and TISSIER M., 1978. Evaluation of nutrient requirements for gestation, p. 235-238. In: JARRIGE R. (Ed.): L'Alimentation des ruminants. INRA Publications, Versailles (FRANCE).

SAUVANT D. and MORAND-FEHR P., 1977. Evaluation des besoins énergétiques de la chèvre (Evaluation of energy requirements of goats). (Unpubl.).

SAUVANT D., 1981. Alimentation énergétique des caprins (Energy feeding in goats), Vol. 1, p. 55-79. In: MORAND-FEHR P., BOURBOUZE A. and DE SIMIANE M. (Eds.): Nutrition and systems of goat Feeding. Symposium International, Tours (FRANCE), May 12-15, 1981, ITOVIC-INRA, Paris (FRANCE).

SAUVANT D., MORAND-FEHR P. and BAS P., 1983. L'intérêt des lipides dans les aliments concentrés, observations chez la chèvre (Advantage of lipid supplies in concentrate feeds for goats), pK1 - K17. In: CAAA: Quels aliments concentrés pour les vaches fortes productrices de lait ? ADEPRINA, Paris (FRANCE).

SAUVANT D., MORAND-FEHR P. and BAS P., 1984. Facteurs favorisant l'état de cétose chez la chèvre (Factors increasing cetosis risks in goats), p. 369-378. In: YVORE P. and PERRIN G. (Eds.): Goat diseases. INRA Publications, Versailles (FRANCE).

SAUVANT D., LEGENDRE D., TERNOIS F. and MORAND-FEHR P., 1987a. Indirect quantification of lipomobilization in goats at the onset of lactation, (Poster). 4th Intern. Conference on Goats, March 8-13, 1987, Brasilia (BRAZIL).

SAUVANT D., HERVIEU J., GIGER S., TERNOIS F., MANDRAN N. and MORAND-FEHR P., 1987b. Influence of dietary organic matter digestibility on goat nutrition and production at the onset of lactation. Ann. Zootech., 36: 335-336 (Abst.).

SAUVANT D., MANDRAN N., HERVIEU J., TERNOIS F. and GIGER S., 1988. La sécrétion des acides gras du lait en fonction de caractéristiques de la ration et de l'état nutritionnel chez la chèvre au démarrage de la lactation (Effect of diets and nutritional status of goats in early lactation on milk fatty acids secretion). Reprod. Nutr. Dévelop., 28: 177-178.

SENGAR O.P.S., 1980.Indian Research on protein and energy requirements. J. Dairy Sci. 63: 1655-1670.

SINGH S.N. and SENGAR O.P.S., 1978. Investigation on milk and meat potential of Indian goats. Final Tech. Rep. P.L. 480, Res. Proj. n° A7 - AH - 18, RBS Coll., Bichpuri (INDIA).

SKJEVDAL T., 1974. Potatoes and swedes in the diet of ruminants. III - Studies in lactating dairy goats. Agric. Univ. Norway, Dep. Animal Prod. Report n° 169, 42 pp.

STOHMAN et al., 1868. Cité par HAENLEIN G.P.W., 1980.

WILKINSON J.M. and STARK B.A., 1987. The Nutrition of goats, p. 91-106. In: HARESIGN W. and COLE D.J.A. (Eds.): Recent advances in animal nutrition. Butterworths (UK).

Chapter 7

ENERGY NUTRITION IN GROWING GOATS

M.R. SANZ SAMPELAYO, P. BAS and P. SCHMIDELY

INTRODUCTION

A survey of available information related to energy nutrition in growing goats is given in this chapter. The fact there were no specific data for goats on this subject has led to the assumption that the energy requirements of goat kids could be considered as similar to those of lambs.

The management of energy nutrition of an animal species during growth requires knowledge of its specific requirements which will depend on its genetic potential as well as on the type of growth required. On the other hand, it is necessary to know and take into account the nutritive value of feeds with which such needs will be fulfilled. To that end, the most interesting information is obtained from trials where an energy balance has been established. The energy retention may be calculated by a comparative slaughter method or estimated by calorimeter techniques. The great advantage of the slaughter method is the possibility of keeping the experimental animals more or less, under conditions similar to those of a normal production. The calorimeter techniques even if separated from the practical conditions have the advantage of obtaining successive measurements on the same animals. Information from feeding trials as well as from comparative slaughter methods and calorimeter techniques are now available for growing goats. All these results will be referred to and compared when possible, with data obtained in other young ruminants.

USE OF ENERGY FOR MAINTENANCE

Energy requirement for maintenance

The energy requirement for maintenance is theoretically that necessary for supporting body functions, normal activity and body temperature in the absence of any gain or loss of tissue. Maintenance is commonly measured as the metabolizable energy (ME) corresponding to zero energy retention (MEm) and it is usually derived from regression analyses of data from experiments in which several levels of feeding are used. This requirement was estimated by the comparative slaughter method for Saanen kids fed with goat milk (JAGUSH et al., 1983) and for Granadina kids fed with goat milk and a milk replacer (SANZ SAMPELAYO et al., 1988b), during the first 23 and 30 days of life, respectively. The weaning period can modify not only the efficiency in the utilization of the new energy resources, but also the level of energy intake. The maintenance requirement in Granadina kids was estimated after carrying out an early and progressive weaning within 31-45 days of age. The energy balance was established using the comparative slaughter method just after weaning, from 46 to 60 days of age (MUÑOZ HERNANDEZ, 1984). This energy requirement of 5 month old Granadina goats has been estimated by indirect calorimetric assays (AGUILERA et al., 1985). The values obtained from all these experiments are given in Table 1.

Table 1

UTILIZATION OF METABOLIZABLE ENERGY (ME) FOR MAINTENANCE IN THE YOUNG GOAT, ME REQUIREMENTS (MEm; KJ/kg$^{.75}$/day) AND ME EFFICIENCIES (km)

Breed	Period	MEm		k m	References
		At energy retention = 0	At weight gain = 0		
Saanen	Milk feeding*	458	483	-	(1)
Granadina	Milk feeding*	444	362	0.73	(2) (3)
Granadina	Milk feeding**	427	362	0.54	(2) (3)
Granadina	Just after weaning	557	459	0.78	(3)
Granadina	Ruminant period	427	-	0.75	(4)

* : Goat milk; ** : Milk replacer.
(1)   JAGUSH et al., 1983; (2) SANZ SAMPELAYO et al., 1988b; (3) MUÑOZ HERNANDEZ, 1984;
(4)   AGUILERA et al., 1985.

At zero energy retention, the average MEm value estimated in animals receiving goat milk and milk replacer was equal to 443 kJ ME/kg W$^{.75}$/day. This value proves to be similar to those obtained in lambs (WALKER and NORTON, 1970) and in young calves (VERMOREL et al., 1979) fed with milk or milk replacers. The energy requirement for maintenance is also considered as the ME intake when weight gain is zero. At zero empty body weight gain (EBWG) for goat milk and milk replacer fed Granadina kids, the estimated value MEm was equal to 362 kJ ME/kg W$^{.75}$/day, which is different from that estimated at zero energy retention (SANZ SAMPELAYO et al., 1988b). However, for Saanen kids this energy requirement estimated at both zero energy retention and zero bodyweight gain (BWG) was more similar, 482 vs 458 kJ ME/kg W$^{.75}$/day (JAGUSCH et al., 1983). It is well known that a situation of zero energy retention does not necessarily correspond to zero weight gain. In the animal fed at maintenance a lipid mobilization for a higher protein retention could take place (CLOSE and FOWLER, 1983).

Efficiency of utilization of ME for maintenance

Maintenance processes mainly need ATP. In monogastric and pre-ruminant animals the ME of glucose is the most valuable for the synthesis of ATP, while fat ME is 5 % less and amino acid ME 10 - 20 % less valuably (VAN ES, 1979). In Granadina kids, the fasting heat production was estimated as the heat production at zero ME intake (MUÑOZ HERNANDEZ, 1984). The estimated values represented 0.75 and 0.54 of the maintenance requirements for goat milk and milk replacer fed animals, respectively. Although the error involved in the method for estimating the fasting heat production could partly explain the obtained values, they are lower than those obtained (about 0.80) for pre-ruminant lambs and calves fed with liquid diets (WALKER and JAGUSCH, 1969; NEERGAARD, 1979). These results could be in agreement with those of TANABE and KAMEOKA (1976) who reported that kids are able to use free fatty acids as an energy source.

Due to rumen fermentations, ME in ruminants is about 10 % less efficient than in monogastric animals because about 10 % is lost as heat. However, the volatile fatty acids and amino acids absorbed have approximately the same potential for ATP synthesis (VAN ES, 1979).

Just after weaning the efficiency of ME for maintenance was estimated in Granadina kids to be equal to 0.78 (MUÑOZ HERNANDEZ, 1984). For 5-month old animals of the same breed, weighing between 13.1 and 21.9 kg, AGUILERA et al. (1985) reported an efficiency of ME use for maintenance equal to 0.75.

USE OF ENERGY FOR GROWTH

Voluntary feed intake

The energy ingested by animals during their growing period is an important factor of their growth capacity. As far as young goats are concerned, there are only a few studies where the intake capacity and intake regulation have been analysed (FEHR and SAUVANT, 1974; LU et al., 1987; BAS, 1989; SCHMIDELY, 1988, unpublished results) and SANZ SAMPELAYO, 1988, unpublished). From the above mentioned studies it is possible to conclude that in the first period of life, the abomasum development could be the first regulatory factor and that between 2 - 4 weeks, the maximum feed intake starts to be efficiently regulated by the energy density (BAS, 1988, unpublished; SANZ SAMPELAYO, 1988, unpublished).

The level of intake is often assumed to be a function of metabolic weight and feeding scales are frequently expressed as a fixed multiple of energy requirement for maintenance. For young goats, as indicated by MORAND-FEHR et al. (1982), the level of dry matter intake whether related to metabolic weight or not depends largely on the growth stage. In Granadina kids fed with a milk replacer, the level of voluntary feed intake was a linear function of metabolic weight. This value for the first month of life was equal to 903 kJ ME/kg $W^{.75}$/day (SANZ SAMPELAYO, 1988, unpublished). For Alpine and Saanen kids, the maximum level of intake was reached between 2 - 3 weeks of age, which corresponds to 1 574 and 1 330 kJ ME/kg $W^{.75}$/day, respectively for goat milk and milk replacer fed animals (BAS, 1988, unpublished ; SCHMIDELY, 1988, unpublished).

Energy requirement for growth

The total requirement of ME for maintenance and growth established some time ago have been revised during the last ten years. These values were obtained from feeding experiments and only in relation to ruminant animals (NRC, 1981; RAJPOOT et al., 1981; SAUVANT, 1981; MORAND-FEHR et al., 1982; LU et al., 1987). SAUVANT (1981) reported that the total requirement for maintenance and growth in young goats was around 837 kJ ME/kg $W^{.75}$/day, indicating an absence of specific and accurate data depending on breed, age, weight, etc. The NRC (1981) indicated an energy requirement for weight gain equal to 30.3 kJ ME/g gain. Results ranging between 21.5 and 45.2 kJ ME/g weight gain, were reported by MORAND-FEHR et al. (1982) from work on young animals of different breeds and of varying potential, under various production conditions and using different methodologies. LU et al. (1987) reported that the growth requirement of Alpine and Nubian young goats within 4 - 8 months of age was equal to 37.7, 59.0 and 57.4 kJ ME/g gain, respectively, for animals fed diets with 12.8, 11.6 and 10.3 MJ ME/kg dry matter. For Granadina kids fed with goat milk and a milk replacer, this growth requirement was equal to 13.1 and 14.8 kJ ME/g gain (SANZ SAMPELAYO et al., 1988a).

Efficiency of utilization of ME for growth

The efficiency of ME utilization for energy retention (kpf) has been estimated in Granadina kids fed with goat milk and milk replacer (SANZ SAMPELAYO et al., 1988b).

The values were different according to the feeding type, 0.73 and 0.58, respectively for goat milk and milk replacer, and confirms the findings of WALKER and JAGUSCH (1969), WALKER and NORTON (1970), NEERGAARD (1979) and DEGEN and YOUNG (1982) for milk or milk replacer fed pre-ruminant animals. A very low value for kpf was found by JAGUSCH et al. (1983) in Saanen kids. These authors reported that the ME of goat milk was used for energy retention with an efficiency of 0.45 and they indicated that this value could reveal the persistence of highly thermogenic brown fat or a nutritional deficiency in the goat milk used, but not an energy or protein deficiency (JAGUSCH et al., 1983).

From multiple regression equations in which ME is the dependent variable and lipid and protein accretions the independent variables, it is possible to estimate the energy cost for protein and fat deposition and the partial efficiencies of utilization of ME for protein and fat retention. These values were estimated with the comparative slaughter method and for Saanen kids fed with goat milk (JAGUSCH et al., 1983) and Granadina kids fed goat milk a milk replacer (SANZ SAMPELAYO et al., 1988b) (Table 2). The partial efficiency for fat deposition was lower than that for protein retention, contrary to what happens normally (FOWLER et al., 1980; CLOSE and FOWLER, 1983). However, KIELANOWSKI (1965) reported that the partial efficiencies for protein and fat deposition in milk fed lambs were 0.80 and 0.63, respectively. FOWLER et al. (1980) indicated according to KIELANOWSKI's suggestion (1965), that the partial efficiency for protein retention may be a high value as estimated in young goats. At the same time, FOWLER et al.(1980) and CLOSE and FOWLER (1983) reported that high efficiencies of protein accretion are usually associated with low fat deposition efficiencies what may be a result of nutrient partitioning towards protein retention versus lipid storage at this stage of growth. We think that a considerable amount of additional information is required to explain these results.

Table 2

UTILIZATION OF METABOLIZABLE ENERGY (ME) FOR GROWTH IN YOUNG GOATS.

Breed	Feeding	(A)	(B)	(C)	(D)	(E)	(F)	(G)
Saanen	Goat milk	-	0.45	28.8	76.3	0.83	0.52	(1)
Granadina	Goat milk	13.1	0.73	26.2	61.2	0.91	0.65	(2)
Granadina	Milk replacer	14.8	0.58	30.5	69.8	0.78	0.57	(2)

(A) MEg : ME requirements for growth (kJ/g gain); (B) kpf : ME efficiency for growth;
(C) kJ/g protein, and (D): kJ/g fat : ME costs for protein and fat deposition;
(E) kp and (F) kf : ME efficiency for protein and fat retention.
(G) Bibliographic source : (1) JAGUSCH et al., 1983; (2) SANZ SAMPELAYO et al., 1988b.

FACTORS AFFECTING GROWTH REQUIREMENTS

Changes in body composition during growth

As the animal increases in weight, the physical and chemical composition of the body changes. The nature of these changes depends on animal factors such as genotype but also on nutritional factors. To describe the overall changes which occur during growth, it is very useful to establish relationships between the different components of the gain. These aspects were analysed in detail by VERMOREL (1975) in relation to pre-ruminant lambs and calves. Similar data are not yet available for young goats. However, due to the interest for this matter, we include in this section some very partial data referring to Saanen, Granadina and Alpine kids during their milk feeding period. The values for Saanen and Granadina kids are shown in Table 3.

Table 3

ENERGY COST AND COMPOSITION OF WEIGHT GAIN IN YOUNG GOATS

Breed	Period (days)	Feeding	Weight (kg)	Gain (g/day)	ME (MJ/day)	ER (MJ/day)	Protein (g/day)	Fat (g/day)
Saanen	2-23	Goat milk	5.2	162.5	4.10	1.58	26.9	24.4
Granadina	2-15	Goat milk	3.8	101.2	2.73	1.13	14.7	19.6
Granadina	2-15	Milk replacer	3.5	74.6	2.43	0.76	12.7	11.3
Granadina	16-30	Goat milk	6.4	161.8	4.53	1.94	22.9	35.1
Granadina	16-30	Milk replacer	5.4	130.5	3.68	1.32	16.1	23.6

For Saanen breed : data from animals ingesting two times or more the energy for maintenance (JAGUSCH et al., 1983); for Granadina breed : data from animals ingesting two times or more the energy for maintenance (SANZ SAMPELAYO et al., 1988b).

VERMOREL (1975) reported that in lambs the quantity of protein retained per day is related to growth rate because of the quantity of water retained during protein deposition. The fat stores estimated in g/day or g/kg gain show a significant relation with the energy retained because of its high energy density. Therefore, within the same animal breed and age group, the quantity of energy retained per kg gain increases with increasing fat deposits. The differences between breeds reflect different fattening capacities. When the animal grows, the protein content in the gain decreases to a minimum level while the fat increases so that the animals become fatter and fatter. The ME per kg gain varies according to age and type of feed depending on the use of energy for growth.

BAS and MORAND-FEHR (1987), BAS (1990) and SCHMIDELY (1988, unpublished) analyzed the ME costs for growth in Alpine kids fed with goat milk or different milk replacers. Lower growth costs were observed after ingestion of a milk replacer with a lower energy density,which was associated to a poorer fattening and a better protein utilization.

Energy sources

The value of feeds as energy sources,has been discussed by VAN ES (1979). The quantity of protein together with the energy necessary for its utilization will determine the growth rate. The remaining energy will be stored as fat which will depend on the origin of this energy, as well as the metabolic pathways involved in fat and glucose ME utilization for fat deposition.

The ME value of feeds may be estimated from its gross or digestible energy through a metabolizability factor which depends on the type of feeds considered. In the same way and taking into account the km, kpf and the level of production, the net energy (NE) value can be obtained from the ME one. Goat milk and two milk replacers with different fat contents (24.8 vs. 32.6 percent) were evaluated in Granadina kids (SANZ SAMPELAYO, 1988, unpublished) as energy source. The digestibility was greatest for goat milk followed by the milk replacer, with the lowest fat content. The digestibility and metabolizability of gross energy as well as the ME (MJ/kg) values for the three feeds were: 98.1, 91.4 and 86.8; 95.4, 88.6 and 84.1 and 22.7, 20.2 and 20.2, respectively. The metabolizability data were similar to those obtained in calves fed with milk or milk replacer (NEERGAARD, 1979). It must be noted that the metabolizability/digestibility ratio was very similar for the three diets, about 0.97. This result shows that digestibility is the main factor of variation in the ME value of the feeds. For Granadina kids just after weaning given a diet based on lucerne hay and concentrate, the ME and metabolizability values were equal to 11.3 MJ/kg and 61.9 %, respectively (MUÑOZ HERNANDEZ, 1984). For ruminant animals of the Granadina breed, at 5 months of age, a diet based on barley, sunflower meal and lucerne had a digestible and metabolizable energy equal to 13.1 and 11.0 MJ/kg, respectively (AGUILERA et al., 1985). DEGEN and YOUNG (1982) reported that for weaning lambs, the ME content of a diet similar to those of young goats was 11.6 MJ/kg.

PRACTICAL RECOMMENDATIONS
FOR ADEQUATE ENERGY NUTRITION

MORAND-FEHR et al. (1982) reported that goat kids make suitable use of milk replacers, with growth and development depending on nutritive value. The fat quantities added to the milk powder, the type of carbohydrates used and the carbohydrate/fat ratio are the most important factors that influence the energy value of milk replacers (see Chapters 21 and 24). In this section we shall only mention that the optimum fat content of milk replacers for goat kids to obtain good fattening will depend on the animal breed. BAS and MORAND-FEHR (1987) and BAS et al. (1987) gave Alpine kids different milk replacers with different fat contents (16, 18 and 27 %). The authors found similar growth as well as a higher fattening in kids fed the milk replacer with the highest fat content. The adipose subcutaneous tissue was not affected. SANZ SAMPELAYO et al. (1988b) gave Granadina kids during their first month of life, goat milk or a milk replacer (33.5 vs. 24.8 % of fat), the fattening obtained being higher with goat milk. In Granadina kids receiving either goat milk or a milk replacer with similar fat content (33.5 vs. 32.6), from birth to 8 - 12 kg liveweight, the same authors (SANZ SAMPELAYO et al., 1988a) obtained carcasses with the same fattening state as assessed by dissection and sampling of subcutaneous adipose tissue.

The most important suggestion to guarantee proper energy supplies during weaning, is the utilization of a high quality hay or a suitable concentrate with a high energy density (see Chapter 22). After weaning, the appropriate allowances of energy for young goats become similar to those for adult animals (see Chapter 23). MORAND-FEHR et al. (1987) reported that very starch rich feeding has to be avoided in 5 - 7 months old female kids to prevent excess fattening that can negatively affect the milk production within the first year. So, if the hay quality is appropriate, the intake of concentrate must be limited to no more than 100 - 200 g/day. According to the energy recommendations suggested by different authors from 1979 till now, the energy requirements depend on the growth rates for similar weights and/or ages. We therefore present recommended allowances for the first five months of life which significantly depend on the breed used (Table 4).

Table 4

RECOMMENDED ALLOWANCES OF METABOLIZABLE ENERGY (ME) FOR YOUNG GOATS
(Data with a safety margin of 5 % above the mean value of total requirements)

Age (day)	Mean weight (kg)	Weight gain(g/day)	ME content of feeds(MJ/kg)	Recommended ME (kJ/day)	Breed
0- 30	5-6	150	23.26*	3850	Granadina[a]
0- 30	5-6	120	20.85**	3260	Granadina[a]
21- 45	7.5-12.2	200-300	17.88**	6950	Saanen and Alpine[a]
45- 60	6-10	100	11.25***	3700	Granadina[c]
60- 90	12-16	100-160	***	6045	Alpine and Norwegian[d]
90-150	16-25	100-140	***	7560	Alpine and Norwegian[d]

(*) Goat milk; (**) Milk replacer; (***) solid food; (a) SANZ SAMPELAYO et al. (1988b); (b) BAS (1988, unpublished) and SCHMIDELY (1988, unpublished; (c) MUÑOZ HERNANDEZ (1984); (d) MORAND-FEHR et al. (1982) and SKJEVDAL (1982).

78

SUMMARY

A survey of available information related to energy nutrition in growing goats is given. Maintenance requirements of 443, 557 and 427 kJ ME/kgW$^{.75}$/day for the milk feeding period just after weaning and for the ruminant period, respectively, have been proposed. For growth, these requirements are rather heterogeneous according to breed, age and methodologies ranging between 13.1 and 14.8 kJ ME/g weight gain for milk fed animals and between 21.5 and 59.0 kJ ME/g weight gain for ruminant animals fed with different diets. The efficiency of utilization of metabolizable energy for maintenance ranged from 0.54 to 0.73 for the milk feeding period, 0.78 just after weaning and 0.75 for the ruminant period. For growth the efficiency ranges from 0.45 to 0.73 for milk fed kids. Energy costs of protein and fat deposition have been estimated for milk and milk replacer fed animals resulting in low values for protein deposition from 26.2 to 30.3 kJ ME/g and high ones for fat retention, from 61.2 to 76.3 kJ ME/g. Dietary energy density appears to be the primary factor affecting the voluntary feed intake related to metabolic weight. The most important factors determining the energy use in these animals are the breed and feeding conditions.

Keywords: Growing goat, Energy requirements, Maintenance, Growth, Efficiency of energy utilization, Energy sources.

RÉSUMÉ

L'information disponible sur l'utilisation de l'énergie par le chevreau est passée en revue. On en a déduit les besoins d'entretien de 443, 557 et 427 kJ ME/kg P$^{0.75}$/jour, respectivement pendant la période lactée, juste après le sevrage et pendant la période où le chevreau est ruminant. Les besoins de croissance sont plutôt hétérogènes et dépendent de la race, de l'âge et des méthodologies utilisées; leurs valeurs se situent de 13.1 à 14.8 kJ EM par g de gain de poids pour les animaux nourris au lait et de 21.5 à 59.0 kJ EM par g de gain de poids pour les animaux ruminants recevant différents régimes. L'estimation de l'efficacité de l'utilisation de l'énergie métabolisable pour l'entretien est de 0.54 à 0.73 pendant la période lactée, de 0.78 juste après le sevrage et de 0.75 pour la période suivante. Pour la croissance, l'efficacité se situe entre 0.45 et 0.73 pour les chevreaux nourris au lait. Les coûts énergétiques des dépôts de protéines et de graisses ont été estimés pour les animaux recevant du lait de la mère ou de remplacement. Les valeurs sont faibles pour les dépôts de protéines de 26.2 à 30.3 kJ EM/g et élevées pour les dépôts de graisses de 61.2 à 76.3 kJ EM/g. La densité énergétique du régime paraît être le facteur primaire limitant la consommation alimentaire rapportée au poids métabolique. Les facteurs les plus importants qui déterminent l'utilisation énergétique chez ces animaux sont la race et la nature du régime.

REFERENCES

AGUILERA J.F., LARA L., PRIETO C. and MOLINA E., 1985. Energy requirements for maintenance in goats of Granadina breed, p. 283-289. Proc. Intern. Symp. about Goat Exploitation in Arid Zones, Dec. 9-13, 1985. Fuerteventura, Canarias Islands (SPAIN).

BAS P. and MORAND-FEHR P., 1987. Effect of goat milk or milk replacer intake on growth and carcass quality of kids, Vol. 2, p. 1470 (Abst.). 4th Intern.Conf. on Goats, March 8-13, 1987, Brasilia (BRAZIL).

BAS P., MORAND-FEHR P., SCHMIDELY P. and HERVIEU J., 1987. Effect of dietary lipid supplementation on pre-and post-weaning growth and fat deposition in kids. Ann. Zootech., 36: 339 (Abst.).

BAS P., 1990. Influence of weaning age on growth, body composition and lipid metabolism of Alpine male kids. Thèse de Docteur Ingénieur, Paris 6, (FRANCE).

CLOSE W.H. and FOWLER V.R., 1983. Energy requirements of pigs, p. 159-174. In: HARESING W. (Ed.): Recent Advances in Animal Nutrition. Butterwords, London (UK).

DEGEN A.A. and YOUNG B.A., 1982. Intake and energy retention and heat production in lambs from birth to 24 weeks of age. J. Anim. Sci., 54: 353-362.

FEHR P.M. and SAUVANT D., 1974. Effets séparés et cumulés du nombre de repas et de la température du lait sur les performances des chevreaux de boucherie (Effects of meal number and milk temperature on kid performance). Ann. Zootech., 23: 503-518.

FOWLER V.R., FULLER M.F., CLOSE W.H. and WHITTEMORE C.T., 1980. Energy requirements for the growing pig, p. 151-156. In: MOUNT L.E. (Ed.): Energy Metabolism of Farm Animals. EAAP No 26. Butterwords, London (UK).

JAGUSCH K.T., DUGANZICH D.M., KIDD G.T. and CHURCH S.M., 1983. Efficiency of goat milk utilization by milk-fed kids. N-Z. J. Agric. Res., 26: 443-445.

KIELANOWSKI J., 1965. Estimates of the energy costs of protein deposition in growing animals, p. 13-20. In: BLAXTER K.L. (Ed.): Energy Metabolism of Farm Animals. EAAP No 11. Academic Press, London (UK).

LU C.D., SAHLU T. and MARCOS FERNANDEZ J., 1987. Assessment of energy and protein requirements for growth and lactation in goats, Vol. 2, p. 1229-1247. 4th Intern. Conf. on Goats, March 8-13, 1987, Brasilia (BRAZIL).

LU C.D., POTCHOIBA M.J. and SAHLU T., 1987. Effect of dietary energy density and protein level on growth in dairy goats, Vol. 2, p. 1387-1388 (Abst.). 4th Intern. Conf. on Goats, March 8-13, 1987, Brasilia (BRAZIL).

MORAND-FEHR P., HERVIEU J., BAS P. and SAUVANT D., 1982. Feeding of young goats, p. 90-104. 3rd Intern. Conf. on Goat Production and Disease, Jan. 10-15, 1982, Tucson, Arizona (USA).

MORAND-FEHR P., SAUVANT D. and BRUN-BELLUT J., 1987. Recommandations alimentaires pour les caprins (Nutrient allowances for goats). Bull. Tech. CRZV Theix, INRA, (70), 213-222.

MUÑOZ HERNANDEZ F.J., 1984. Ensayos de metabolisme en ganado caprino desde el nacimiento hasta la etapa de rumiante. Lactancia artificial (Experiments on young goat metabolism from birth to ruminant states). Doctoral Thesis, Veterinary Faculty, University of Cordoba (SPAIN).

NATIONAL RESEARCH COUNCIL, 1981. Nutrient Requirement of Goats: Angora, Dairy and Meat Goats in Temperate and Tropical Countries. National Academy Press, Washington, D.C. (USA).

NEERGAARD L., 1979. Influence of specially extracted soya meal on nitrogen and energy metabolism in the preruminant calf, p. 43-47. In: MOUNT L.E. (Ed.): Energy Metabolism of Farm Animals. EAAP No 26. Butterwords, London (UK).

RAJPOOT R.L., SENGAR O.P.S. and SINGH S.N., 1981. Energy and Protein in Goats, Vol. 1, p. 101-124. In: MORAND-FEHR P., BOURBOUZE A. and DE SIMIANE M. (Eds.): Nutrition and systems of goat feeding. Symposium International, Tours (FRANCE), May 12-15, 1981, INRA-ITOVIC, Paris (FRANCE).

SANZ SAMPELAYO M.R., LARA L., GIL EXTREMERA F. and BOZA J., 1988a. Carcass quality of the Granadina breed goat kid under intake of a specific milk replacer. Seminar of FAO subnetwork on Goat Nutrition and Feeding, Oct. 3 - 5, 1988, Potenza (ITALY).

SANZ SAMPELAYO M.R., MUÑOZ F.J., GUERRERO J.E., GIL EXTREMERA F. and BOZA J., 1988b. Energy metabolism of the Granadina breed goat kid. Use of goat milk and a milk replacer. J. Anim. Physiol. Anim. Nutr., 59: 1-9.

SAUVANT D., 1981. Alimentation énergétique des caprins (Energy nutrition in goats), Vol. 1, p. 55-59. In: MORAND-FEHR P., BOURBOUZE A. and DE SIMIANE M. (Eds.): Nutrition and systems of goat feeding. Symposium International, Tours (FRANCE), May 12-15, 1981, INRA- ITOVIC, Paris (FRANCE).

SKJEVDAL T., 1982. Nutrient requirements of dairy goats based on Norwegian Research, p. 105-108. Proc. 3rd Intern. Conf. on Goat Production and Disease, Jan. 10-15, 1982. Tucson, Arizona (USA).

TANABE S. and KAMEOKA K., 1976. Effect of feeding a carbohydrate-free diet on the growth and metabolism of preruminant kids. Br. J. Nutr., 36: 47-59.

VAN ES A.J.H., 1979. Evaluation of the energy value of feeds: overall appreciation, p. 15-24. In: PIGDEN W.J., BALCH C.C. and GRAHAM N. (Eds.): Standardization of Analytical Methodology for feeds. Ottawa (CANADA).

VERMOREL M., 1975. Le métabolisme énergétique du veau et de l'agneau préruminants (Energy metabolism in preruminant calf and lambs). Les industries de l'alimentation animale, 1: 1-12.

VERMOREL M., BOUVIER J.C. and GEAY Y., 1979. Energy utilization by growing calves. Effects of age, milk intake and feeding level, p. 49-53. In: MOUNT L.E. (Ed.): Energy Metabolism of Farm Animals. EAAP No 26. Butterwords, London (UK).

WALKER D.M. and JAGUSCH K.T., 1969. Utilization of the metabolizable energy of cow's milk by the lamb, p. 187-193. In: BLAXTER K.L., KIELANOWSKI J. and THORBER G. (Eds.): Energy Metabolism of Farm Animals. EAAP No 12. Oriel Press Limited, Newcastle (UK).

WALKER D.M. and NORTON B.N., 1970. The utilization of energy by the milk-fed lambs, p. 125-128. In: CHURCH A. and WENK C. (Eds.): Energy Metabolism of Farm Animals. EAAP No 13. Juris Orion-Verlag, Zürich (SWITZERLAND).

Chapter 8

PROTEIN NUTRITION AND REQUIREMENTS OF ADULT DAIRY GOATS

J. BRUN-BELLUT, J.E. LINDBERG and M. HADJIPANAYIOTOU

INTRODUCTION

During the past ten years extensive efforts have been made to increase the understanding of nitrogen utilization in goats. In contrast to sheep, for which a large amount of data is available, corresponding information on goats is more limited. The purpose of this article is to review the most recent data on nitrogen metabolism in goats,and from these to derive estimates of nitrogen requirements, and finally to identify relevant areas for future research.

Different research teams of the European goat sub-network have contributed data : Dijon (France), Grangeneuve (Switzerland), Nancy (France), Nicosia (Cyprus), Paris (France) and Uppsala (Sweden). The studies cover a wide range of aspects of nitrogen metabolism, such as protein degradation in the rumen, rumen microbial protein synthesis, rumen degraded protein requirements of microbes, nitrogen recycling, efficiency of utilization of digested protein for milk protein synthesis, and body nitrogen mobilization.

Accurate results are needed to improve estimates of nitrogen requirements, which in turn may be used as a basis for recommendations (e.g. ARC, NRC, PDI) (JARRIGE et al., 1978). In particular, values are needed for rumen microbial efficiency, and nitrogen requirements of different tissues, such as the udder. To verify estimated nitrogen requirements, it is necessary to know the actual level of nitrogen utilization by the animal. Therefore, in this review the data have been grouped into three main sections : Nitrogen metabolism in the rumen, nitrogen requirements, and indicators of nitrogen status in dairy goats.

NITROGEN UTILIZATION IN THE RUMEN

Rumen degradation and outflow

There have been suggestions in the literature that goats have more "efficient" digestion than sheep (see Chapter 5). It is possible that aspects of rumen metabolism could be different amongst the two species, however, there is very little concrete evidence of major differences. Comparative studies on rumen metabolism in sheep and goats indicate only minor differences in the concentration of volatile fatty acids (acetate, propionate, butyrate) and ammonia (ANTONIOU and HADJIPANAYIOTOU, 1985). In addition, in studies of rumen degradability of feedstuffs using the nylon bag technique, no significant differences were found between goats and sheep (HADJIPANAYIOTOU et al., 1988). The extent of rumen digestion is determined by the rate of outflow of undigested feed particles from the rumen. In goats, as well as in sheep and cattle, the retention time of feed particles in the rumen decreases as feed intake increases (LINDBERG, 1988; HADJIPANAYIOTOU et al., 1988; NAJAR, 1988). HADJIPANAYIOTOU et al. (1988) compared rumen outflow rates of small feed particles in dry and lactating goats and sheep using Cr-mordanted soybean meal. They found no significant differences in rumen outflow rates in the two species at the production levels studied. The experiments conducted by HASNA (1989) and BRUN-BELLUT et al. (unpubl.) showed a liquid outflow rate ranging between 2.6 and 8.0 % per hour for goats with no significant difference between the means for dry and lactating goats (4.7 % and 4.2 % per h, respectively) or if they received high or low levels of rumen degraded protein in the diet (3.7 and 5.2 % per hour, respectively).

Rumen microbial protein synthesis

It is generally assumed that the synthesis of microbial protein per unit of degraded organic matter in the rumen is of the same magnitude in goats, sheep and cattle. Very few direct measurements have been conducted with goats.

The microbial crude protein (MCP) flow can be estimated from the appearance of microbial RNA in the duodenal digesta. Using this method, LAURENT (1985) measured duodenal microbial flows in lactating dairy goats ranging from 15 to 20 g N per day. This corresponded to 98 and 180 g MCP/kg Digested Organic Matter (DOM), and 151 to 303 g MCP/kg Rumen Degraded Organic Matter (RDOM). Except for a maize silage diet, where dry goats had a lower DOM than sheep and a higher microbial efficiency (109 g/kg DOM for sheep and 159 g/kg DOM for goat) there were no significant differences observed between sheep and goats (LAURENT, 1985). In dry goats fed ammonia and ethylendiamine treated straw and maize silage, also using microbial RNA as a marker, similar although more variable efficiencies were reported (HASNA, 1989). HALBOUCHE (1989), using diaminopimelic acid (DAPA) as the microbial marker, measured microbial flows which ranged from 26.4 g N per day in ad libitum fed dry goats. For lactating goats the efficiencies ranged from 180 to 225 g MCP/kg DOM. Using ^{15}N labelling, BRUN-BELLUT et al. (unpubl.) measured efficiencies in microbial protein synthesis from 106 to 160 g MCP/kg DOM in dry and lactating goats fed a range of different diets. This respectively corresponded to 186 and 160 g MCP/kg RDOM. The range of efficiencies found appeared to be related to ration quality and to the level of rumen degradable protein in the diet.

The MCP flow can also be estimated indirectly from regressions on nitrogen excretion in the faeces (BRUN-BELLUT, 1986 ; GIGER, 1987) and from the urinary excretion of allantoin (LAURENT et al., 1983 ; LAURENT, 1985 ; LINDBERG, 1985).

Using a large number of nitrogen balance measurements (280 in each experiment) and the regression :

Fecal Nitrogen = a (Non degraded Nitrogen) + b(DOM) + c(Non digestible Organic Matter or Crude fiber),

BRUN-BELLUT (1986) and GIGER (1987) obtained a value of 30 g (σ = 0.3 g) of undigestible microbial protein per kg DOM. If MCP digestibility fell within the range of 0.7 to 0.8, the MCP flow was between 100 and 130 g/kg DOM. Assuming a conversion factor for MCP to allantoin of 0.04 (LAURENT, 1985 ; LINDBERG, unpubl.) it can be calculated than in lactating dairy goats fed a range of diets at varying intakes (LINDBERG, 1985) the efficiency of MCP was on average 175g per kg DOM (σ = 26).

It appears that most of the values given above for the efficiency of MCP in goats are within the same range as those given for sheep and cattle (ARC, 1984 ; NRC, 1985). Microbial protein synthesis is permitted by available organic matter and nitrogen which are degraded in the rumen. Some of the observed differences in microbial protein synthesis per kg of DOM could result from differences between the kinetics of the degradation of organic matter in the rumen.

Nitrogen recycling

Nitrogen for rumen microbial growth can come from degraded feed protein and non-protein nitrogen, as well as from recycled nitrogen via the rumen wall and the saliva. Many authors have shown that nitrogen recycling can be quantitatively important. VARADY et al. (1979) estimated that 20 g of urea was recycled daily to the rumen of sheep receiving hay and barley diets. EGAN et al. (1984) have shown large variations in the amount of recycled nitrogen with different diets in sheep. There have been few measurements conducted on goats, although HOUPT and HOUPT (1968) indicated that the rumen epithelium of goats and seemed to have a higher permeability to urea than that of the sheep.

Studies conducted by BRUN-BELLUT et al. (unpubl.) indicated that in goats, recycled urea ranged between 0.4 and 9.7 g/day, and varied with the nitrogen intake and physiological status of the animals. Correcting for the loss of ammonia, the net recycled N urea varied between -7 and +3.5 g/day.

Net nitrogen recycling can be estimated as the difference between nitrogen intake and duodenal nitrogen flow. For lactating goats which were receiving between 23 and 39 g nitrogen intake daily, the net nitrogen recycling ranged from +22.6 g N to -2.5 g N per day (BRUN-BELLUT, 1986). It thus appears that this net recycling could be very high (up to 43 % of duodenal nitrogen flow). To estimate the level of RDP which could be replaced by recycled nitrogen, the effect of different levels of RDP on digestibility and milk yield was estimated in lactating goats in two experiments (BRUN-BELLUT et al., unpubl.). When over 80 g RDP/kg DOM was available, the digestibilities of organic matter and crude fiber wer not affected by RDP level. At 67 g RDP/kg DOM, digested organic matter and milk yield were reduced. However between 80 and 108 g RDP/kg DOM, it is possible that microbial synthesis could be affected without a corresponding change in digestibility. These preliminary results seem to indicate that recycling could provide 20 % of the RDP requirement of the animal. If the level of RDP was decreased by 35 %, digestibility was not affected, but microbial synthesis was depressed.

NITROGEN REQUIREMENTS

Maintenance requirements

Maintenance nitrogen requirements should cover irreversible losses in faeces, urine, hair and scurf. Maintenance nitrogen requirements can be estimated by two different methods: the factorial method, or from nitrogen balance data. In the factorial method the minimal loss of nitrogen is estimated from the endogenous urinary nitrogen (EUN), the metabolic fecal nitrogen (MFN) and from losses in hair and scurf. Using nitrogen data the maintenance requirements are assumed to be equivalent to the quantity of ingested or digested nitrogen at which zero nitrogen balance is obtained. Both dry dairy goats and castrated male goats have been used to obtain these estimates which are reported in Table 1.

The factorial method

Endogenous Urinary Nitrogen (EUN) has been defined as circulating nitrogen which originates from turnover catabolism of amino acid of tissue proteins and which is excreted in the urine (THEWIS, 1974). It can be estimated either by direct measurement of nitrogen excretion at zero nitrogen intake, or by regression analysis. The validity of using feeds or diets low in nitrogen to estimate EUN has been questioned, due to inhibition of rumen metabolism (ARC, 1984). BLANCHART et al. (1980) estimated EUN from the regression :

$$\text{Urinary Nitrogen} = a \times \text{Digested Nitrogen} + \text{EUN}$$

Table 1

MAINTENANCE NITROGEN REQUIREMENTS IN GOATS

AUTHORS	YEAR	NUMBER OF GOATS Nbr. periods	TYPE	WEIGHT (kg)	N. REQUIREMENTS FACTORIAL BALANCE (1)		EUN (2)	MFN (3)
Majumdar	1960	6	DF	23.8	1.48	2.47	0.11	0.41
Khouri	1974	6x4	LF			2.0		
Mba et al.	1975	8	CM	7.2		0.8-2.1		
Itoch et al.	1978		CM			2.12		
Kurar and Mugdal*	1979		CM			2.57		
Blanchart et al.	1980	23x6	CM	50	1.6	1.7	0.11	
Davendra	1980		DF		1.14		0.13	0.22
Rajpoot et al.	1980	6x2	CM	36			0.12	0.43
Sengar	1980		LF	35	1.4			
				25		3.4		
Akinsoyinu	1981		DF			2.2		
Balasubramanya*	1981		CM			2.5		
Mugdal and Singh*	1981	12x2	CM	18	1.5			
Rajpoot	1981	27	DF	12	2.6			
Reynolds	1981	4	CM	30	1.9	2.0		
Kurar and Singh*	1982		CM			3.9		
Brun-Bellut et al.	1984	216	LF	48		2.0		
Chandra and Kurar	1984	12	DF	55		3.5		
Cisuk and Lindberg	1985	185	LF	50		2.2		
						0.31***		
Brun-Bellut	1986	281	LF	50		2.2**		
Giger	1987	154	LF	55		3.0**	0.17	0.20
Hadjipanayiotou	1988	12	DF	65		2.1		

(*)   cited by Chandra and Kurar (1984); (**) PDI; (***) AATN
(1)   g DCP / kg W$^{.75}$;   (2)   g N / kg W$^{.75}$;   (3) g N / 100g DM

in 6 castrated male goats receiving 23 different diets (from straw to maize silage plus soyabean meal). The EUN value obtained was 113 mg N/kg W$^{.75}$. Similarly, BRUN-BELLUT (1986) obtained an EUN of 100 mg N/kgW$^{.75}$ in lactating goats.

GIGER (1987) expanded the regression model to include PDIE, PDIN-PDIE, milk nitrogen and nitrogen retention. Using this regression on lactating goats, this author estimated EUN to be 171 mgN/kg.W$^{.75}$. The latter value is significantly higher than in dry goats and castrated male goats. The physiological status of goats is more complex during lactation ; so the regression models did not allow an accurate estimate of EUN.

Metabolic fecal nitrogen (MFN): MASON (1969) showed that fecal nitrogen was composed of undigested feed nitrogen (UDN) and non dietary fecal nitrogen (undigested bacterial cells, epithelial cells, mucus, enzymes, and bile). JARRIGE et al. (1978) and VAN SOEST (1982) divided fecal nitrogen into three parts: UDN, undigestible microbial nitrogen and endogenous nitrogen which comes from desquamated intestinal cells.

MFN has been estimated indirectly from variations in fecal nitrogen. Two types of equations have been proposed for calculating MFN. The first of these was (MASON, 1969):

Fecal nitrogen = a x Ingested or Digested nitrogen + MFN

Several authors have suggested that MFN is correlated with dry matter intake (DMI). To be able to compare different estimates of MFN, it is necessary to calculate MFN as a function of DMI. MFN values of 0.22 g/100 g DMI (DEVENDRA, 1981), 0.41 g N/100 g DMI (MAJUMDAR, 1960) and 0.50 g N/100 g DMI (REYNOLDS, 1981) have been reported.

The second type of equation used to estimate MFN was (JARRIGE et al., 1978 ; VAN SOEST, 1982):

Fecal Nitrogen = a x Nitrogen intake + b x DOM + c x undigested crude fiber

In this equation, MFN was equal to c x undigested OM (UDOM) or c x undigested crude fiber (UDCF) (BRUN-BELLUT et al., 1984). A value of 12 g MFN/kg undigested crude fiber was obtained for lactating dairy goats which corresponds to 0.11 g/100 g DMI (BRUN-BELLUT et al., 1984a). This value was very low compared with estimates obtained by others (Table 1). GIGER (1987), using lactating goats, proposed a model where both the different nitrogen and fiber fractions were utilized, she estimated MFN to be 0.2 g/100 g DMI.

These results confirmed that DMI and $W^{.75}$ were not good estimators of MFN and that the use of UDCF or UDOM, if these variables were not correlated with N intake or DOM, would be better for estimating the quantity of MFN. However, when such equations were used, very small estimates of endogenous losses in feces were obtained. This may have occurred if RDP was not sufficient for microbial synthesis, in which case endogenous nitrogen which reached the rumen would be utilized and thus the part of fecal microbial protein which came from endogenous nitrogen would not be accounted for in MFN. Nitrogen losses via the skin must also be accounted for in estimates of whole - body nitrogen losses. The minimum daily loss for dairy goats in metabolic crates (BRUN-BELLUT, 1986) was estimated to be 0.02 g N/kg $W^{.75}$. No published results exist for animals living in cold or hot climates.

The minimal N required to maintain the animal can be calculated from the above information.

Minimal N losses = MFN + EUN + skin losses.

= 0.15 g/kg $W^{.75}$ + 0.11 g/kg $W^{.75}$ + 0.02 g/kg $W^{.75}$

= 0.28 g N kg $W^{.75}$.

This is equivalent to 1.75 g CP/kg $W^{.75}$.

It should be noted that this minimal nitrogen loss is not a constant and will thus vary with diet composition and feed intake, as well as the physiological status of the subject.

A principal problem is the conversion of the estimated minimal nitrogen loss to a nitrogen requirement in useful, practical units. In the DCP system, requirements are calculated as the sum of EUN and skin plus hair losses ; MFN is considered as a part of the undigested nitrogen in the feces. However, in more recent protein evaluation systems such as PDI (JARRIGE, 1978), ARC (1984) and NRC (1985), the true intestinal digestibility has been used to estimate the value of diet ; thus MFN must be considered in the nitrogen losses and added to EUN and skin losses to determine nitrogen requirements. To transform nitrogen losses to nitrogen requirements, NRC (1985) has proposed a coefficient to represent the efficiency of conversion of digested protein to EUN. However, this coefficient was not used for MFN and skin losses. If the coefficient proposed by NRC for cattle is used, the nitrogen requirements for maintenance may be calculated:

$$0.02 + \frac{EUN}{0.67} + MFN = 2.1 \text{ g DCP/kg } W^{.75}$$

Nitrogen balance

As mentioned above, maintenance nitrogen requirements have also been estimated by regression analysis where variations in nitrogen retention were explained by nitrogen intake, digested nitrogen, and absorbed amino acid nitrogen.

Using this approach HADJIPANAYIOTOU (1988) estimated maintenance nitrogen requirements of dry non-pregnant goats to be 3.52 g CP/kg $W^{0.75}$ and 1.75 g DCP/kg $W^{.75}$ based on pooled N balance date. It was also shown that the estimates of maintenance nitrogen requirements varied substantially with the level of nitrogen intake. GIGER (1987) using 49 N balance measurements on dry goats, estimated maintenance nitrogen requirements from regressions of digested nitrogen intake and PDI intake on nitrogen retention. Values of 3.37 g PDI/kg $W^{.75}$ (PDI: Digested Protein in Intestine, JARRIGE et al., 1978) and 4.37 g DCP/kg $W^{.75}$ were found.

Requirements during lactation

Nitrogen requirements for lactation have been estimated from experiments where different levels of protein have been fed, and from models of adjustment based on nitrogen balance data. HADJIPANAYIOTOU (1986) found that for goats in mid-lactation, producing 2.5 kg of fat corrected milk, a diet containing 16 % CP was sufficient to maintain production. At the end of lactation, 10 % crude protein in the diet was enough to cover the requirements. Using a regression model on 281 nitrogen balance periods, BRUN-BELLUT et al. (1984a) obtained an estimated maintenance requirement of 2.25 g DCP/kg $W^{.75}$ or 2.20 g PDI/kg $W^{.75}$. This corresponded to an efficiency for milk protein production of 0.78 for the DCP system and 0.74 for the PDI system. GIGER (1987), using 154 nitrogen balance periods in dairy goats, obtained maintenance requirements of 3.69 g DCP/kg $W^{.75}$ or 3.02 g PDI/kg $W^{.75}$, with an efficiency for milk protein production of 0.54 in the DCP system and 0.56 in the PDI system. The discrepancies amongst these results and those cited above, might be explained if some of the animals were in negative energy balance, and hence some amino acids were irreversibly lost to energy metabolism. The model utilized (GIGER, 1987) did not exclude animals in negative nitrogen balance. This would result in an overestimate of nitrogen requirements and an underestimate of the efficiency of transformation of digested nitrogen to milk protein nitrogen.

CISZUK and LINDBERG (1985), using 185 nitrogen balance periods in dairy goats, estimated maintenance requirements to be 0.229 g N/kg $W^{.75}$ with efficiencies for total protein production at peak yield of 0.54 in the DCP system and 0.71 in the AAT-PBV system. At low nitrogen intakes the utilization of digested nitrogen and AAT was close to 0.65 and 0.75 respectively (CISZUK and LINDBERG, 1985) giving estimates of maintenance requirements of 2.19 g DCP and 1.94 g AAT/kg $W^{.75}$.

As indicated in the studies presented above, the variations in efficiencies of converting digested nitrogen to milk nitrogen could be explained by the quality of protein. In 1980 CISZUK showed that with the same amount of digested nitrogen, goats which received soyabean meal and urea had a lower milk nitrogen efficiency than goats which received soyabean meal and fish meal (0.53 vs 0.72). A similar effect was also shown by CISZUK and LINDBERG (1988) with diets containing increasing amounts of urea and fish meal. In dairy cows the possible lack of specific amino acids for milk protein synthesis has been discussed in great detail (TAMMINGA and OLDHAM, 1980). No recently published data on amino acid requirements for dairy goats appears to exist. Another explanation for the variations in efficiency could be the lack of energy and the utilization of protein as an energy source.

Nitrogen mobilization

It appears likely that part of the body protein pool can be mobilized during short periods of protein and energy deficiencies, to cover protein requirements for high priority processes. One such process is the synthesis of milk. Of a total of 280 nitrogen balance measurements, GIGER (1987) found 76 incidences of negative nitrogen balance in lactating goats during the first part of lactation. On average, the mobilization of nitrogen, estimated from the nitrogen balance data, amounted to 26 g/week. BRUN-BELLUT et al. (1984a) found that negative nitrogen balances occured mainly during the second and third weeks of lactation. In the latter study, the average nitrogen mobilization was 55 g/ 2 weeks (maximum value: 155 g). Using a regression model, BRUN-BELLUT et al. (1984a) estimated the efficiency of milk protein production from body nitrogen reserves to be 0.70. Using this efficiency it can be calculated that on average 230 g of milk protein could be produced from mobilized body protein during the first three weeks of lactation.

DETERMINATION OF NITROGEN STATUS OF GOATS

In practical feeding situations it is difficult to determine if a given diet is able to cover the animal's requirements for a known level of production. There are situations where the feeding value of a diet is unknown and again others where the feeding value is known but not the animal's requirements. If protein intakes are above requirements nitrogen is lost from the body mainly through urinary excretion. For these reasons indicators are useful to show the relationship between nitrogen intake and nitrogen requirements.

Lactating goats

Milk Urea Level

It has been shown (BRUN-BELLUT et al., 1983) that milk urea level (MUL) was correlated strongly with the utilization of digested nitrogen for maintenance and milk protein yield. MUL was also correlated with the level of urinary nitrogen ($r = 0.64$; $n = 281$) or urinary urea ($r = 0.72$). At the same time, MUL was correlated with the difference between digested nitrogen and milk nitrogen yield (N Excess), $r = 0.57$, and the level of RDP in diet (in PDI system: PDIN-PDIE), $r = 0.43$:

$$MUL = 0.285 + 0.003 (PDIN-PDIE) + 0.113 \text{ N Excess} - 0.007 \text{ N balance}.$$

$$(r = 0.77, n = 281)$$

BRUN-BELLUT et al. (1983) suggested that if MUL was above 300 mg/l there was an excess of RDP, or of digested protein, or an insufficiency of dietary energy. In the other hand, if MUL was below 300 mg/l, this indicated either an insufficiency of RDP or of digested protein (Table 2).

Urinary Allantoin

The purine catabolite allantoin is excreted in the urine of ruminants when nucleic acids are metabolised. It appears that a major part of the allantoin excretion is derived from digested nucleic acids synthetized by rumen microorganisms (LAURENT, 1985 ; LINDBERG, 1985). For goats, LAURENT et al. (1983), and LINDBERG (1985) have shown that urinary allantoin excretion is strongly correlated with the quantity of DOM. A regression equation which explained variations of urinary allantoin yield was proposed (LAURENT, 1985).

$$\text{Urinary Allantoin} = 0.003 \text{ DOM} + 0.02 (PDIN-PDIE)/UFL - 0.06 \text{ Milk Nitrogen Balance}$$

$$(r = 0.71, n = 189)$$

Table 2.

MILK UREA LEVEL AND URINARY ALLANTOIN EXCRETION

Urinary allantoin (g/d)		MUL (mg/l) 280	320
	lack of RDP but N recycling is enough	O K	RDP excess or DCP excess
4.5	lack of RDP	lack of RDP and fermented energy	lack of fermented energy
0			

DOM and the level of RDP in the diet explained 44 % of the variance of urinary allantoin yield. Only 6 % of the variance was explained by milk nitrogen and nitrogen balance. In goats, 40 to 50 % of digested RNA was excreted in the urine as allantoin (LAURENT, 1985 ; LINDBERG, unpubl.). LAURENT (1985) presented one equation to estimate the amount of microbial protein which was synthesized in the rumen.

$$\text{Microbial N} = \frac{\text{Allantoin Nitrogen}}{0.5 \times 0.8} \times \frac{\text{Total Nitrogen in bacteria}}{\text{N RNA}}$$
$$(\text{g/day})$$

The simultaneous utilization of MUL and urinary allantoin excretion may give a better description of the nitrogen status of lactating goats than either parameter alone (Table 2, BRUN-BELLUT et al., 1984b). It would be of great value if the amount of urinary allantoin excreted was related to milk or blood allantoin levels, as the latter are easier to obtain. Unfortunately, no relation was found between blood or milk allantoin levels and urinary allantoin excretion (BRUN-BELLUT, 1986).

Another possible means of estimation of urinary allantoin excretion is the use of the allantoin/creatinine ratio in a spot sample of urine (LINDBERG, 1985). This ratio could be used alone, provided a constant diurnal excretion ratio of allantoin and creatinine, and definition of allantoin / creatinine levels for different diets and levels of production.

Dry goats

In non-lactating goats, MUL must be replaced by blood or plasma urea levels (PUL). In lactating goats there were very high correlations between MUL and PUL measured simultaneously:

MUL = 0.86 PUL + 149   (r = 0.92, n = 144; BRUN-BELLUT et al., 1987)

PUL was also strongly correlated with ingested nitrogen, and RDP level (GIGER et al., 1985):

PUL = 0.240 + 0.140 ingested N/kg $W^{.75}$ + 0.0106 (PDIN-PDIE)   (r = 0.76, n = 187)

It is however, important to realize that there could be large variations in the PUL levels during the day. Both the kind of diet fed and the feeding practice have an influence on the PUL during the course of the day (BAS et al., 1980 ; BRUN-BELLUT et al., 1987). It appears that the maximum PUL occurs between two and three hours after feeding, while the minimum PUL occurs just before feeding. Thus if PUL was used to estimate the nitrogen status of goats, the factors mentioned above must be defined for each situation.

CONCLUSION

The variation in estimates of nitrogen requirements, expressed as crude protein and digestible crude protein, underlines the importance of accurately describing feed protein quality.

Most results concerning nitrogen requirements for maintenance were lower than 2.5 g DCP/kg $W^{.75}$. Researchers who found values higher than 3 g/kg $W^{.75}$ utilized high protein diets, and used models which do not tolerate excess nitrogen, or used animals in negative energy balance. Based on present information, maintenance nitrogen requirements for goats are in the region of 2.5 g/kg $W^{.75}$. This value is lower than those previously suggested by different authors for sheep than those previously suggested by different authors for sheep and goats. Possible changes in body composition could account for this difference.

Milk protein efficiency values were the same as proposed by TAMMINGA and OLDHAM (1980). A recommendation of 0.66 for milk protein efficiency appears to be adequate based on current knowledge.

It appears that further studies should be undertaken to allow a better description of rumen metabolism in goats. Topics such as rumen microbiology, rumen degradability and rumen recycling of nitrogen require more attention. In addition, further estimates of rumen microbial protein synthesis, and factors affecting this process, are needed.

Direct estimates of basal nitrogen requirements in dairy goats are lacking, as well as specific amino acid requirements for production.

Further increases in our understanding of nitrogen metabolism in goats will allow optimization of production for the benefit of dairy goat producers.

SUMMARY

The nitrogen requirements of dairy goats have been extensively studied during the past ten years. Using factorial methods and nitrogen balance data, different research teams of the European Goat Subnetwork have estimated the nitrogen requirements for maintenance and production of dry and lactating dairy goats. Maintenance nitrogen requirements were in the range 1.9 to 3.7 g DCP/kg $W^{.75}$ in different studies, while the efficiency of milk protein production from digested protein was close to 0.70. Rumen microbial protein synthesis was estimated to range from 100 to 160 g MCP/kg DOM. It appeared that levels of urea in milk or plasma and urinary allantoin excretion could be utilized to estimate the nitrogen status of goats.

Keywords : Dairy goat, Nitrogen Requirements, Nitrogen utilization, Microbial protein synthesis, Nitrogen recycling.

RÉSUMÉ

Les besoins azotés des chèvres laitières ont été amplement étudiés pendant les dix dernières années. A l'aide des méthodes factorielles et des résultats de bilans azotés, différentes équipes de recherches du sous réseau caprin européen ont estimé les besoins azotés d'entretien et de production des chèvres laitières taries et en lactation. Les besoins azotés d'entretien se situent entre 1,9 et 3,7 g MAD/kg $P^{0,75}$ selon les résultats de différentes études alors que l'efficacité de la production de protéines du lait à partir des protéines absorbées est proche de 0,70. On estime que la synthèse des protéines microbiennes du rumen est de l'ordre de 100 à 160 g/kg MOD. Les niveaux d'urée dans le lait et l'excrétion urinaire d'allantoïne peuvent apparemment être utilisés pour estimer l'état azoté des chèvres.

REFERENCES

Agricultural Research Council, 1980. The nutrient requirements of ruminant livestock, Commonwealth Agricultural Bureaux, Slough (UK).

Agricultural Research Council, 1984. The nutrient requirements of ruminant livestock. Commonwealth Agricultural Bureaux, Slough (UK).

AKINSOYINU A.O., 1981. Protein metabolism and requirements, Vol. 1, p. 127-136. In: MORAND-FEHR P., BOURBOUZE A. and DE SIMIANE M. (Eds.): Nutrition and systems of goat feeding. Symposium International, Tours (FRANCE), May 12-15, 1981, INRA-ITOVIC (FRANCE).

ANTONIOU T. and HADJIPANAYIOTOU M., 1985. The digestibility of five roughages offered alone or with concentrate, by sheep and goats. J. Agric. Sci. (Camb.) 105:663-671.

BAS P., ROUZEAU A. and MORAND-FEHR P., 1980. Variations diurnes et d'un jour à l'autre de la concentration de plusieurs paramètres sanguins chez la chèvre en lactation (Diurnal and daily variations in blood parameter concentrations in lactating goats). Ann. Rech. Vet. 11: 409-420.

BLANCHART G., BRUN-BELLUT J. and VIGNON B., 1980. Comparaison des ovins aux caprins quant à l'ingestion, la digestion et la valeur alimentaire de diverses rations (Comparison of level of intake, digestion and nutritive value of various diets in sheep and goats). Reprod. Nutr. Dévelop., 20: 1731.

BRUN-BELLUT J., 1986. Détermination des besoins azotés de la chèvre en lactation (Determination of nitrogen requirements in lactating goats). Thèse INPL, Nancy (FRANCE).

BRUN-BELLUT J., LAURENT F. and VIGNON B., 1983. Taux d'urée du lait et utilisation de l'azote par la chèvre laitière (Urea content in milk and nitrogen utilization in lactating goats), Vol. 2, p. 165-168. In: 4th International Symposium Protein Metabolism and Nutrition, Clermont-Ferrand (FRANCE), INRA Publ., Versailles (FRANCE).

BRUN-BELLUT J., BLANCHART G. and VIGNON B., 1984a. Détermination des besoins azotés de la chèvre en lactation (Determination of nitrogen requirements in lactating goats). Ann. Zootech., 33: 171-186.

BRUN-BELLUT J., LAURENT F. and VIGNON B., 1984b. Taux d'urée du lait, allantoine urinaire, témoins de la nutrition azotée chez la chèvre en lactation (Urea content in milk and allantoin content in urine: parameters estimating of nitrogen nutrition in lactating goats). Can. J. Anim. Sci., 64: 281.

BRUN-BELLUT J., ABDULAHAD N. and VIGNON B., 1987. Effet du jeûne et de la fréquence de distribution des repas sur les variations de l'urémie et du taux d'urée du lait chez la chèvre laitière (Effect of fasting and number of daily meals on the variations of uremia and milk urea content in dairy goats). Reprod. Nutr. Dévelop. 27: 275-276.

CHANDRA S. and KURAR C.K., 1984. Protein requirements of Beetal goats for maintenance. Indian J. Anim. Sci. 54: 189-193.

CISZUK P., 1980. Nitrogen balance and digestibility in lactating goats on rations with varied nitrogen and energy sources. Report 53, Swedish Univ. Agric. Sci., Dep. Anim. Nutr., Uppsala (SWEDEN).

CISZUK P.and LINDBERG J.E., 1985. Total nitrogen retention in lactating goats in relation to digested nitrogen and estimated absorption of amino acids. Acta Agric. Scand. Suppl. 25: 163-176.

CISZUK P. and LINDBERG J.E., 1988. Responses in feed intake, digestibility and nitrogen retention in lactating dairy goats fed increasing amounts of urea and fish meal. Acta Agric. Scand., 103-118.

DEVENDRA C., 1980. The protein requirements for maintenance of indegenous Kambing Katjang goats in Malaysia. Nutr. Abst. Rev. 52: 2347 (Abst.).

EGAN J.K., BODA K. and VARADY J., 1984. Regulation of nitrogen metabolism and recycling, Sept. 10-14, 1984. 6th Intern. Symp. Ruminant Nutrition, BANFF (CANADA).

GIGER S., SAUVANT D., DORLEANS M. and HERVIEU J., 1986. Variations de l'urémie en fonction de la quantité et de la nature des matières azotées alimentaires chez la chèvre en lactation (Effects of level and kind of dietary protein on uremia in lactating goats). Reprod. Nutr. Dévelop. 27: 201-202.

GIGER S., 1987. Influence de la composition de l'aliment concentré sur la valeur alimentaire des rations destinées au ruminant laitier (Effects of concentrate conposition on nutritive value of diets for dairy ruminants). Thèse de docteur-ingénieur. INA-PG, Paris (FRANCE).

HADJIPANAYIOTOU M., 1986. Studies on the response of lactating Damascus goats to dietary protein. J. Anim. Physiol. Anim. Nutr. 57: 41-52.

HADJIPANAYIOTOU M., 1988. Nitrogen requirements of Damascus goats during maintenance. Miscellaneous reports 32: 1-7.

HADJIPANAYIOTOU M., KOUMAS A., GEORGHIADES E. and HADJIDEMETRIOU D., 1988. Studies on degradation and outflow rate of protein supplements in the rumen of dry and lactating Chios ewes and Damascus goats. Anim. Prod. 46: 243-248.

HALBOUCHE M., 1989. Flux d'azote dans le tube digestif des caprins (Nitrogen flow in goat digestive tractus). Thèse INPL, Nancy (France).

HASNA D., 1989. Estimation de la valeur azotée des aliments soumis à un traitement technologique (Estimation of nitrogen value of technologically traited feeds). Thèse INPL, Nancy (FRANCE).

HOUPT T.R. and HOUPT K.A., 1968. Transfer of urea across the rumen wall. Amer. J. Physiol. 214: 1296-1303.

ITOCH M., HARYY T., TANO R. and IWASAKI K., 1978. Maintenance requirements of energy and protein for castrated japanese native goats. Nutr. Abst. Rev. 49: 142 (Abst.).

JARRIGE R., JOURNET M. and VERITE R., 1978. L'azote (Nitrogen), p. 89-128. In: JARRIGE R. (Ed.): Alimentation des ruminants. INRA publ., Versailles (FRANCE).

KHOURI S., 1974. Contribution à l'étude de l'évaluation de l'efficacité nutritive des protéines tannées chez le ruminant (Evaluation of nutritive efficiency of formaldehyde treated protein in ruminants). Thèse, Univ. Sci. et Tech. du Languedoc, Montpellier (FRANCE).

KURAR C. and SINGH S.P., 1982. Protein requirements for maintenance of crossbred goats. Indian J. Dairy Sci. 35: 85-87.

LAURENT F., BLANCHART G. and VIGNON B., 1983. Excretion de l'allantoïne urinaire chez la chèvre laitière (Urinary excretion of allantoin in dairy goats), Vol. 2, p. 333-337. 4th International Symposium Protein Metabolism and Nutrition. INRA publ., Versailles (FRANCE).

LAURENT F., 1985. Flux d'acide ribonucléique dans le tube digestif de petits ruminants (Flow of ribonucleic acid in the digestive tractus of small ruminants). Thèse INPL, Nancy (France).

LINDBERG J.E., 1985. Urinary allantoin excretion and digestible organic matter intake in dairy goats. Swedish J. Agric. Res. 15: 31-37.

LINDBERG J.E., 1988. Retention time of small particles and of water in the gut of dairy goats fed at different levels of intake. J. Anim. Physiol. Anim. Nutr. 59: 173-181.

MAJUMDAR B.N., 1960. Studies on goat nutrition. 1- Minimum protein requirements of goats for maintenance. Endogenous urinary nitrogen and metabolic fecal nitrogen excretion studies. 2- Digestible protein requirements for maintenance from balance studies. J. Agric. Sci. (Camb.) 34: 329-340.

MASON V.C., 1969. Some observations on the distribution and origin of nitrogen in sheep feces. J. Agric. Sci. (Camb.) 73: 99-111.

MBA A.U., EGBUIWE C.P. and OYENUGA V.A., 1975. Nitrogen balance studies with Red Sokoto Maradi goats for the minimum protein requirements. E. A. Agric. For. J. 10: 285.

MUGDAL V.D. and SINGH N., 1981. Nitrogen excretion and protein requirements, Vol. 1, p. 142-151. In: MORAND-FEHR P., BOURBOUZE A. and DE SIMIANE M. (Eds.): Nutrition and systems of goat feeding. Symposium International, Tours (FRANCE), May 12-15, 1981, INRA-ITOVIC, Paris (FRANCE).

NAJAR T., 1988. Effet de différents facteurs zootechniques sur le transit digestif du ruminant. Expérimentations sur chèvres laitières (Effects of various zootechnical factors on digestive transit in ruminants. Experiments in dairy goats). Thèse INA-PG, Paris (FRANCE).

National Research Council, 1985. Ruminant nitrogen usage. National Academic Press, Washington D.C. (USA).

RAJPOOT R.L., 1981. Energy and protein in goat nutrition, Vol. 1, p. 101-124. In: MORAND-FEHR P., BOURBOUZE A. and DE SIMIANE M. (Eds.): Nutrition and systems of goat feeding. Symposium Internernational, Tours (FRANCE), May 12-15, 1981, INRA-ITOVIC, Paris (FRANCE).

RAJPOOT R.L., SENGAR O.P.S. and SINGH S.N., 1980. Goats: protein requirements for maintenance. Intern. Goat Sheep Res. 1: 182-189.

REYNOLDS I., 1981. Nitrogen metabolism in indigenous Malawi goats. J. Agric. Sci. 96: 347-351.

SENGAR O.P.S., 1980. Indian research on protein and energy requirements of goats. J. Dairy Sci. 63: 1655-1670.

TAMMINGA S. and OLDHAM J.D., 1980. Amino acid utilization by dairy cows. Concept of amino acid requirements. Livest. Prod. Sci. 7: 453-463.

THEWIS A., 1974. Métabolisme de l'azote chez le ruminant (Nitrogen metabolism in ruminants). Ann. Gembloux 80: 139-153.

VAN SOEST P.J., 1982. Nutritional ecology of the ruminant, p. 374. O. & B. Books Inc., Corvallis, Oregon (USA).

VARADY J., TASHENOV K.T., BODA K., FEJES J. and KOSTA K., 1979. Endogenous urea secretion into the sheep gastro-intestinal tract. Physiol. Bohemoslavaca 28: 551-559.

Chapter 9

PROTEIN NUTRITION AND REQUIREMENTS
OF GROWING GOAT

M. HADJIPANAYIOTOU, J. BRUN-BELLUT and J. A. LINDBERG

INTRODUCTION

The increasing cost of protein supplements has enhanced research to determine optimal protein levels for cows (SATTER and ROFFLER, 1975) and sheep (PAPAS, 1977a, b; GONZALES et al., 1979; HADJIPANAYIOTOU, 1982a). Information on the nutrient requirements of goats and the nutritive value of feedstuffs for goats are limited. Because of similarities of the digestive system of domestic ruminants, the nutritive value of feedstuffs for goats has been assumed to be equal to that of sheep and cattle.

It has been pointed out, however, that the digestive efficiency of goats varies according to breed and type, location and dietary protein level (HUSTON, 1978). Rather recently, an internationally prepared standard for feeding goats has been issued under the sponsorship of the US National Research Council (NRC, 1981). Furthermore, a number of papers on nitrogen nutrition of goats has been presented at the International Symposium on goats held in Tours (France), Tucson (Arizona - USA) and Brasilia (Brazil). It is widely accepted however, that digestible crude protein (DCP) is an inadequate measure of the protein value of feeds and it is superseded by systems that take account of degradation of protein in the rumen, of the efficiency of microbial protein synthesis and the supply of dietary protein to the small intestine (ARC, 1980, 1984; BURROUGHS et al., 1975; VERITE et al., 1979).

Research work conducted in some institutions participating in the European research network on goats was mainly concentrated on the collection of data that will enable us to use the new approach for feeding growing goats. More specifically, the rate and extent of degradation of protein supplements in the rumen, factors affecting their degradation (outflow rate, chemical treatment) and the effects of protein source and level on the performance of growing goats were studied. The present paper will review data on nitrogen nutrition of growing goats obtained from institutions participating in the European sub-network on goats and compare our data with those obtained from other places and with other species of animals.

PROTEIN IN GROWING GOAT

Goat meat has high protein and low calorie content (Table 1). HAENLEIN (1986) postulated that such composition suggests favourable production economics because it is cheaper nutritionaly to produce lean. Body composition changes as animals grow older and heavier. The general trend is an increase in fat content with increasing age and fattening, and a reduction in the moisture and protein content of meat. In the studies of GAILI et al. (1972) goats had less chemical fat, more moisture and protein in their meats than sheep. Cold carcass weight and weight of intestinal and kidney fat as a percentage of slaughter weight were increased as kids grew older and heavier (ECONOMIDES, unpubl.). Unfortunately there are no comparative data on protein content of empty body weight of lambs and kids offered the same diets and slaughtered at the same mature weight. The impression therefore, that goat is a lean species might originate from the fact that it has a very low content of subcutaneous fat. The higher the feeding level, especially that of concentrates, the more rapid the deposition of fat, and consequently the faster the growth (FEHR et al., 1976) and the larger the protein deficiency of the diet relative to energy supply (MORAND-FEHR, 1981).

In general it is not profitable to allow the animal to deposit more fat than is needed for obtaining the desired meat quality. To attain this it would be important to have a good knowledge of the total quantities of protein and energy which will be produce, so that diets can be fed which suit the requirements of the animal. In kids like other ruminants, non-carcass parts may comprise 40 to 50 % of their weight. The nitrogen and energy content of these non-carcass parts cannot be neglected. However, information on their composition is scarce. Furthermore, most of the information on rates of synthesis of protein comes from rats, rabbits, sheep and men. There is an urgent need therefore, to produce more such data on goats to improve their gross feed efficiency.

Table 1

EDIBLE PROTEIN AND CALORIC CONTENT OF MEATS
(after Mc DOWELL and BOWE, 1977)

Species	Protein (g/kg)	Calories (mcal/kg)
Goat	183	2.341
Mutton	144	3.720
Pork	138	4.369
Beef	155	3.601
Dairy cattle	179	2.918
Veal	191	1.945
Poultry	210	2.194

In line with cattle and sheep, when energy intake is kept at a constant level, nitrogen retention is closely related to nitrogen intakes (LINDBERG, 1989). Nitrogen accretion has priority over fat deposition in general, and more so in a young animal with a high protein requirement. This also shows that nitrogen can be retained despite substantial weight losses. In general, there is a change in protein and fat content of empty-body-weight gain with advancing maturity (ARC, 1980). This was also shown in kids that retained 1.21 mg N/kJ GE at 2-6 weeks of age compared with 0.89 mg N/kJ GE at 9-13 weeks of age (LINDBERG, 1989).

PROTEIN REQUIREMENTS

Relatively limited work has been conducted in this area. SENGAR (1980) reported that protein requirement is a function of energy intake and the protein requirements of growing goats varied from 4.7 to 4.3 g DCP/kg $W^{.75}$. SKJEVDAL (1982) based on Norwegian experiments recommended 40, 60, 80, 100, 90 g DCP/day for kids weigning (liveweight at the end of period) 8, 12, 17, 27, 35 or 45 kg and of 0-6, 6-12, 12-18, 18-34, 34-44 and over 44 weeks of age, respectively. French recommendations (MORAND-FEHR et al., 1982) for different age periods (birth to 30 days, 30-60 days, 60-90 days, 90 to 120 days, 120-150 days, 150 to 180 days and 180-210 days) liveweight (6.5, 11.5, 16.3, 20.7, 24.5, 27.6, 30 kg average liveweight in mid period) and growth rates (165, 165, 155, 140, 115, 90 and 70 g/day) are 19.7, 12.7, 9.5, 7.6, 6.2, 5.2 or 4.7 g DCP/kg $W^{.75}$, respectively. French recommendations are lower than Norwegian recommendations. In a recent growth trial (LU et al., 1987), average protein requirement for growth was observed to be 0.82, 0.62 and 0.50 g CP/g of gain in growing dairy goat kids fed 15, 13 and 11 % CP diet *ad libitum*, respectively. SCHMIDELY et al. (1988) gave 5.88 g DCP/kg $W^{.75}$ estimated from the regression equation of digestible CP (DCP) on retained protein. The NRC (1981) recommended that 0.28 g CP or 0.195 g DCP was required for per g bodyweight gain. Without considering physiological stage, dietary energy concentration, rate of growth or protein degradation in the rumen, this value is comparable to 0.30 - 0.36 g CP/g of gain (NRC, 1985) and to 0.20 - 0.38 g CP/g of gain (ARC, 1980) for growing sheep.

Although the DCP has been widely used in many goat studies, CP is considered less biased in defining protein requirements. In an attempt to better define protein requirements for ruminants the UK and French councils on nutrient requirements proposed a new system based on ruminal degradable protein and undegradable protein (ARC, 1980; VERITE et al., 1979). For the establishment of a similar system for goats there is an urgent need for the determination of various parameters required (efficiency of microbial protein synthesis in the rumen, absorption rate of protein and amino acids in the small intestine, rate and extent of degradation of dietary protein in the rumen).

The basal N requirement (211 mg/kg $W^{.75}$) of growing kids estimated by one of us (LINDBERG, 1989) was slightly lower than the N requirement of lactating goats (229 mg/kg $W^{.75}$) (CISZUK and LINDBERG, 1985). The basal N requirement of the kids was comparable with basal urinary N losses from giving N-free liquid to calves (194 mg/kg $W^{.75}$ BLAXTER and WOOD, 1951) and lambs (190 mg/kg $W^{.75}$; WALKER and FAICHNEY, 1964). Considerably higher total N excretion (350 mg N/kg $W^{.75}$) values were obtained in cows, steers and lambs sustained on intragastric infusion of nutrients (ORSKOV et al., cited by ARC, 1984).

Differences in nitrogen requirements of growing goats obtained by various research workers might be ascribed to breed differences, nitrogen source and levels used, growth rates, source and level of dietary energy, body condition, body composition, sex of the animals and methods used in determining requirements.

RESPONSE TO DIETARY PROTEIN LEVELS

Growth rate of kids increased with increasing protein content (10.9, 14.7 and 16.2 %) of concentrated feeds offered *ad libitum* to Damascus kids weaned from their dams at 76 days of age (LOUCA and HANCOCK, 1977). Similarly, MAVROGENIS et al. (1979) reported that both the growth and feed utilization of kids fed *ad libitum* were higher with high protein diets up to 7 months of age. In the same study, male kids responded positively to increased protein level in the diet, whereas female response was only marginal, presumably because males have larger mature size and as a result deposit more protein and less fat than females for a longer period of time (ANDREWS and ORSKOV, 1970). Increasing (11.3, 16.0 and 20.9 % CP) the protein content of the diet resulted in significantly greater liveweight gains in Australian Cashmere goat, although the improved growth could be largely attributed to increased intake rather than to enhanced feed efficiency (ASH and NORTON, 1987a).

In the studies of LOUCA and HANCOCK (1977) and MAVROGENIS et al. (1979) a positive response of kids to increasing levels of dietary CP up to 180 g/kg DM was noted. In a subsequent study (HADJIPANAYIOTOU, 1982b) the performance of male and female Damascus kids offered a concentrate mixture with 180 or 200 g CP/kg feed DM over the period 60 to 102 days of age was studied. In the latter study, the final weight, weight gain, feed intake and feed to gain ratio of kids fed the 18 % or 20 % CP concentrate mixtures were similar. Decrease of dietary CP from 18 % to 16 % during the period 103 to 156 days of age resulted in higher feed to gain ratios compared to those on the 18 % or 20 % CP. Positive response to higher (193 vs 163 g CP/kg feed dry matter) level has also been found and in local Cyprus kids (Table 2). Compared with lambs, research work at the

Cyprus Agricultural Research Institute (HADJIPANAYIOTOU, 1987b) demonstrated that dietary CP concentration for early weaned Damascus kids (52 days of age) and Chios lambs (35 to 42 days of age) to attain maximum growth rates is similar (180 g CP/kg DM) up to 90 to 100 days of age. Thereafter, maximum growth rates can be obtained in male lambs even when the dietary protein concentration is reduced to 160 g CP/kg, but in kids, the dietary protein should be maintained at 180 g CP/kg DM until 140 days of age. This might be partly due to the fact that kids consume less feed (ECONOMIDES, 1986) than lambs (80 vs 96 g DM/kg $W^{.75}$), and at the same physiological age (100-150 days) have less weight relative to their mature body size.

Table 2.

THE EFFECT OF PROTEIN LEVEL ON THE PERFORMANCE OF CYPRUS LOCAL KIDS
(HADJIPANAYIOTOU, 1987a)

	Low protein		High protein		
	male	female	male	female	S D
No. of animals	14	15	16	13	-
Initial weight (kg)	14.2	12.5	14.8	12.2	2.82
Final weight (kg)	19.3	16.1	21.9	16.5	3.61
Weight gain (kg/day)	0.134	0.089	0.177	0.106	0.038
Concentrate intake (kg/day)	0.98		0.85		-

The percentage of lean and fat in the "best end neck" joint (7th to 10th rib inclusive) was not significantly altered with increased dietary CP of Damascus kids (HADJIPANAYIOTOU, 1982b). This was in line with the data of ASH and NORTON (1987b) where there was no effect of dietary CP concentration on the chemical composition of the empty body weight gain or on carcass tissue distribution of Australian Cashmere goat. However, this is at variance with the data of ANDREWS and ORSKOV (1970) with lambs who reported a marked decrease in fat content of carcass with increased dietary protein concentration.

RESPONSE TO DIFFERENT PROTEIN SUPPLEMENTS

The use of diets of high CP content results in high concentration of rumen NH_3-N which cannot be efficiently utilized by rumen micro-organisms, and NH_3 is excreted in the urine. One way to improve ruminal protein output is to protect dietary protein from microbial degradation and/or select among protein supplements of low degradability.

Feeding kids formaldehyde (HCHO) treated groundnut cake resulted in higher growth rates and better feed conversion efficiencies compared to the control (MUDGAL and SENGAR, 1981). Contrary, in our studies (HADJIPANAYIOTOU, unpubl.) protection of soybean meal with HCHO (3 kg 40 % HCHO/t soybean meal) did not improve daily bodyweight gain and/or feed to gain ratio in either early weaned lambs or kids offered pelleted (5 mm cubes) barley grain based (75 to 85 kg) concentrates (190 CP/kg DM) *ad libitum* along with 100 g of lucerne hay daily (Table 3). The lack of difference between the control and the treated group obtained in our study might be ascribed to the high concentrate feeding system which induces low rumen pH values that might lower rumen degradation values (LOERCH, 1985) of untreated soybean meal and the relatively adequate supplies of rumen undegradable and degradable protein from the control group. A major problem encountered in many studies has been decreased CP digestibility associated with HCHO treatment. In our study however, although treatment decreased protein and DM degradation in the forestomachs of goats (HADJIPANAYIOTOU, unpubl.) apparent DM and CP digestibilities of the finished diets were not affected by treatment (control: DM: 0.83, CP: 0.77; treatment: DM: 0.81, CP: 0.76).

Table 3

THE EFFECT OF FISH MEAL (FM) AND OF FORMALDEHYDE (HCHO) TREATMENT OF SOYBEAN MEAL (SB) (FSB) ON THE PERFORMANCE OF CHIOS MALE LAMBS (L) AND DAMASCUS KIDS (K)

| | Concentrate mixture | | | | | | | | |
| | Soyabean meal (SB) | | HCHO SB (FSB) | | Fish meal (FM) + SB (FMSB) | | FM + FSB | | |
Species	L	K	L	K	L	K	L	K	SD
No. of animals	32	18	13	10	33	18	13	10	-
Daily weigth gain (g)	353	275	348	300	352	317	364	318	48
Feed intake (g/day) :									
- concentrate	1216	1113	1078	1124	1214	1085	1168	1040	-
- concentrate + hay	1316	1213	1178	1224	1314	1184	1268	1140	-
Feed/gain ratio :									
- concentrate	3.65	4.06	3.62	3.75	3.64	3.44	3.80	3.27	-
- concentrate + hay	4.09	4.14	3.90	4.08	3.97	3.49	4.07	3.58	

Protein sources (Fish meal: FM vs Soyabean meal: SB) resulted in different gain responses in lambs than kids. Male kids in contrast to lambs grew faster and had better feed conversion efficiency on the FM supplemented diet than those on the SB supplemented diet (Table 3). This difference might be ascribed to differences in rumen degradable / undegradable protein and/or amino acid requirements between the two species. Just after weaning, FM gave better results for feed intake and weight gain than soybean meal and in the study of TANABE et al. (1975). Under the conditions prevailing in our studies feeding formaldehyde-treated soybean meal to lambs and kids cannot be justified. This is so concerning feeding FM to female kids and male and female lambs. With regard to feeding FM to fast growing male kids, the incremental revenue resulting from better feed conversion efficiency from feeding FM should be compared with the incremental cost associated with the higher cost of FM before including fish meal in the diets of kids.

Rainfed leguminous grains (HADJIPANAYIOTOU et al., 1985) are good sources of protein (260-313 g/kg DM) and can be used as protein supplements in concentrated feeds. Growth rate, feed intake and feed to gain ratio of kids was similar when soybean meal was replaced partly or completely by common vetch or broad beans grains (Table 4). Kids on field beans supplemented diets performed equally well with those on soybean meal and in the study of STAUB (1974).

Non protein nitrogen has also been used in diets of growing goats. Indeed, urea fed goats tended to grow faster than goats fed groundnut meal (MBA et al., 1974) or groundnut/palm kernel meals (MBA et al., 1975). Furthermore, the groundnut cake fed goats required more protein than urea-fed goats for each g of bodyweight gain.

Table 4

THE EFFECT OF REPLACEMENT OF SOYBEAN MEAL (SB) BY BROAD BEANS (BB) OR
COMMON VETCH GRAIN (VG) ON THE PERFORMANCE OF FATTENING KIDS OFFERED
CONCENTRATE (185 g CP/kg DM) *AD LIBITUM*
(KOUMAS and ECONOMIDES, 1987)

Diet	Protein supplement* Kids					
	S B	SB+BB	B B	SB+VG	VG	S E
Initial weight (kg)	17.75	17.71	17.33	17.71	17.58	1.30
Final weight (kg)	35.09	36.86	33.84	34.07	35.88	2.36
Feed intake (kg) :						
- concentrate	67	70	66	67	70	-
- Lucerne hay	7	7	7	7	7	-
Feed / gain ratio :						
- concentrate	3.89	3.65	4.04	4.12	3.80	-
- concentrate + hay	4.27	4.02	4.47	4.54	4.19	-

(*) 70 days on test.

PROTEIN DEGRADATION IN THE RUMEN

Information on protein degradation in the rumen is required by the new protein evaluation systems that have been proposed to replace DCP or CP in feeding standards (BURROUGHS et al., 1975; ROY et al., 1977; VERITE et al., 1979; ARC, 1980, 1984). The extent and rate of degradation of dietary protein in the rumen is affected by the outflow of small particles from the rumen. Increasing the outflow rate of small particles from the rumen has been shown to result in a significant decrease of DM and CP degradability of protein supplements (ORSKOV et al., 1983). Effective DM and CP degradabilities of soybean and fish meals at 0.05 and 0.08 per h outflow rate were similar for Chios sheep and Damascus goats (HADJIPANAYIOTOU et al., 1988).

Feeding level, proportion of roughage to concentrate, particle size of the basal feed, environmental temperature, feeding frequency and stage of production are known to alter the rumen liquid turn-over rate (ARC, 1984). Studies conducted in Cyprus with mature Chios sheep and Damascus goats have demonstrated that outflow rate is affected by the level of feed intake (HADJIPANAYIOTOU et al., 1988), the proportion of roughage to concentrate, but not by the stage of production (dry vs lactation) (HADJIPANAYIOTOU and HADJIDEMETRIOU, unpubl.). further studies with fast growing Chios lambs and Damascus kids of similar age (approx. 105 days) and at the same level and type of feeding (*ad libitum* feeding of concentrates along with 80 g DM of lucerne hay/head/day) showed no difference between species for outflow rate of small particles from the rumen (Table 5).

Table 5.

OUTFLOW RATE OF SOYBEAN MEAL FROM THE RUMEN
OF CHIOS LAMBS AND DAMASCUS KIDS

	Lambs	Kids
No. of animals	14	15
Feed intake (g DM/kg body weight)	35.1	34.4
Outflow rate (% per h)	5.95	6.11

CONCLUSIONS

Information on protein nutrition of growing goats is very limited. Published data on N-requirements of growing goats differ significantly. Under certain conditions however, the NRC (1981) recommendations for goats are comparable to those of ARC (1980) and NRC (1985) for sheep. There is an urgent need for more data on the protein and energy composition of non-carcass and carcass components of growing goats for a more efficient feed utilization. Early weaned lambs and kids require similar dietary (180 g CP/kg DM) protein concentration up to 100 days of age, whereas thereafter, maximum growth in lambs but not in kids can be obtained with lower dietary CP levels. There are no differences in outflow rate of protein supplements from the rumen of lambs and kids. Many protein supplements can replace each other from the diets of growing lambs and kids. Positive response to fish meal has been reported by male kids but not lambs.

SUMMARY

Studies on nitrogen nutrition of growing goats are limited. Goats had less chemical fat and more moisture and protein in their meats than sheep. Protein degradation and outflow rate of protein supplements in the rumen was similar in sheep and goats. Male kids, in contrast to lambs, grew faster and had better feed conversion efficiency on the fish meal supplemented diet than on the soybean meal diet. Broad beans and vetch grain replaced successfuly soybean in diets of growing kids. Data on the effect of formaldehyde treatment of protein supplements on kid performance are contradictory.

Growth rate of early weaned (52 days of age) kids increased with increasing dietary protein concentration up to 180 g CP/kg feed DM.

Published data on N-requirements of growing goats differ significantly. However, without considering physiological stage, growth rate, dietary energy density and microbial degradation in the rumen, the National Research Council (NRC) recommendations for goats (0.28 g CP/g bodyweight gain) are comparable to those of the Agricultural Research Council (0.20 - 0.38 g CP/g bodyweight gain) and NRC (0.30 - 0.36 g CP/g body weight gain) for sheep.

Keywords: Growing goat, Protein requirements, Protein level, Protein source, Growth.

RÉSUMÉ

La nutrition azotée des caprins en croissance a été étudiée de façon limitée. La viande de caprins contient moins de lipides et plus d'eau que la viande de mouton. La dégradation protéique et le transit des protéines alimentaires dans le rumen sont comparables chez les caprins et les ovins. Les chevreaux mâles, à la différence des agneaux, croissent plus rapidement et ont une meilleure efficacité alimentaire quand ils reçoivent comme source azotée de la farine de poisson que lorsqu'ils ingèrent du tourteau de soja. Les graines de fève et de vesce peuvent remplacer favorablement le soja dans les rations destinées aux chevreaux en croissance. Les résultats relatifs aux effets du traitement à la formaldéhyde des compléments azotés sont contradictoires.

La vitesse de croissance des chevreaux sevrés précocement (à 52 jours) augmente avec l'accroissement du taux protéique dans le régime jusqu'à 180 g de MAT par kg de MS.

Les résultats publiés sur les besoins azotés des chevreaux en croissance diffèrent significativement. Cependant, sans tenir compte du stade physiologique, de la vitesse de croissance, de la densité énergétique du régime et de la dégradation microbienne dans le rumen, les recommandations du National Research Council (NRC) pour les caprins (0.28 g MAT par g de gain de poids) sont comparables à celles de l'Agricultural Research Council (0.20 - 0.38 g de MAT par g de gain de poids) et à celles du NRC pour l'agneau (0.30 - 0.36 g de MAT par g de gain de poids).

REREFENCES

ARC, 1980. Nutrient Requirements of Ruminants, p. 121-181. Commonwealth Agricultural Bureaux, Slough (UK).

ARC, 1984. The Nutrient Requirements of Ruminant Livestock. Suppl. No 1, p. 45. Commonwealth Agricultural Bureaux, Slough (UK).

ANDREWS R.P. and ORSKOV E.R., 1970. The nutrition of the early weaned lamb. II - The effect of dietary protein concentration, feeding level and sex on body composition at two live weights. J. Agric. Sci. (Camb.), 75: 19-26.

ASH A.J. and NORTON B.W., 1987a. Studies with the Australian Cashmere goat. I - Growth and digestion in male and female goats given pelleted diets varying in protein content and energy level. Aust. J. Agric. Res., 38: 957-969.

ASH A.J. and NORTON B.W., 1987b. Studies with the Australian Cashmere goat. II - Effects of dietary protein concentration and feeding level on body composition of male and female goats. Aust. J. Agric. Res., 38: 971-982.

BLAXTER K.L. and WOOD W.A., 1951. The nutrition of the young Ayrshire calf. I - The endogenous nitrogen and basal energy metabolism of the calf. Brit. J. Nutr., 5: 11-25.

BURROUGHS W., NELSON D.K. and MERTENS D.R., 1975. Protein physiology and its application in the lactating cows. The metabolizable protein feeding standard. J. Anim. Sci., 41: 933-944.

CISZUK P. and LINDBERG J.E., 1985. Total nitrogen retention in lactating goats in relation to digested nitrogen and estimated absorption of amino acids. Acta Agric. Scand., 25 (suppl.): 163-176.

ECONOMIDES S., 1986. Comparative studies of sheep and goats: milk yield and milk composition and growth rate of lambs and kids. J. Agric. Sci. (Camb.), 106: 477-484.

FEHR P.M., SAUVANT D., DELAGE J., DUMONT B.L. and ROY G., 1976. Effect of feeding methods and age at slaughter on growth performance and carcass characteristics of entire young male goats. Livest. Prod. Sci., 3: 183-194.

GAILI E.S.E., GHANEM U.S. and MUKHTAR A.M.S., 1972. A comparative study of some carcass characteristics of Sudan desert sheep and goats. Anim. Prod., 14: 351-357.

GONZALEZ J.J., ROBINSON J.J. and MAC HATTIE I., 1979. The effect of source and level of dietary protein on milk production in ewes. Anim. Prod., 28: 453 (Abst.).

HADJIPANAYIOTOU M., 1982a. Protein levels for Chios lambs given high concentrate diets. Ann. Zootech., 31: 269-278.

HADJIPANAYIOTOU M., 1982b. Protein levels for early weaned Damascus kids on high-concentrate diets. Techn. Bull., 43, Agric. Res. Inst., Nicosia (CYPRUS).

HADJIPANAYIOTOU M., 1987a. Feed evaluation, nitrogen requirements and the effect of supplementary feeding on the performance of small ruminants in Cyprus, p. 319-334. In: Isotope aided studied on livestock productivity in Mediterranean and North African Countries. IAEA, Vienna (AUSTRIA).

HADJIPANAYIOTOU M., 1987b. Intensive feeding systems for goats in the Near East, Vol. 2, p. 1109-1141. 4th Intern. Conf. on Goats, March 8-13, 1987, Brasilia (BRAZIL).

HADJIPANAYIOTOU M., ECONOMIDES S. and KOUMAS A., 1985. Chemical composition, digestibility and energy content of leguminous grains and straws grown in a Mediterranean region. Ann. Zootech., 34: 23-30.

HADJIPANAYIOTOU M., KOUMAS A., GEORGHIADES E. and HADJI- DEMETRIOU D., 1988. Studies on degradation and outflow rate of protein supplements in the rumen of dry and lactating Chios ewes and Damascus goats. Anim. Prod., 46: 243-248.

HAENLEIN F.W., 1986. Production of Goat Meat. Farm Animals, 1: 21-27.

HUSTON J.E., 1978. Forage utilization and nutrient requirements of the goat. J. Dairy Sci., 61: 988-993.

KOUMAS A. and ECONOMIDES S., 1987. Replacement of soybean meal by broad bean or common vetch grain in lamb and kid fattening diets. Techn. Bull., 88. Agric. Res. Inst., Nicosia (CYPRUS).

LINDBERG J.E., 1989. Nitrogen metabolism and urinary excretion of purines in goat kids. Brit. J. Nutr., 61: 309-321.

LOERCH S.C., 1985. Effects of slowly degradable protein sources on performance of feedlot cattle under various feeding systems. Nutr. Rep. Int., 32: 1229-1240.

LOUCA A. and HANCOCK J., 1977. Genotype by environment interactions for postweaning growth in the Damascus breed of goat. J. Anim. Sci., 44: 927-931.

LU C.D., SAHLU T. and FERNANDES J.M., 1987. Assessment of energy and protein requirements for growth and lactation in goats, Vol. 2, p. 1229-1247. 4th Intern. Conf. on Goats, March 8-13, 1987, Brasilia (BRAZIL).

MAVROGENIS A.P., ECONOMIDES S., LOUCA A. and HANCOCK J., 1979. The effect of dietary protein levels on the performance of Damascus kids. Techn. Bull., 27, Agric. Res. Inst., Nicosia (CYPRUS).

Mc DOWELL R.E. and BOVE L., 1977. The goat as a producer of meat. Cornell Int. Agr. Mimeogr., 56, Dept. Anim. Sci., Cornell Univ., Ithaca, NY (USA).

MBA A.U;, AKINSOYINU A.O. and OLUBAJO F.O., 1974. Studies on comparative utilization of urea and groundnut cake rations by West African dwarf goats. I - N-balance and growth. Nigerian J. Anim. Prod., 1: 209-216.

MBA A.U., EGBUIWE C.P. and OYENUGA E.A., 1975. Nitrogen balance studies with Red Sokoto (Maradi) goats for the minimum protein requirements. East Afr. Agric. For. J. 40: 285-291.

MORAND-FEHR P., 1981. Growth, p. 253-283. In: C. GALL (Ed.): Goat Production. Academic Press, New York (USA).

MORAND-FEHR P., HERVIEU J., BAS P. and SAUVANT D., 1982. Feeding of young goats, p. 90-104. 3rd Intern. Conf. on Goat Production and Disease. Jan. 10-15, 1982, Tucson, Arizona (USA).

MUGDAL V.D. and SENGAR S.S., 1981. Effect of feeding treated and untreated protein on the growth rate pattern, nutrients utilization and body composition in kids, Vol. 1, p. 180-193. In: MORAND-FEHR P., BOURBOUZE A. and DE SIMIANE M. (Eds): Nutrition and systems of goat feeding. Symposium International, Tours (FRANCE), May 12-15, 1981, ITOVIC-INRA (FRANCE).

NRC, 1981. Nutrient requirements of domestic animals. No. 15, Nutrient requirement of goats. National Academy Press, Washington, DC (USA).

NRC, 1985. Nutrient requirements of sheep. Sixth revised edition. National Academy Press, Washington, DC (USA).

ORSKOV E.R., HUGHES-JONES M. and ELIMAN M.E., 1983. Studies on degradation and outflow rate of protein supplements in the rumen of sheep and cattle. Livest. Prod. Sci., 10: 17-24.

PAPAS A., 1977a. Protein requirements of lactating Chios ewes. J. Anim. Sci., 44: 672-679.

PAPAS A., 1977b. Protein requirements of Chios sheep during maintenance. J. Anim. Sci., 44: 665-671.

ROY J.H.B., BALCH C.C., MILLER E.I., ORSKOV E.R. and SMITH R.H., 1977. Calculation of N-requirement for ruminants from nitrogen metabolism studies, p. 126-129. In: TAMINGA S. (Ed.): Symposium on protein metabolism and nutrition. EAAP, Publ. n° 22. PUDOC, Wageningen (THE NETHERLANDS).

SATTER L.D. and ROFFLER R.E., 1975. Nitrogen requirement and utilization in dairy cattle. J. Dairy Sci., 58: 1219-1237.

SCHMIDELY Ph., MORAND-FEHR P. and BAS P., 1988. Estimation of nitrogen requirement for maintenance and growth in kids treated or not with anabolic agents. Seminar of FAO subnetwork on Goat Nutrition and Feeding, Oct. 3-5, 1988. Potenza (ITALY).

SENGAR O.P.S., 1980. Indian research on protein and energy requirements of goats. J. Dairy Sci. 63: 1655-1670.

SKJEVDAL T., 1982. Nutrient requirements of dairy goats based on Norwegian research, p. 105. 3rd Intern. Conf. on Goat Prod. and Disease. Jan. 10-15, 1982, Tucson, Arizona (USA).

STAUB G., 1974. Utilisation comparée de la féverole graine à différentes sources azotées sur chevreaux (Comparison of *Faba vicia* bean and other nitrogen resources utilization in growing kids). Mémoire de fin d'études, ENSSAA, Dijon (FRANCE).

TANABE S., HARYU T. and TANO R., 1975. The effect of dietary protein concentration on the growth of kids. Bull. Nat. Inst. Anim. Ind., (29), 59-67.

VERITE R., JOURNET M. and JARRIGE R., 1979. A new system for protein feeding of ruminants; The PDI system: Livest. Prod. Sci., 6: 349-367.

WALKER D.M. and FAICHNEY G.J., 1964. Nitrogen balance studies with the milk-fed lamb. II - The partition of urinary nitrogen and sulphur. Brit. J. Nutr., 18: 201-207.

Chapter 10

MINERAL NUTRITION OF GOATS

J. KESSLER

INTRODUCTION

Knowledge of mineral metabolism in the goat and especially knowledge of the major and trace mineral requirements is still fragmentary. However, various publications on this subject demonstrate that the amount of available information increases steadily (HAENLEIN 1980, 1987; KESSLER, 1981; LAMAND, 1981; NRC, 1981; MBA, 1982). The growing interest in goat feeding in countries with big goat populations is instrumental in this matter. Furthermore, the use of goats as model ruminants is becoming more and more important in physiological trials. It is the aim of the present review to summarize knowledge of the mineral metabolism of the goat. Supplementary information for cattle and sheep is not given except where this is absolutely necessary. Instead, reference is made to the literature relevant to those subjects (ARC, 1980; UNDERWOOD, 1981; NRC, 1985; MERTZ, 1987; NRC, 1988).

According to today's knowledge, 23 major and trace minerals are essential to goats (table 1). A deficiency of these elements causes metabolism disorders which can only be prevented or cured by supplying the lacking elements. In the following chapter those essential elements are discussed which are of practical importance to daily goat feeding. Relevant literature is cited for the other elements (NRC, 1980; MERTZ, 1987).

Table 1

ESSENTIAL MINERALS

Major minerals		Trace minerals	
Calcium	(Ca)	Iron	(Fe)
Phosphorus	(P)	Iodine	(I)
Magnesium	(Mg)	Copper	(Cu)
Potassium	(K)	Manganese	(Mn)
Sodium	(Na)	Zinc	(Zn)
Chlorine	(Cl)	Cobalt	(Co)
Sulfur	(S)	Molybdenum	(Mo)
		Selenium	(Se)
		Chromium	(Cr)
		Tin	(Sn)
		Vanadium	(V)
		Fluorine	(F)
		Silicon	(Si)
		Nickel	(Ni)
		Arsenic	(As)
		Lead	(Pb)

MAJOR ELEMENTS

Functions and metabolism

About 99 % of the calcium (Ca) is stored in the skeleton and in the teeth. In the bones calcium acts as the main structural element and also serves as a relatively easily available Ca-reserve which is important for the maintenance of Ca-balance in the organism (homeostasis). The remaining one percent of the Ca is in the extracellular fluids (blood plasma and tissue fluid) and in the soft tissue. Calcium plays a dominant role in conducting nerve impulses, in muscle contraction and in the coagulation of blood (transformation of prothrombin to thrombin). The main absorption site for Ca is the proximal small intestine; both active and passive (diffusion) transport occur. Regulation of Ca-metabolism is mainly effected by the parathyroid hormone, calcitonin and metabolically active vitamin D.

Some 75 - 85 % of the total phosphorus (P) content of the organism is found in the formation of the skeleton in the form of small, flat crystals (hydroxylapatite). The P-containing molecule adenosin-triphosphate plays a key role in the energy transport and storage. P is also component of numerous enzymes involved in carbohydrate, fat and protein metabolism. Another function of P is the buffering of the body fluids. Furthermore, it is important in cell division and as a component of nucleic acids (carriers of genetic information). P is mainly absorbed in the distal part of the small intestine. Its transport through the gut wall takes place passively, depending on the level and form of P-intake, and actively, depending on the P-requirement. Homeostasis of P-metabolism is closely connected with that of Ca-metabolism and is indirectly regulated by the same hormones.

About 60 - 70 % of the magnesium (Mg) occurs in the bones, whereas the remaining 30 - 40 % is found in the soft tissues and in the extracellular body fluids. Many enzymes involved in the carbohydrate, fat and protein metabolism are activated by Mg-ions. Moreover, Mg plays an important role in neuromuscular activity. The main absorption site of Mg is the rumen. In contrast to Ca and P there are no indications that Mg-metabolism is controlled or regulated by hormones. The kidney is of great importance for Mg-excretion. According to the level of intake of Mg, the kidney excretes a larger or smaller quantity of Mg via the urine.

Whereas potassium (K) is mainly intracellular, sodium (Na) is mainly extracellular. The roles of the two electrolytes are the maintenance of osmotic pressure in body fluids, the regulation of the acid-base metabolism of the organism and the maintenance of potentials across cell membranes (neuromuscular excitation). K and Na seem to be absorbed all along the digestive system. Regulation of K- and Na-metabolism is chiefly based on the hormone aldosterone. Amounts of both K and Na which are not retained are mainly excreted in the urine.

Requirements

The following estimates of the mineral requirements of goats are based on the factorial method. As various data are summarized elsewhere (KESSLER, 1981) we intend primarily to describe new results.

There have been few measurements of the inevitable minimal faecal and urinal losses, or the net minimum endogenous requirement for calcium, in goats. By analogy with the requirement of sheep it is assumed that the minimum requirement of goats is 20 mg Ca/kg LW per day. On the basis of experiments with West African dwarf goats made by ADELOYE and AKINSOYINU (1984, 1985) and by AKINSOYINU (1986) the estimates for the maintenance requirement for phosphorus are 27 mg for the growing animal, 12 mg for the adult female and 19.2 mg/kg LW per day for the castrated male animal. Inevitable minimal endogenous losses of 30 mg P/kg LW per day were found by KODDEBUSCH (1988) in lactating German Saanen goats. The latter value is taken in consideration when calculating the requirements. There are no new data for Mg. Therefore, 3.5 mg/kg LW per day is still accepted. There are no data available for the inevitable faecal and urinary K- and Na-losses in the goat. The requirement for K and Na is therefore assumed to be 50 mg and 15 mg/kg LW per day, respectively, as in sheep.

The net requirement for pregnancy in the goat mainly results from the mineral composition of the kid at its birth. On the basis of investigations made by LÜDKE (1971) and especially from the recent results obtained by PFEFFER and KEUNECKE (1986) with Saanen goats an average content of 11.5 g Ca, 6.6 g P, 0.3 g Mg, 2.1 g K and 1.7 g Na per kg empty body weight for the new born kid is to be expected.

The net requirement for lactation is determined by the mineral content of the milk. Among the European breeds such as Saanen and Alpine it varies only a little. Mean values per kg milk are 1.25, 1.0 and 0.14 g for Ca, P and Mg respectively (KESSLER, 1981; LOEWENSTEIN, 1982; BRENDEHAUG and ABRAHAMSEN, 1987). There are, however, big differences among the various African and Asian breeds. Depending on the breed the values per kg milk are between 0.88 and 2.04 g for Ca, 0.65 - 1.24 g for P and 0.13 - 0.87 g for Mg (PARKASH and JENNESS, 1968; AKINSOYINU and AKINYELE, 1979; SAWAYA et al., 1984; GNAN and ERABTI, 1985). According to GUEGUEN and BARLET (1978) and KESSLER (1981, 1987) the milk of Saanen goats and that of Alpines contains 2.1 g K and 0.4 g Na per kg. Compared to these values the K- and Na- contents (1.7 g and 0.6 g per kg milk) are somewhat different in African and Asian breeds.

Up to now the net requirements for body growth in the goat have had to be derived from that in sheep. The investigations made by PFEFFER and KEUNECKE (1986) now allow specific estimates for the goat. According to these authors mean values per kg weight gain, in the weight range birth to 39 kg LW, are: 10.7 g for Ca, 6.0 g for P and 0.4 g for Mg. In the weight range up to 31 kg LW the values for K and Na are 2.4 g and 1.6 g. In the weight range of 32 - 39 kg the quantities of K and Na retained fall each to 0.4 g.

In cases of an adequate or an elevated Ca intake, 25.1 or 36.9 g per day, MARAVAL et al. (1984) estimated the coefficient of true absorption of Ca in lactating goats at 18.0 or 20.2 %, respectively. Comparisons of these values with the coefficient of apparent absorption of Ca found in various experiments (KESSLER, 1981) suggest that the mentioned values are somewhat low. A true availability of 30 % seems to be more realistic. This is also confirmed by HOVE (1984) who observed - according to the Ca-intake - values of 20 - 43 % in lactating goats. P seems to be quite well used by the goat. ADELOYE (1987) found a coefficient of apparent absorption of about 55 % and a coefficient of true absorption of 90 % in lactating and non-lactating West African dwarf goats. About 37 % of the dietary P is derived from monosodium phosphate. In his investigation KODDEBUSCH (1988) also estimated the true availability of dietary P at 90 %. Compared to cattle and sheep (GUEGUEN et al., 1987) these values seem to be high. In the present recommendations a true availability of 65 % is assumed. There are no research results available for the true absorption of Mg by the goat. Great variations (18 to 57 %) have been observed in the apparent absorption (KESSLER, 1981). Considering the various influences, a coefficient of true absorption of about 20 % could be appropriate. The true availability of K and Na cannot be estimated but very roughly. 90 % for K and 80 % for Na might be adequate.

On the basis of the data given in table 2 the daily recommended mineral allowances for the goat are shown in table 3. The Ca- and P-values for maintenance correspond to the absolute NRC recommendations (1981) for goats with low activity. There is an obvious difference between the present estimates of Ca-supply for milk production and those of the NRC: The reason is that the NRC-values for the true availability for Ca (68 %) are too high. However, there are no differences between the NRC and the present data as far as the recommended P-supply per kg milk is concerned.

Table 2.

VALUES USED
IN CALCULATING RECOMMENDED DAILY MINERAL ALLOWANCES FOR GOATS

	Ca	P	Mg	K	Na
Net minimum endogenous requirement (mg/kg LW/day)	20	30	3.5	50	15
Net requirement for pregnancy (g/kg fœtus)	11.5	6.6	0.3	2.1	1.7
Net requirement for lactation (g/kg milk)	1.25	1.0	0.14	2.1	0.4
Net requirement for body growth (g/kg LW)	10.7	6.0	0.4	2.4 (0.4)*	1.6 (0.4)*
Coefficient of true absorption (%)	30	65	20	90	80

(*) : above 32 kg LW.

Table 3.

RECOMMENDED DAILY MINERAL ALLOWANCES FOR THE DAIRY GOAT (55 kg LW)

	Daily requirements (g)				
	Ca	P	Mg	K	Na
Maintenance	3.7	2.5	1.0	3.1	1.0
Gestation (incl. maintenance)					
1st - 3rd month (a)	3.7	2.5	1.0	3.1	1.0
4th - 5th month (b)	7.9	3.6	1.2	3.4	1.2
Lactation (incl. maintenance)					
1 kg (milk/d)	7.9	4.0	1.7	5.4	1.5
2 kg "	12.1	5.5	2.4	7.8	2.0
3 kg "	16.3	7.0	3.1	10.1	2.5
4 kg "	20.5	8.5	3.8	12.4	3.0
5 kg "	24.7	10.0	4.5	14.8	3.5
6 kg "	28.9	11.5	5.2	17.1	4.0

(a) In addition, per kg milk: 4.2 g Ca, 1.5 g P, 0.7 g Mg, 2.3 g K and 0.5 g Na.
(b) Liveweight gain 1.5 kg/week.

Deficiency and excess

The best known metabolic disorder in dairy goats related to calcium is the hypocalcemic puerperal paresis (milk fever). According to investigations in cattle and goats, milk fever is associated with high Ca-excretion in the milk at the beginning of the lactation and a disturbed Ca-homeostasis (JÖNSSON, 1979; BARLET, 1984). Anorexia, atony of the gastro-intestinal tract, ataxia or paresis and subnormal temperature are - according to OVERBY's and ODEGAARD's (1980) observations on 40 dairy goats - the clinical symptoms. In general, milk fever in goats occurs one week before to three weeks after parturition. Cases of hypocalcemic parturient paresis have been reported from various countries such as Norway, U.K., France, Chile, etc. (OVERBY and ODEGAARD, 1980; NAUS, 1982; VERSCHUERE, 1982; PAYNE, 1983). The number of cases varies greatly between the different countries. There are few indications of the efficacy of prophylactic treatments such as low Ca-supply before parturition, oral or parenteral administration of vitamin D and of its metabolites as well as pour-in of easily soluble Ca-gels (MERRALL, 1985).

Like cattle, goats can also fall sick with calcinosis when eating certain plants, e.g. golden oat-grass (Trisetum flavescens [L.] P.B.; KESSLER, 1982). This is caused by the compound 1.25 dihydroxy-cholecalciferol (glycosidic-bound form) found in yellow oat-grass (RAMBECK and ZUCKER, 1985), which can lead to an uninhibited Ca-absorption from the intestine. Goats with calcinosis show atypical symptoms such as emaciation, lower milk yield, locomotor disorders and enhanced abdominal respiration. The necropsy reveals, among other things, calcification of the thoracic aorta, of lungs and kidneys as well as of the digestive tract. As is reported by WANNER et al. (1986) the calcinogenic effects of yellow oat-grass in goats can vary, depending on the mode of fodder preservation. Yet, some other factors also play a role, as numerous investigations show (SIMON et al., 1978; STARK, 1979).

Phosphorus deficiency in growing goats is indicated by a reduced feed intake and a reduced growth rate as well as by a stiff gait and by frequent lying (LÜDKE and HENNIG, 1971). Oestrus and conception do not seem to be negatively influenced. A dietary P-excess is said to stimulate the formation of urinary calculi (SATO and OMORI, 1977; BELLENGER et al., 1981; BARLET, 1984). Also goats can be protected from this disorder if the dietary Ca:P ratio is greater than 2.5:1 or if the ration is supplemented with NH_4Cl (UNANIAN DIAS E SILVA et al., 1982). According to McDOWELL (1985) P-deficiency is the most prevalent mineral deficiency of the grazing ruminant in warm climates.

Magnesium deficiency, manifested in its extreme form as hypomagnesaemic tetany, occurs sporadically in goats (LITTLEDIKE et al., 1983). Inducing factors are an inadequate Mg-supply and low absorption of Mg, caused by - among other things - high ammonium and potassium contents in the rumen. Stress of any kind are also risk factors. MACKENZIE (1980) describes the symptoms of hypomagnesaemia as follows: anxiety, uncontrolled movement, staggering, collapse, convulsion and death.

According to SCHELLNER (1972/73b) the effects of a sodium deficiency are lower milk yield in adult goats and reduced growth in kids. Furthermore, as is the case in cattle the sodium deficiency symptoms such as pica, restlessness and a rough, dull coat can also be observed in goats. A sodium excess - 2 to 3 % added dietary salt in a complete mixed diet with free choice water - seems to have a negative influence on the body weight gains in growing goats (IVEY et al., 1986). Whereas a K-deficiency in goats is rather exceptional, an excess of K in ordinary goat rations occurs quite often. According to KESSLER (1987) high K-contents in the ration can negatively influence Mg-utilization by goats.

TRACE ELEMENTS

Functions and metabolism

Most of the body iron (Fe) is bound to hemoglobin and to a smaller extent to myoglobin. Iron is mainly stored in the liver, in the spleen and in the bone marrow in the form of ferritin and hemosiderin. Fe is a component of various enzymes such as cytochrome oxidase and peroxidases. Iron is important in the binding and transport of molecular oxygen and in resistance to infection. Dietary Fe is absorbed in the duodenum and transported in the blood by the transferrin. As Fe can only be excreted by the body to a limited extent, regulation of homeostasis primarily occurs at the level of absorption (mucosal block).

Copper (Cu) is an important constituent or activator of various enzymes such as cytochrome oxidase, monoamino-oxidase and phenol-oxidase. These enzymes take part in the respiratory chain, in the metabolism of sulphur amino-acids and in the formation of the hair pigment melanin. Lysil oxidase is known to be essential for maintaining the integrity of connective tissue (lungs, bones etc.). Cu also seems to be necessary for the biosynthesis of hemoglobin and for the maturation of the erythrocytes. Cu is chiefly absorbed in the small intestine, but absorption has also been observed in the stomach and large intestine. Absorption of copper is influenced by many endogenous factors (e.g. digestive juices) and exogenous factors (feed ingredients etc.). Certain investigations indicate that - especially in cases of low Cu-levels - the absorption rate is also determined by the requirement of the animal. In the plasma Cu chiefly exists in the ceruloplasmin. Excessive Cu is mainly stored in the liver.

Cobalt (Co) is required by the rumen microorganisms for the synthesis of vitamin B12. In the form of vitamin B12 it is also of importance to the animal. Vitamin B12 takes part in nitrogen, carbohydrate and fat metabolism. Moreover, there are connections with the process of hematopoiesis.

As a constituent of the thyroid hormones T3 and T4, iodine (I) plays a particular role in metabolism. With 70 - 80 % of the total I-content of the animal the thyroid gland is the tissue or organ with the highest I-concentration. During the foetal and juvenile periods I influences the differentiation, growth and development of cells. In adult animals I influences basal metabolism and fertility. In ruminants I-absorption occurs in the rumen and to a lesser extent in the small intestine. The kidney is the main organ of excretion. There is a close relation between I-excretion in the urine, the I-content of the plasma and the uptake of I by the thyroid gland.

Manganese (Mn) plays a role in metabolism as an activator of enzymes (glycosyltransferase, phosphoenol pyruvate carboxykinase etc.) and as a constituent of metalloenzymes (arginase, pyruvate carboxylase etc.). Mn also influences bone metabolism, oestrus and hemoglobin synthesis. There is also a probable connection between Mn and the immunity system. Absorption of Mn takes place all along the small intestine. Only 1 - 4 % of the dietary Mn is absorbed. The ruminant does not seem to have effective reserves of Mn.

As a constituent of enzymes (carboanhydrase, alkaline phosphatase, dehydrogenases etc.) and of compounds such as insulin, zinc (Zn) plays a role in energy, protein and mineral metabolism. Furthermore, there are close relations between Zn and the vitamin A and various hormones (growth hormone, prolactin). The best known function of Zn is in the formation and regeneration of skin and hair cells. Zn is mainly absorbed in the small intestine. There seems to exist an interdependance between absorption rate, requirement and Zn-status. The largest proportion of the non-retained Zn is excreted in the faeces.

Selenium (Se) is an integral constituent of the Se-dependent enzyme glutathion peroxidase (GSH-Px) and plays an important role in fat metabolism. Whereas vitamin E - as an antioxidant - hinders the formation of lipid peroxides. Peroxides are reduced by GSH-Px to less noxious hydroxy acids. Furthermore, Se is believed to influence fertility and resistance to infection. Se is mainly absorbed in the small intestine and in the caecum. There seems to be a certain adaptation of Se-absorption to Se-supply, especially in cases of Se-deficient rations. Orally administered Se is excreted in the faeces, if administered otherwise it is excreted in the urine.

Molybdenum (Mo) concentrations are high in the skeleton, in the liver and in the kidney. Like most other trace elements Mo is part of various enzymes such as xanthine oxidase/dehydrogenase and nitrate reductase. Thus the main roles of Mo, namely participation in the formation of uric acid and in the catabolism of nitrate, are known. Mo is mainly absorbed in the small intestine and excreted in the faeces and less in urine. There are close interactions between sulphur and its metabolites and Mo. The absorption of Mo-anions, for example, is obstructed by the presence of sulphate anions.

Requirements

There are three methods for estimating the trace mineral requirements of dairy goats: balance trials, the dose-effect interrelations and factorial methods. Balance trials are not very effective in study of trace mineral metabolism and the factorial method causes various technical problems. Therefore, data now available mainly stem from dose-effect trials.

There are no data available for the exact iron requirement of the goat. According to LAMAND (1984) the requirement is 30 mg/kg DM. WILKINSON and STARK (1987) suggest that 40 mg/kg DM will meet the requirement.

According to the investigations on Cu-deficiency in goats made by ANKE (1974) a Cu-supply of 8 mg/kg DM meets the requirement. LAMAND (1984) fixes the lower limit at 7 mg and recommends 10 mg/kg DM, provided that there are no interactions with other feed components (Ca, S, Mo, Fe). According to HUMPHRIES et al. (1987) 10 mg might be too high for the non-ruminating kid. These workers presume that cases of Cu-toxicities observed in Angora kids were caused by the administration of a milk replacer intended for calves and containing 10 mg Cu per kg DM.

The goat seems to be less prone to cobalt deficiency than the sheep (CLARK et al., 1987); daily weight gains were not depressed by the intake of grass with a Co-content of less than 0.04 mg/kg DM. MGONGO et al. (1984) reported, however, that a Co-supply of less than 0.01 mg/kg DM did cause deficiency symptoms in East African short-horned goats. 0.1 mg Co/kg DM should certainly meet the Co-requirement of the goat.

HENNIG et al. (1980) and GROPPEL et al. (1981) observed reduced feed intake, growth depression and reproductive disorders in growing and adult goats fed a diet containing 0.05 mg iodine/kg DM compared to a ration with 0.42 mg. LAMAND (1981) reports cases of goitre in kids whose mothers were given a diet containing less than 0.3 mg I/kg DM in the feed. On the basis of these results, and in comparison to other ruminants, the I-requirement of goats can - depending on performance - be fixed at 0.4 - 0.6 mg/kg DM. If the ration contains larger amounts of Brassicas, of nitrate or of cyanogenetic glycosides (BATH et al., 1979) the I-supply needs to be raised to 1.8 mg/kg DM.

According to ANKE and GROPPEL (1970) a manganese level of 20 mg/kg DM does not entirely meet the Mn-requirement of the goat. Therefore 60 mg/kg DM are recommended (HENNIG, 1972) for the goat, as for the other ruminants. WILKINSON and STARK (1987) suggest a Mn-concentration of 40 mg/kg DM to be fully adequate. According to today's knowledge of Mn-metabolism this value might be more realistic than the higher level.

NEATHERY et al. (1973) observed obvious zinc-deficiency symptoms in sexually mature, adult goats which received a diet with 4 mg zinc/kg feed. No pathological alterations, however, were observed in the control animals given 40 mg Zn/kg. Zn deficiency also occurred in primiparous goats fed a ration containing 24 mg Zn/kg DM (SCHELLNER, 1972/73a). In investigations with 9 months old female goats, CHHABRA and ARORA (1985) compared rations containing 15 and 65 mg Zn with or without a supplement of 1800 IU vitamin A. Except for the group with 65 mg Zn and 1800 IU vitamin A, all treatments showed more or less obvious metabolism disorders. On the basis of these results 50 mg/kg DM should meet the Zn-requirement of the goat.

110

During the last years many experiments on the selenium metabolism of the goat were carried out (ALLEN and MILLER, 1981; HUSSEIN et al., 1985; GUBLER, 1986; SZILAGYI et al., 1987). Yet, there exist only estimates for the Se-requirement of the goat. In general a Se-level of 0.1 mg/kg DM is supposed to be adequate.

As ANKE et al. (1978) reported a molybdenum allowance of < 0.06 mg does not meet the requirement of growing, pregnant and lactating goats. On the basis of results obtained in numerous trials with goats in different physiological periods, ANKE et al. (1985) recommend a minimal Mo-content of 0.1 mg/kg DM in the ration.

Table 4 summarizes the trace element requirement per kg feed DM of the goat.

Table 4.

RECOMMENDED TRACE ELEMENT ALLOWANCES FOR THE GOAT

	Recommended dietary allowances (mg/kg DM)
Fe	30 - 40
Cu	8 - 10
Co	0.1
I	0.4 - 0.6
Mn	40
Zn	50
Se	0.1
Mo	0.1

Deficiency and excess

Iron deficiency mainly occurs in kids in the form of alimentary iron-deficiency anaemia and - as BOSS and WANNER (1977) reported - it is often observed between the 20th and 60th day of life. Among other methods parenteral administration of iron dextran is successfully used in prophylactic treatments (WANNER and BOSS, 1978). Iron deficiency results, for example, from a heavy infection with gastrointestinal strongyles and has also been described in older goats (KESSLER et al., 1981). The reason for this anaemia may be that as a result of the blood-suckling activity of the parasites, iron metabolism is disturbed by parasite toxins and erythropoesis is inhibited.

Copper deficiency in kids seems to occur in various regions. Cases of swayback or enzootic ataxia have been reported from Australia (SEAMAN and HARTLEY, 1981), England (INGLIS et al., 1986), The Netherlands (WOUDA et al., 1986), New Zealand (MERRALL, 1985) and from Switzerland (VON BEUST et al., 1983). WOUDA et al. (1986) described the symptoms as follows: ataxia, loss of postural control, spasticity of the hind limbs and muscular weakness, often progressing to permanent recumbency. On the basis of post mortem examinations the same authors assume that - as in sheep - genetic differences are responsible for the susceptibility of goats to Cu-deficiency; African dwarf goats seem to be more susceptible than other breeds. Illthrift, anaemia, change of coat colour, fertility disorders and other symptoms can be observed in Cu-deficient adult goats. Whereas a Cu-intake of more than 15 mg/kg DM in the ration is toxic for the sheep, the toxic limit for the cow is at 100 mg. As is shown by DIECKHOFF (1986) in his review article on copper, the goat seems to be less susceptible to chronic Cu-poisoning than the sheep. According to LAMAND (1981) the toxicity limit for goats is approximately

30 mg/kg DM. SOLAIMAN et al. (1988), however, could not observe any apparent signs of toxicity in growing Nubian goats receiving orally up to 600 mg Cu-sulphate per day in gelatine capsules during several weeks. HUSSEIN (1985) also did not find any signs of chronic copper poisoning in adult goats of the Swedish land race when orally administered copper sulphate capsules containing 15 mg Cu/kg LW on five days per week during seven weeks. However, if for three months before and during the period of the Cu-administration an intramuscular injection of selenium/vitamin E (0.6 mg Se and 30 mg vitamin E per 10 kg LW) was given, the goats showed typical Cu-poisoning symptoms. As already stated elsewhere, a Cu-content of 10 mg/kg DM in milk replacers seems to be the critical limit for the non-ruminating kid (HUMPHRIES et al., 1987).

In his investigation on thyroid status in cobalt and vitamin B12 deficient goats MGONGO et al. (1981) was able to experimentally induce a Co-deficiency in East African Short-horned goats which received a ration with < 0.01 mg Co/kg DM during 15 weeks. The observed deficiency symptoms were macrocytic and normochromic anaemia, loss of body-weight and irregular oestrus cycles (MGONGO et al., 1981; MGONGO et al., 1984). A field case of Co-deficiency in a 7 months old female goat was reported by BRAIN (1983). Yet, CLARK et al. (1987) assume that the goat is less susceptible to Co-deficiency than the sheep. Thus the intake of pasture grass with 0.035 mg Co/kg DM during 4 months did not lead to a significant difference in live weight gain in growing Angora goats compared to control animals, receiving additional Co doses parenterally.

As in other livestock the typical symptom of iodine-deficiency in the kid is thyroid hypertrophy accompanied by slow growth and reduced vitality. I-deficiency in adult goats results in reduced feed intake, reduced conception rate and increased abortion rate (HENNIG et al., 1980; GROPPEL et al., 1981). Furthermore, the incidence of hairlessness in newborn kids increases. Primary or secondary I-deficiency in goats seems to occur on all continents of the world. Comprehensive investigations on I-deficiency in grazing goats were made for example in Australia (CAPLE et al., 1985). Goitre in adult goats of the Murciana breed and of its descendants when fed Brassicas were reported by DE LAS HERAS et al. (1984) from Spain.

According to ANKE (1974), fertility is especially influenced negatively by an extreme manganese deficiency (1.9 mg/kg feed; silent heat, lower non return rate, increased abortion rate, unbalanced sex ratio). High rearing losses, kids with leg deformations (carpometacarpal joint), paralyses and nervous disturbances are further consequences. A normal conception rate, but a lower Mn-content in the milk was found by SCHELLNER (1972/73a) when feeding a ration with an average Mn-content of 21 mg/kg DM to primiparous goats.

MILLER et al. (1964) were the first research workers to experimentally induce a zinc deficiency in goats. In comparison to the control animals, the adult male goats fed various rations poor in Zn showed a lower feed intake and reduced growth rates. Moreover, they displayed typical symptoms such as a listless, weak and unthrifty appearance, dull rough hair coat, loss of hair and/or keratinized skin, very small testicles etc. Similar results were obtained by NEATHERY et al. (1973) with Zn-contents of 4.1 mg/kg in the ration. A field case of Zn-deficiency in a herd of Pygmy goats is reported by NELSON et al. (1984) from the USA. Feed analyses in European and tropical regions (McDOWELL, 1985) suggest that marginal Zn-deficiency in goats may occur more often than is commonly known. As in other ruminants the goat seems to be relatively resistant to high Zn-levels. The first symptoms of chronic Zn-poisoning (decrease in haematocrit, hemoglobin and erythrocyte values and increase in Zn-content of plasma and hair) were observed by KESSLER et al. (1983) in 8-months old male Saanen goats fed Zn-contaminated maize (> 1 800 mg Zn/kg DM) during a period of 4 weeks.

Numerous reports suggest that selenium/vitamin E deficiency often occurs, especially in kids (TONTIS, 1984; MBWIRIA et al., 1986). Illthrift and white muscle disease (nutritional muscular dystrophy) are the most important symptoms. According to observations made by TONTIS (1984) the myocardial form of the white muscle disease occurs less frequently than the skeletal form. The injection of Se/vitamin E and the administration of a Se/vitamin E-containing drench proved to be effective as a prophylactic treatment (GUBLER, 1986). In pregnant and lactating goats a Se-deficiency could be experimentally induced by feeding a ration with 0.04 mg Se/kg DM (SZILAGYI et al., 1987). As far as we know there are no indications that Se/vitamin E deficiency occurs in adult goats in practice. Specific vitamin E-selenium effects on goat reproduction have been claimed, however, but there has been no factual scientific evidence to support this (MERRALL, 1985). As is shown by levels of glutathione peroxidase in the blood, Se-excretion in the milk and of Se-storage in the hair, organic Se is better utilized than inorganic Se by the goat (KESSLER, 1988). This has to be taken into consideration especially in connection with Se-toxicity. PATHAK and DATTA (1984) describe the symptoms of a Se-poisoning, in local goats in Assam, as follows: loss of appetite and body weight, depression, constipation followed by diarrhoea, polydipsia, polyuria, respiratory distress and subnormal rectal temperature.

As in cattle and sheep molybdenum deficiency in goats hardly appears to occur in the fields. Experimentally induced Mo-deficiency in goats results in depressed growth, lower conception rate and in more frequently occurring abortions (ANKE et al., 1985). However, Mo-excess is more important and is associated with secondary Cu-deficiency. There are great differences between the various ruminant species in tolerance to Mo (NRC, 1980; ANKE et al., 1985). According to the latter authors the goat seems to be particularly Mo-tolerating. Yet, the mechanism of this tolerance is said not to be connected with Cu-metabolism.

Besides the problems of requirement, deficiency and excess, the question of the utilization of the various chemical mineral compounds by the goat arises. As far as we know there do not exist any systematic data. Therefore the results obtained with cattle and sheep as well as with laboratory animals have to be used (DRESSLER, 1971; PEELER, 1972; JIMENEZ, 1980; KRATZER and VOHRA, 1986).

SUMMARY

Based on relevant literature, today's knowledge of the mineral nutrition of goats (major and trace minerals) is presented. Functions and metabolism of the minerals are discussed at first. Then, the daily mineral requirements of the goat are estimated with the factorial method (major elements) or with the dose-effect interrelations (trace elements). A discussion of possible effects of a mineral excess or a mineral deficiency on the metabolism in the goat terminates this review.

Keywords : Goat, Major elements, Trace elements, Function, Metabolism, Requirement, Deficiency, Excess.

RÉSUMÉ

Les connaissances actuelles concernant l'approvisionnement des caprins en minéraux essentiels (éléments majeurs et oligo-éléments) sont présentées à partir des résultats provenant de la bibliographie ayant trait à ce sujet. Les fonctions et le métabolisme de chaque élément sont brièvement décrits. Les apports journaliers recommandés en minéraux pour les caprins sont ensuite estimés par la méthode factorielle pour les éléments majeurs et par la relation dose-effet pour les oligo-éléments. Ensuite, l'auteur discute des conséquences éventuelles d'une carence ou d'un excès en minéraux sur le métabolisme des caprins.

REFERENCES

ADELOYE A.A. and AKINSOYINU A.O., 1984. Phosphorus: An estimation of requirements by the female West African dwarf goat. J. Anim. Prod. Res., 4 : 121-129.

ADELOYE A.A. and AKINSOYINU A.O., 1985. Phosphorus requirement of the young West African dwarf (Fouta djallon) goat for maintenance and growth. Nutr. Reports Intern., 32 : 239-244.

ADELOYE A.A., 1987. Monosodium phosphate as phosphorus source for the goat. Nutr. Reports Intern. 35 : 775-781.

AKINSOYINU A.O., 1986. Minimum phosphorus requirement of the dwarf goat for maintenance. Trop. Agric. (Trinidad), 63 : 333-335.

AKINSOYINU A.O. and AKINYELE I.O., 1979. Major elements in milk of the West African dwarf goats as affected by stage of lactation. J. Dairy Res., 46 : 427-431.

ALLEN J.C. and MILLER W.J., 1981. Transfer of selenium from blood to milk in goats and noninterference of copper with selenium metabolism. J. Dairy Sci., 64 : 814-821.

ANKE M. and GROPPEL B., 1970. Manganese deficiency and radioisotope studies on manganese metabolism, p. 133-136. In : MILLS, C.F. (Ed.) : Trace element metabolism in animals, Edinburgh and London (UK).

ANKE M., 1974. Die Bedeutung der Spurenelemente für die tierischen Leistungen. Tag. Ber. Akad. Landwirtsch., Wiss. Berlin, 132 : 197-218.

ANKE M., GRÜN M., PARTSCHEFELD M. and GROPPEL B., 1978. Molybdenum deficiency in ruminants, p. 230-233. In : KIRCHGESSNER M. (Ed.) : Trace element metabolism in man and animals. 3 - Institut für Erährungsphysiologie, München (GERMANY).

ANKE M., GROPPEL B. and GRÜN M., 1985. Essentiality, toxicity, requirement and supply of molybdenum in human and animals, p. 154-157. In : MILLS C.F., BREMNER I. and CHESTERS J.K. (Eds.) : Trace elements in man and animals. TEMA 5 - Commonwealth Agricultural Bureaux, Slough (UK).

ARC, 1980. The nutrient requirements of ruminant livestock. Commonwealth Agricultural Bureaux, Slough (UK).

BELLENGER C.R., RUTAR A.J., ILKIW J.E. and SALAMON S., 1981. Urolithiasis in goats. Austr. Vet. J., 57 : 56.

BARLET J.P., 1984. Calcium, phosphore, troubles du métabolisme phosphocalcique (Calcium, phosphorus and disorders relative to metabolism of calcium and phosphorus), p. 393-398. In : YVORÉ P. and PERRIN G. (Eds.) : Les maladies de la chèvre. Colloque international, Oct. 9-11, 1984, Niort (FRANCE).

BATH G.F., WENTZEL D. and VAN TONDER E.M., 1979. Cretinism in Angora goats. J. South Afr. Vet. Ass., 50 : 237-239.

BOSS P.H. and WANNER M., 1977. Das Blutbild der Saanenziege. Schweiz. Arch. Tierheilk., 119 : 111-119.

BRAIN L.T.A., 1983. Cobalt deficiency in a young goat. Goat Vet. Soc. J., 4 : 45.

BRENDEHAUG J. and ABRAHAMSEN R.K., 1987. Trace elements in bulk collected goat milk. Milchwissenschaft, 42 : 289-290.

CAPLE I.V., AZUOLAS J.K. and NUGENT G.F., 1985. Assessment of iodine status and thyroid function of sheep and goats kept under pastoral conditions, p. 609-613. In : MILLS C.F., BREMNER I. and CHESTERS J.K. (Eds.): Trace elements in man and animals. TEMA 5, Commonwealth Agricultural Bureaux, Slough (UK).

CHHABRA A. and ARORA S.P., 1985. Effect of Zn deficiency on serum vitamin A level, tissue enzymes and histological alterations in goats. Livest. Prod. Sci., 12 : 69-77.

CLARK R.G., MANTLEMAN L. and VERKERK G.A., 1987. Failure to obtain a weight gain response to vitamin B12 treatment in young goats grazing pasture that was cobalt deficient for sheep. New Zealand Vet. J., 35 : 38-39.

DE LAS HERAS M., GARCIA DE JALON J.A., GARCIA MARIN J.F., GALLEGO M., BADIOLA J.J., y BASCUAS J.A., 1984. Bocio congenito en cabras asociado al consumo de coles. Med. Vet., 1 : 41-47.

DIECKHOFF H.-J., 1986. Vergiftung der Haussäugetiere mit dem Schwermetall Kupfer unter besonderer Berücksichtigung der Wiederkäuer (eine Literaturstudie). Diss. Tiergesundheitsamt der Landwirtschaftskammer, Hannover (GERMANY).

DRESSLER D., 1971. Mineralische Elemente in der Tierernährung. Eugen Ulmer, Stuttgart (GERMANY).

GNAN S.O. and ERABTI H.A., 1985. The composition of Libyan goat's milk. The Austr. J. Dairy Techn. Dec. : 163-165.

GROPPEL B., HENNIG A., GRÜN M. and ANKE M., 1981. Untersuchungen zum Jodstoffwechsel. 2. Mitteilung. Arch. Tierernährung. 31 : 153-164.

GUBLER D., 1986. Vergleichende Untersuchung verschiedener Formen der Vitamin E - Selen - Supplementierung beim Kleinwiederkäuer. Dissertation Institut für Tierzucht der Universität, Bern (SWITZERLAND).

GUEGUEN L. and BARLET J.P., 1978. Besoins nutritionnels en minéraux et vitamines de la brebis et de la chèvre (Mineral and vitamin requirements of ewes and goats), p. 19-37. 4e Journées de la recherche ovine et caprine, ITOVIC-SPEOC, Paris (FRANCE).

GUEGUEN L., DURAND M. and MESCHY F., 1987. Apports recommandés en éléments minéraux majeurs pour les ruminants (Major elements allowances for ruminants). Bull. Tech. CRZV Theix., (70), 105-112.

HAENLEIN G.F.W., 1980. Mineral nutrition of goats. J. Dairy Sci. 63 : 1729-1748.

HAENLEIN G.F.W., 1987. Mineral and vitamin requirements and deficiencies, 1249-1266. 4th Intern. Conf. on Goats, March 8-13, 1987, Brasilia (BRAZIL).

HENNIG A., 1972. Mineralstoffe, Vitamine, Ergotropika. VEB, Berlin (GERMANY).

HENNIG A., GROPPEL B., ANKE M. and GRÜN M., 1980. Untersuchungen zum Jodstoffwechsel. Arch. Tierernährung, 30 : 695-705.

HOVE K., 1984. Intestinal radiocalcium absorption in the goat: Measurement by a double-isotope technique. Brit. J. Nutr., 51 : 145-156.

HUMPHRIES W.R., MORRICE P.C. and MITCHELL A.N., 1987. Copper poisoning in Angora goats. Vet. Rec., 121 : 231.

HUSSEIN K.S.M., 1985. Copper toxicity in small ruminants interaction of selenium and the role of metallothionein. Sveriges Lantbruksuniversitet, Uppsala (SWEDEN).

HUSSEIN K.S.M., JONES B.-E.V. and FRANK A., 1985. Selenium copper interaction in goats. Zbl. Vet. Med. A., 32 : 321-330.

INGLIS D.M., GILMOUR J.S. and MURRAY I.S., 1986. A farm investigation into swayback in a herd of goats and the result of administration of copper needles. Vet. Rec., 118 : 657-660.

IVEY D.S., TEH T.H., ESCOBAR E.N. and SAHLU T., 1986. Salt requirement for growing dairy goats. J. Dairy Sci., 69 (suppl. 1) : 112.

JIMENEZ A.A., 1980. Availability of minerals for ruminants. Feedstuffs, 52 : 16.

JÖNSSON G., 1979. Aetiologie und Prophylaxe der Gebärparese des Rindes. Uebers. Tierernährg., 7: 193-216.

KESSLER J., 1981. Eléments minéraux majeurs chez la chèvre - Données de base et apports recommandés (Major and trace elements in goats. Basic data and recommended allowances), Vol., 1, p. 196-209. In : MORAND-FEHR P., BOURBOUZE A. and DE SIMIANE M. (Eds.) : Nutrition and systems of goat feeding. Symposium International, Tours (FRANCE), May 12-15, 1981, INRA-ITOVIC, Paris (FRANCE).

KESSLER J., WANNER M. and PFISTER K., 1981. Einfluß einer starken Verwurmung (Magen-Darm-Strongyliden) auf Mineralstoffstoffwechsel, ausgewählte haematologische und klinisch-chemische Parameter sowie auf die Entwicklung bestimmter Organe beim wachsenden Kleinwiederkäuer. Zeitschr. Tierphys. Tierern. und Futtermittelkde, 46 : 170-181.

KESSLER J., 1982. Goldhaferbedingte Kalzinose beim Kleinwiederkäuer. Mitt. für die Schweizerische Landwirtschaft, 30 : 179-184.

KESSLER J., WANNER M. and RIHS T., 1983. Estimation of the toxicity of zinc-contaminated maize for the goat, p. 27-39. In : BOKORI J. (Ed.) : Trace elements in animal nutrition. Budapest (HUNGARY).

KESSLER J., 1987. Effect of a high dietary potassium content on Mg-, K- and Na-metabolism in the lactating goat. Ann. Zootech., 36 : 329 (Abst.).

KESSLER J., 1988. In Press.

KODDEBUSCH L., 1988. Untersuchung zur Verwertung von Phosphor verschiedener Herkünfte bei laktierenden Ziegen. Diss. Institut für Tierernährung, Rheinische Friedrich-Wilhelms-Universität, Bonn (GERMANY).

KRATZER F.H. and VOHRA P., 1986. Chelates in nutrition. CRC Press, Florida (USA).

LAMAND M., 1981. Métabolisme et besoins en oligo-éléments (Trace element metabolism and requirements), Vol. 1, p. 210-217. In : MORAND-FEHR P., BOURBOUZE A. and DE SIMIANE M. (Eds.) : Nutrition and systems of goat feeding. Symposium International, Tours (FRANCE), May 12-15, 1981, INRA-ITOVIC, Paris (FRANCE).

LAMAND M., 1984. La pathologie liée aux carences en oligoéléments (Pathology relative to trace element deficiencies), p. 415-423. In : YVORÉ P. and PERRIN G. (Eds.) : Les maladies de la chèvre. Colloque international, Oct. 9-11, 1984, Niort (FRANCE).

LITTLEDIKE E.T., STUEDEMAN J.A., WILKINSON S.R. and HORST R.L., 1983. Grass tetany syndrome, p. 173-195. In : FONTENOT J.P. (Ed.) : Role of magnesium in animal nutrition. Proc. John Lee Pratt Intern. Symp., Blacksburg, Virginia (USA).

LOEWENSTEIN M., 1982. Dairy goat milk and factors affecting it, p. 226-236. 3rd Intern. Conf. on Goat Production and Diseases, Jan. 10-15, 1982, Tucson (USA).

LÜDKE H., 1971. Phosphor und Fruchtbarkeit bei Milchkühen und Ziegen. Wissenschaftliche Zeitschrift der Karl-Marx-Universität, Leipzig, 20 : 482-486.

LÜDKE H. and HENNIG A., 1971. Der Einfluß des Phosphors auf das Wachstum und die Fortpflanzungsleistung von Ziegen. Monatshefte für Veterinärmedizin, 26 : 890-893.

MACKENZIE D., 1980. Goat husbandry. Faber and Faber, London (UK).

MARAVAL B., LAURENT F. and VIGNON B., 1984. Effect of diet composition on Ca and Mg utilization by lactating goats. Paper N 5b.16. 35th Annual Meeting of the EAAP, Aug. 6-9, 1984, The Hague (NETHERLANDS).

MBA A.U., 1982. Mineral nutrition of goats in Nigeria, p. 109-112. 3rd Intern. Conf. on Goat Production and Disease, Jan. 10-15, 1982, Tucson (USA).

MBWIRIA S.K., DICKINSON J.O. and BELL J.F., 1986. Blood selenium concentrations of sheep and goats from selected areas of Kenya. Trop. Anim. Hlth Prod., 18 : 159-165.

MAC DOWELL L.R., 1985. Nutrition of grazing ruminants in warm climates. Academic Press Inc., San Diego, California (USA).

MERTZ W., 1987. Trace elements in human and animal nutrition. Academic Press Inc., San Diego, California (USA).

MERRALL M., 1985. Nutritional and metabolic diseases, p. 126-151. Proc. of a course in goat husbandry, Palmerson North (NEW ZEALAND).

MGONGO F.O.K., GOMBE S. and OGAA J.S., 1981. Thyroid status in cobalt and vitamin B12 deficiency in goats. Vet. Rec., 109 : 51-53.

MGONGO F.O.K., GOMBE S. and OGAA J.S., 1984. The influence of cobalt/vitamin B12 deficiency as a "stressor" affecting adrenal cortex and ovarian activities in goats. Reprod. Nutr. Dévelop., 24 : 845-854.

MILLER W.J., PITTS W.J., CLIFTON C.M. and SCHMITTLE S.C., 1964. Experimentally produced zinc deficiency in the goat. J. Dairy Sci., 47 : 556-559.

NAUS A.H., 1982. Diseases present in Chilean goats (1975-1980), p. 343. 3rd Intern. Conf. on Goat Production and Disease, Jan. 10-15, 1982, Tucson, Arizona (USA).

NEATHERY M.W., MILLER W.J., BLACKMON D.M., PATE F.M. and GENTRY R.P., 1973. Effects of long term zinc deficiency on feed utilization, reproductive characteristics, and hair growth in the sexually mature male goat. J. Dairy Sci., 56 : 98-105.

NELSON D.R., WOLFF W.A., BLODGETT D., LUECKE B., ELY R.W. and ZACHARY J.F., 1984. Zinc deficiency in sheep and goats: three field cases. JAVMA, 184 : 1480-1485.

NRC, 1980. Mineral tolerance of domestic animals. National Academy Press, Washington, DC (USA).

NRC, 1981. Nutrient requirements of goats: Angora, dairy and meat goats in temperate and tropical countries. N° 15, National Academy Press, Washington, DC (USA).

NRC, 1985. Nutrient requirements of sheep. National Academy Press, Washington, DC (USA).

NRC, 1988. Nutrient requirements of dairy cattle. National Academy Press, Washington, DC (USA).

OVERBY I. and ODEGAARD S.A., 1980. Hypocalcemia in goats. Norsk Veterinaertidsskrift, 92 : 21-25.

PARKASH S. and JENNESS R., 1968. The composition and characteristics of goat's milk: a review. Dairy Sci. Abstr., 30 : 67-87.

PATHAK D.C. and DATTA B.M., 1984. The effects of oral administration of sodium selenite on clinical signs and mortality. Indian Vet. J., 61 : 845-846.

PAYNE J.M., 1983. Maladies métaboliques des ruminants domestiques. Le Point Vétérinaire, Maisons-Alfort (FRANCE).

PEELER H.T., 1972. Biological availability of nutrients in feeds: availability of major mineral ions. J. Anim. Sci., 35 : 695-712.

PFEFFER E. and KEUNECKE R., 1986. Untersuchungen über die Gehalte an Protein, Fett und Mineralstoffen im Körper wachsender Ziegen. J. Anim. Physiol. Anim. Nutr., 54 : 166-171.

RAMBECK W.A. and ZUCKER H., 1985. Vitamin-D-Metabolite in der calcinogenen Pflanze Trisetum flavescens (Goldhafer). Z. Tierphysiol., Tierernährg. u. Futtermittelkde, 54 : 95-96.

SATO H. and OMORI S., 1977. Incidence of urinary calculi in goats fed a high phosphorus diet. Jap. J. Vet. Sci., 39 : 531-537.

SAWAYA W.N., KHALIL J.K. and AL-SHALHAT A.F., 1984. Mineral and vitamin content of goat's milk. J. The American Dietetic Assoc., 84 : 433-435.

SCHELLNER G., 1972/73a. Die Wirkung von Zink- und Manganmangel und Zink- und Manganzufütterung auf Wachstum, Milch- und Milchfettleistung und Fruchtbarkeit beim Wiederkäuer. Jrb. Tierernährung und Fütterung, 8 : 137-155.

SCHELLNER G., 1972/73b. Die Wirkung von Natriummangel und Natriumbeifütterung auf Wachstum, Milch- und Milchfettleistung und Fruchtbarkeit bei Ziegen. Jrb. für Tierernährung und Fütterung, 8: 246-259.

SEAMAN J.T. and HARTLEY W.J., 1981. Congenital copper deficiency in goats. Austr. Vet. J., 57 : 355-356.

SIMON U., DANIEL P., HÄNICHEN T., DIRKSEN G., 1978. Kalzinogene Wirkung verschiedener Sorten des Goldhafers (Trisetum flavescens [L.] P.B.) bei Schafen. Das wirtschaftseigene Futter, 24 : 209-213.

SOLAIMAN, S.G., QURESHI M.A., DAVIS G. and SPARKS J., 1988. Induced copper toxicity in goats. (a) Toxicity. J. Anim. Sci., 66 (suppl. 1) : 371-372.

STARK H., 1979. Untersuchungen zur lichtabhängigen Bildung und Verteilung von Cholecalciferol im Goldhafer (Trisetum flavescens). Vet. Med. Diss. München (GERMANY).

SZILAGYI M., ANKE M., ANGELOW L., BALOGH I. and SURI A., 1987. Biochemical parameters in animals fed on selenium deficient nutrients. Paper N 5.42. 38th Annual Meeting EAAP, Sept. 27 - Oct. 1, 1987, Lisbon (PORTUGAL).

TONTIS A., 1984. Zum Vorkommen der nutritiven Muskeldystrophie (NMD) bei Zicklein in der Schweiz. Schweiz. Arch. Tierheilk., 126 : 41-46.

UNANIAN DIAS E SILVA M., DIAS FELICIANO A SILVA A.E. and SANTA ROSA J., 1982. Observations on several cases of urolytiasis in goats, p. 348 (Abst.). 3rd Intern. Conf. on Goat Production and Disease, Jan. 10-15, 1982, Tucson, Arizona (USA).

UNDERWOOD E.J., 1981. The mineral nutrition of livestock. Commonwealth Agricultural Bureaux, Slough (UK).

VERSCHUERE B., 1982. Epidémiologie de la pathologie caprine. Enquête dans le département des Deux-Sèvres (Epidemiology of goat pathology. Inquiry in Deux Sevres area). Thèse de Doctorat Vétérinaire, ENV Toulouse (FRANCE).

VON BEUST B.R., VANDERVELDE M., TONTIS A. and SPICHTIG M., 1983. Enzootische Ataxie beim Zicklein in der Schweiz. Schweiz. Arch. Tierheilk., 125 : 345-351.

WANNER M. and BOSS P.H., 1978. Parenterale Verabreichung von Eisendextran an neugeborene Zicklein. Schweiz. Arch. Tierheilk., 120 : 369-375.

WANNER M., KESSLER J., MARTIG J. and TONTIS A., 1986. Enzootische Kalzinose bei Ziege und Rind in der Schweiz. Schweiz. Arch. Tierheilk., 128 : 151-160.

WILKINSON J.M. and STARK B.A., 1987. The nutrition of goats, p. 91-106. In : HARESIGN W. and COLE D.J.A. (Eds.) : Recent advances in animal nutrition. Butterworths, London (UK).

WOUDA W., BORST G.H.A. and GRUYS E., 1986. Delayed swayback in goat kids, a study of 23 cases. The Vet. Quarterly, 8 : 45-56.

Chapter 11

VITAMIN NUTRITION OF GOATS

J. KESSLER

INTRODUCTION

There is a marked lack of information on vitamin metabolism and especially on vitamin requirements in the goat. This is clearly shown in a review article by VÖLKER and STEINBERG (1981). It is therefore not surprising that the results obtained in cattle and sheep experiments often have been used for the goat. Needless to say, the physiological pecularities of the goat must be taken in consideration in making such extrapolations.

FAT-SOLUBLE VITAMINS

Functions and metabolism

The most important provitamin A for the ruminant is beta-carotene. In the intestinal wall beta-carotene can be converted to retinol by cleavage or by stepwise degradation. As some investigations reveal, various microorganisms of the digestive tract are able to synthesize carotene (MAJUMDAR and GUPTA, 1960), yet it is not utilized by the goat. In contrast to cattle there is no beta-carotene in the plasma of the goat. It is not clear whether the beta-carotene has an independant function in relation to fertility (oestrus, ovulation etc.) in goats as in cattle.

Vitamin A, chemically speaking an alcohol, is required for the formation and regeneration of the pigment rhodopsin, which is important for vision. Further physiological functions are the formation and maintenance of the epithelial cells of skin and mucous membranes and participation in the bone metabolism. Vitamin A appears to increase resistance against infections and parasitic diseases. The main absorption site for vitamin A is - often in conjunction with lipids - the upper part of the small intestine. With increasing intake the efficiency of utilization of vitamin A decreases. The main storage organ is the liver, which contains about 75 - 90 % of the reserves.

Vitamin D2 (ergocalciferol) and vitamin D3 (cholecalciferol) are the two most important forms of vitamin D. With few exceptions vitamin D is not found in feeds. Therefore, biosynthesis of the vitamin in the skin, induced by UV-rays, is very important for the animal. Vitamin D takes part in the absorption of Ca and P from the gastrointestinal tract, influences their deposition and resorption in the bone and affects the extent of their excretion. The 1.25-(OH)2-D3 metabolized in the liver and kidney is the metabolically active form of the vitamin D. The absorption site of the orally administered vitamin D is the small intestine. The extent of absorption depends both on the supply and on the requirement of the animal.

Various chemical compounds with vitamin E-effects are called vitamin E. Among these compounds alpha-tocopherol is biologically the most active form. Beside its effects as an antioxidant (see also chapter on "Selenium") vitamin E participates in the regulation of carbohydrate, creatine and muscle metabolism and in glycogen metabolism. It is also important for the stimulation of antibody formation. Vitamin E is mainly absorbed in the small intestine like the other fat-soluble vitamins, but the stomach also is of some importance in absorption. Fat tissue and the liver are the main vitamin E storage sites. Vitamin E and its metabolites are chiefly excreted in the faeces.

Requirements

There are hardly any investigations into the vitamin requirements of goats. As is shown in the review article by VÖLKER and STEINBERG (1981) most data are recommended values derived from cattle and sheep. They must be read with this view in mind.

The question whether the goat - like the dairy cow - requires a specific amount of beta-carotene can not be answered at the present time. There are certainly great differences between the beta-carotene metabolism of the goat and that of cattle. Therefore, it is hardly possible to compare them with each other. The recommended vitamin A-level for the goat is estimated - depending on the method of calculation - at 3 500 to 11 000 IU per day (VÖLKER and STEINBERG, 1981). 250 - 1 500 IU per day is the assessment of the daily vitamin D requirement of the goat. Data on the vitamin E allowance are vague. Depending on the author 5 - 100 mg alpha-tocopherol per day have been said to meet the vitamin E requirement of the goat (Table 1).

Table 1

RECOMMENDED VITAMIN ALLOWANCES OF THE GOAT

	Recommended daily allowances		
Beta-carotene		?	
Vitamin A (IU)	3 500	-	11 000
Vitamin D (IU)	250	-	1 500
Vitamin E (mg)	5	-	100

Deficiency and excess

Vitamin A deficiency in kids results in reduced growth rates and lower resistance to diseases. According to VÖLKER and STEINBERG (1981) the consequences in adult goats are night-blindness, poor reproductive performance, pronounced cornification of epithelial cells in all parts of the body and bone deformities. CALDAS (1961) described the clinical symptoms of spontaneous avitaminosis A in a Brazilian goat herd as follows: diarrhea and epizootic abortions in adult females; diarrhea, lachrymation and opacity of the cornea in the kids. DESHMUKH and HONOMODE (1982) improved the sperm quality of bucks, of an Indian local breed, by injection of beta-carotene or of vitamin A, respectively, if the supply with these elements were short. Sperm quantity and libido, however, were not affected. Primary or secondary vitamin A deficiency is to be expected in tropical regions with extreme dry periods (McDOWELL, 1985). As in most species high levels of vitamin A also lead to metabolic disorders in goats. Decreased spinal fluid pressure was observed in adult male goats receiving 186 000 IU vitamin A/kg diet in the form of retinyl palmitate during a period of 16 weeks (NRC, 1987).

Vitamin D deficiency certainly occurs in goats kept exclusively indoors without vitamin D-supplementation in the ration. Rickets (improper calcification of the organic matrix of bone) in kids and osteomalacia (demineralization of the bones) in adult animals can be said to be the most important deficiency symptoms. As in sheep 25 000 IU/kg feed with an exposure time of < 60 days and 2 200 IU with an exposure time of > 60 days is suggested to be a safe upper limit for vitamin D3 (NRC, 1988). Soft tissue calcification is the best known effect of D-hypervitaminosis.

The importance of a vitamin E deficiency in goats has already been treated in connection with Se (see Chapter 10). It may be added that, unlike Se, vitamin E has a relatively low toxicity. This might be one reason for the absence of publications on the toxic limits for vitamin E in ruminants and especially in goats.

WATER-SOLUBLE VITAMINS

Functions and metabolism

In ruminants with a fully functioning forestomach vitamin B1 is produced by the rumen microorganisms. Accordingly the goat is normally not dependent on an exogenous supply of vitamin B1. As a cofactor of numerous enzymes (transketolase, pyruvate decarboxylase) vitamin B1 fulfils important functions in carbohydrate metabolism. It also seems to be involved in nerve transmission and/or excitation. Free thiamine can be absorbed by the ruminant in the rumen; otherwise its main absorption site is the duodenum. Thiamine absorption appears to be passive at high or pharmacological concentrations and active at low or physiological concentrations.

Like vitamin B1, vitamin B12 (cyanocobalamin) is normally produced by the rumen microorganisms in adequate quantities. As described above, the trace element Co is required by the ruminant for this purpose (see Chapter 10). The functions of vitamin B12 are therefore described under that heading.

Requirements

As already mentioned the adult goat has no specific requirements for the water-soluble vitamins, as these are synthesized by its own microorganisms. According to VÖLKER and STEINBERG (1981), however, the pre-ruminant kid should be given 3 - 8 mg vitamin B1 and 0.02 - 0.05 mg B12 per kg DM in the milk replacer.

Deficiency and excess

Vitamin B1 deficiency in the form of polioencephalomalacia (cerebrocortical necrosis or CCN) has been reported in kids as well as in adult goats (ESPINASSE, 1978; SMITH, 1979; MAXWELL, 1980). According to SMITH (1979) affected animals developed the following symptoms: dullness, blindness, muscle tremors, opisthotonus and in heavy cases the signs progress to recumbency, convulsions with persistent opisthotonus and nystagmus. Vitamin B1-hypovitaminosis is mainly induced by the thiamine antimetabolites (vitamin B1-inactivating thiaminases) in certain plants (pteridophyta, oak bark). Thiamine antimetabolites can also be produced by certain bacteria and fungi (mouldy feed). THOMAS et al. (1987) showed that the ruminal and faecal thiaminase activities are elevated in goats with CCN. A further reason for vitamin B1 deficiency is a high intake of feeds rich in carbohydrates, which tends to increase the vitamin B1 requirement (SMITH, 1979).

SUMMARY

Fat- and water soluble vitamins accomplish important tasks in metabolism in the goat. These tasks and some aspects of the vitamin metabolism are discussed at first. Furthermore, recommendations for the vitamin supply of the goat are made on the basis of today's knowledge. The effects of both a vitamin excess and a vitamin deficiency on the metabolism in the goat are described.

Keywords : Goat, Vitamin, Metabolism, Function, Requirement, Deficiency, Excess.

RÉSUMÉ

Les vitamines liposolubles et hydrosolubles jouent un rôle important dans le métabolisme des caprins. En premier lieu, leurs rôles ainsi que quelques aspects du métabolisme des vitamines sont discutés. Il a été tenté ensuite, à partir des connaissances actuelles, d'émettre des recommandations concernant les apports journaliers en vitamines pour les caprins. Enfin, les conséquences d'une carence ou d'un excès en vitamines sur le métabolisme des caprins sont décrites.

REFERENCES

CALDAS A.D., 1961. Avitaminose A espontânea en caprinos. 0 Biologico. 27 : 266-270.

DESHMUKH G.B. and HONOMODE J., 1982. Effect of vitamin A and carotene on reproductive efficiency of bucks, p. 369 (Abst.). 3rd Intern. Conf. on Goat Production and Disease, January 10-15, 1982, Tucson, Arizona (USA).

ESPINASSE J., 1978. La nécrose du cortex cérébral des ruminants (Necrosis of cerebral cortex in ruminants). World Rev. Anim. Prod. 14 : 49-55.

MAJUMDAR B.N. and GUPTA B.N., 1960. Studies on carotene metabolism in goats. Indian J. Med. Res. 48 : 388-393.

MAXWELL J.A.L., 1980. Polioencephalomalacia in a goat. Aust. Vet. J. 56 : 352.

McDOWELL L.R., 1985. Nutrition of grazing ruminants in warm climates. Academic Press, Inc., San Diego, California (USA).

NRC, 1987. Vitamin tolerance of animals. National Academy Press, Washington, DC (USA).

NRC, 1988. Nutrient requirements of dairy cattle. National Academy Press, Washington, DC (USA).

SMITH M.C., 1979. Polioencephalomalacia in goats. J. American Vet. Med. Assoc. 174 : 1328-1332.

THOMAS K.W., TURNER D.L. and SPICER E.M., 1987. Thiamine, thiaminase and transketolase levels in goats with and without polioencephalomalacia. Austr. Vet. J. 64 : 126-127.

VÖLKER L. and STEINBERG W., 1981. The vitamin requirements of goats - a review, Vol. 1, p. 226-233. In: MORAND-FEHR P., BOURBOUZE A. and DE SIMIANE M. (Eds.): Nutrition and systems of goat feeding. Symposium International, Tours (FRANCE), May 12-15, 1981, INRA-ITOVIC, Paris (FRANCE).

Chapter 12

ETIOLOGICAL ASPECTS OF NUTRITIONAL AND METABOLIC DISORDERS OF GOATS

D. SAUVANT, Y. CHILLIARD and P. MORAND-FEHR

INTRODUCTION

One of the main purposes of research in animal nutrition is to determine requirements for all nutrients, which may be limiting factors in performance and efficiency. For practical feeding, knowledge of nutritional requirements alone is insufficient and must be combined with other data to constitute ultimate recommendations or allowances. Such recommendations are modified values of the requirements and are obtained by taking into account margins of safety and the possibilities of nutrient mobilization. However, the primary basis of overall recommendations depends upon knowledge of nutritional requirements. For goats, only a few reliable and specific data are available (HAENLEIN, 1980b; SENGAR, 1980; SAUVANT, 1981; NRC, 1981). Therefore, the present task is to produce recommendations for practical feeding of goats which will avoid nutritional disorders. Recommendations are generally based upon experimental metabolic studies and clinical observations, which are focused on nutritional disorders.

Since metabolic studies are seldom conducted on dairy goats, this review will be focused on those areas in which scientific research has recently been completed. Two already published reviews on the metabolic and nutritional disorders of goats are available : HAENLEIN (1980a) concerning minerals and HARMEYER and MARTENS (1980), non-protein nitrogen.

GENERAL VIEW OF NUTRITIONAL DISORDERS

Nutritional disorders are primarily a result of a failure in the adjustment of the balance between nutrient input and requirements. These disorders are distinct and specific to a particular nutrient. Imbalances are mainly due to variations in nutrient elimination, storage or *de novo* synthesis. Table 1 presents a general diagram of the origins of production diseases, in relation with these three main metabolic functions. This table describes extreme situations; in fact, one nutrient may be implicated simultaneously in different pathways. The partition of nutrient utilisation among these different pathways varies according to the physiological stage and hormonal status and will differ widely from one nutrient to another.

An excess of any input initiates a process of nutrient elimination or accumulation. Elimination may occur directly, as seen with calcium, or after partial metabolization as with urea produced from excess protein. The accumulation of a nutrient may occur in two situations. It may arise from the insufficient elimination or metabolism of a nutrient for which there is no specific storage, a situation where the nutrient has generally a toxic effect (as ammonia, ketone bodies and some heavy metals),or it may depend on specific storage mechanisms. In the latter case it is important to know the molecules or elements that are stored in their active metabolic form. For instance, fatty acids, phosphorus and calcium are accumulated in adipose or skeletal tissues without transformation, while glucose is metabolized into lipids and is thus no longer available as glucose.

Table 1

GENERAL SCHEME OF THE FUNCTIONS IMPLICATED IN NUTRITIONAL DISEASES

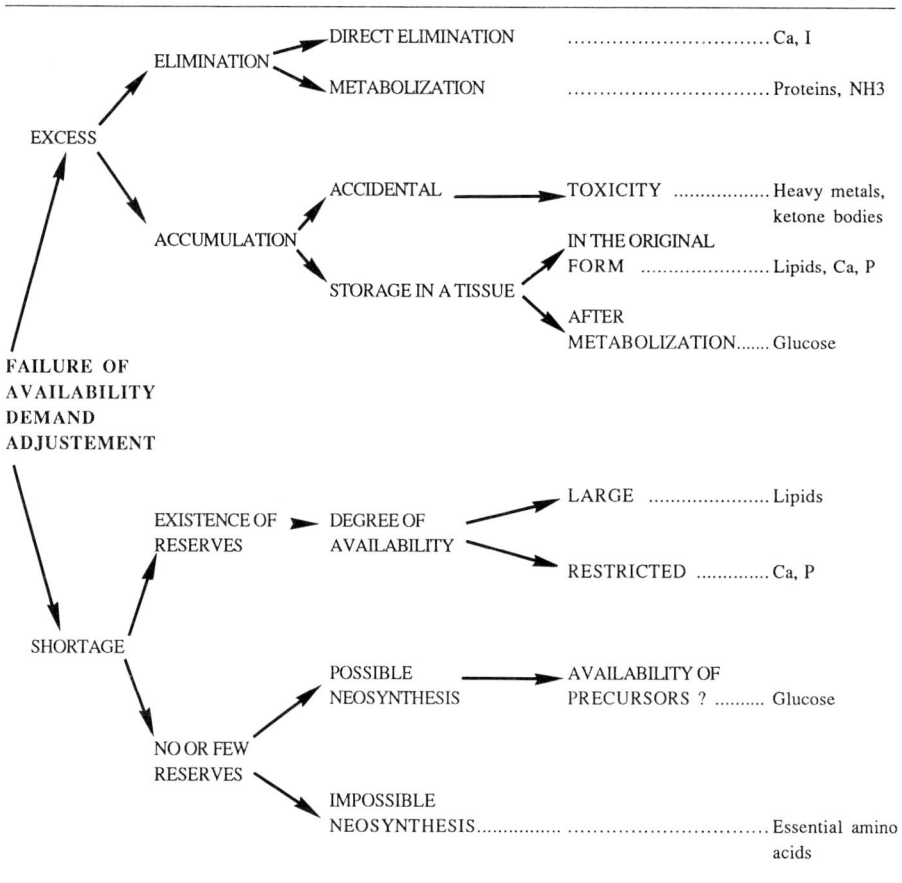

```
                              ┌─ DIRECT ELIMINATION      ............................ Ca, I
              ELIMINATION ────┤
        ┌──                   └─ METABOLIZATION          ............................ Proteins, NH3
EXCESS ─┤
        │                         ACCIDENTAL ──────────→ TOXICITY  ................. Heavy metals,
        │                                                                           ketone bodies
        └── ACCUMULATION ──────┤                         IN THE ORIGINAL
                                                         FORM  ..................... Lipids, Ca, P
FAILURE OF                      STORAGE IN A TISSUE ─────┤
                                                         AFTER
AVAILABILITY                                             METABOLIZATION ....... Glucose
DEMAND
ADJUSTEMENT

                                                          LARGE  ..................... Lipids
          EXISTENCE OF  ──→  DEGREE OF ──┤
          RESERVES            AVAILABILITY
                                                          RESTRICTED  ............. Ca, P

SHORTAGE ─┤                   POSSIBLE  ──────────→  AVAILABILITY OF
                              NEOSYNTHESIS            PRECURSORS ? ......... Glucose
          NO OR FEW ──────┤
          RESERVES            IMPOSSIBLE
                              NEOSYNTHESIS.............. ............................. Essential amino
                                                                                     acids
```

A deficiency in a nutrient may be compensated by mobilization of body reserves. Under such conditions, an animal can support periods of negative balance. For lipids, this adaptive phenomenon may be very efficient. However, the mobilization of these reserves may be associated with metabolic disorders for several reasons :

* a high level of mobilization can lead to diseases due to excesses of metabolites such as ketone bodies;

* a limited capacity for mobilization of existing reserves according to corresponding requirements and gut absorption ability (parturient paresis).

When there is little or no possibility for mobilization of reserves, nutrient shortage leads to a rapid decrease in the performance level. Such a situation is often an etiological factor in various metabolic disorders. However in some situations the possibility of an endogenous synthesis may partly compensate for an insufficient level of input of the exogenous nutrient. But, this depends on precursor availability and is particularly evident in gluconeogenesis.

It is necessary to add a dynamic component to the previous aspects. There are actually several disorders associated with prolonged times of adaptation to enzyme activity, e.g. with the process of ammonia detoxification and intestinal absorption of calcium. In ruminants, such as the dairy goat, sufficient time for adaptation to a new diet is necessary to avoid digestive and metabolic disorders.

ENERGY METABOLISM DISORDERS

Description

The main energy metabolism disorder in goats is due to the accumulation of ketone bodies which become toxic over a certain threshold of concentration in the body fluids. According to PAYNE (1966) about 25 percent of dairy goat mortality in Great Britain, is caused by ketosis which can occur at the end of pregnancy and during the first weeks of lactation. However, mortality rate was lower in the work of MELBY (1987) with Norwegian herds. The incidence of ketosis in goats is associated with both litter size before parturition and high milk production after kidding. Several authors have already described the clinical symptoms and possible treatments of ketosis in goats (KING, 1972; GUSS, 1977), hence they will not be further discussed, in this review which will be focused more on the etiological factors involved.

Ketosis is the consequence of a discrepancy between glucose availability, which is determined by the rate of gluconeogenesis, glucose demand and lipid mobilization. In a situation of intense lipid mobilization associated with a negative energy balance, the nonesterified fatty acids (NEFA) released by adipose tissues invade the whole organism, particularly the liver which then shows signs of steatosis. If glucose precursors and particularly oxaloacetate are not sufficiently available, liver fatty acid oxidation is incomplete and an accumulation of ketone bodies, i.e. aceto-acetate, beta-hydroxybutyrate and acetone. This mobilization is associated with hormonal modifications : growth hormone and probably placental lactogen enhance lipid mobilization while a lack of insulin results in an increase in both fatty liver and ketone body formation (SCHWALM and SCHULTZ, 1975; DEMIGNE et al., 1988).

Influence of the stage in the reproductive cycle

Pregnancy toxemia

During pregnancy the fetuses and products of conception present an exponential growth in weight and requirements. Consequently, during the last weeks before parturition, Alpine goats have a mean live weight gain of 1.5 kg a week, while the total dry matter intake is fairly constant. However, the voluntary intake of dry matter or energy, expressed on a live weight basis, decreases during the last weeks of pregnancy (MORAND-FEHR and SAUVANT, 1978a; see Chapter 3). This discrepancy between input and requirements is associated with a decrease in adipose tissue anabolism (CHILLIARD et al., 1977, 1978) and a progressive increase in lipid mobilization as indicated by plasma non-esterified fatty acid (NEFA) and beta-hydroxybutyrate concentrations (MORAND-FEHR et al., 1977; CHILLIARD, 1985). An increase in lipid mobilization corresponding to enhanced fetal demand for energy and glucose, while glucogenesis is limited by appetite, represents favorable conditions for intense ketogenesis.

Ketosis during lactation

In early lactation the intake capacity of the goat, particularly dairy goats, is not sufficient to satisfy the energy requirements (see Chapter 3). Consequently, the energy balance shows the greatest negative values just after parturition (Figure 1). This situation induces a loss of body weight and a lipomobilization which is maximal during the first days of lactation. This is demonstrated by the plasma concentrations of NEFA, the proportions of stearic and oleic acids in the milk fat, and the milk fat percentage (Figure 2 and CHILLIARD et al., 1987). This lipid mobilization is likely to intensify the degree of liver steatosis as reported in ewes and cows. Liver steatosis may have an unfavorable effect on hepatic metabolism, but until now only limited information is available on this topic (DEMIGNÉ et al., 1988).

During subsequent weeks after parturition there is a progressive decrease in the energy deficit and lipid mobilization (DUNSHEA et al., 1989), associated with an increase in adipose tissue anabolism (CHILLIARD et al., 1979, 1987). However, ketone body levels estimated by the beta-hydroxybutyrate plasma content after one-night fasting, do not reach a maximum immediately after kidding, but 2 to 6 weeks later (Figure 2). This discrepancy in time between the maximum lipid mobilization and ketogenesis is also observed in dairy cows (GARDNER, 1969; FARRIES, 1975; COULON, 1983). Ketogenesis is induced by an increased lactose secretion (Figure 3) and consequently the glucose demand of the udder, if gluconeogenesis becomes insufficient increases ketogenesis. These experimental data agree with clinical observations in cows and goats.

Lipid mobilization constitutes one of the main etiologic factors of ketosis, but the data demonstrate that its level may be high without occurence of ketosis, if gluconeogenesis remains sufficient relative to glucose demand, demonstrating that the determinant factor has the ability to meet glucose requirements. A similar experimental conclusion was drawn by MENAHAN (1966) who induced ketosis in goats by injecting phlorizin, a substance inducing urinary excretion of glucose.

It has sometimes been stated that the dairy goat is less sensitive to lactation ketosis than the dairy cow. In fact no clear experimental evidence of this difference exists between the two species and specific investigations are needed on this subject. However, we disagree with the assumption that there is a lack of fat reserves in goats (NRC, 1981). SKJEVDAL (1979) and CHILLIARD et al. (1981) demonstrated no difference in abdominal and kidney adipose tissue contents between dairy goats and cows, although they are considerable differences in the fat contents in other tissues.

Comparison between the two periods of ketosis risks

Pregnancy toxemia incidence in dairy goats is higher than lactation ketosis, although lipid mobilization estimated by the plasma NEFA content, is larger after kidding. This difference may be because in lactation, unlike in pregnancy, the goat can adapt, by reducing glucose demand and by decreasing the milk production level. HARDWICK et al. (1961) observed a goat milk secretion drop as soon as the plasma glucose level decreased below 0.2 g/l. Otherwise the ratio of lactose to ash may be reduced in the case of lipid mobilization. By contrast in late pregnancy there is no means of adaptation since the fetal glucose demand cannot be sharply reduced without fetal death.

Figure 1

EVOLUTION OF THE REQUIREMENTS AND THE INTAKE OF NET ENERGY
OF ALPINE GOATS DURING EARLY LACTATION
(SAUVANT, 1978)

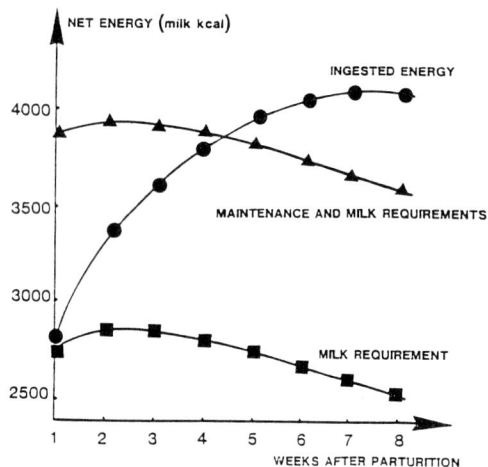

NET ENERGY (milk kcal)

INGESTED ENERGY

4000

MAINTENANCE AND MILK REQUIREMENTS

3500

3000

MILK REQUIREMENT

2500

1 2 3 4 5 6 7 8
WEEKS AFTER PARTURITION

Each point is the mean of 97 animals

Figure 2

LIPOMOBILIZATION PARAMETERS IN GOATS DURING EARLY LACTATION
(SAUVANT, CHILLIARD and MORAND-FEHR, unpubl.)

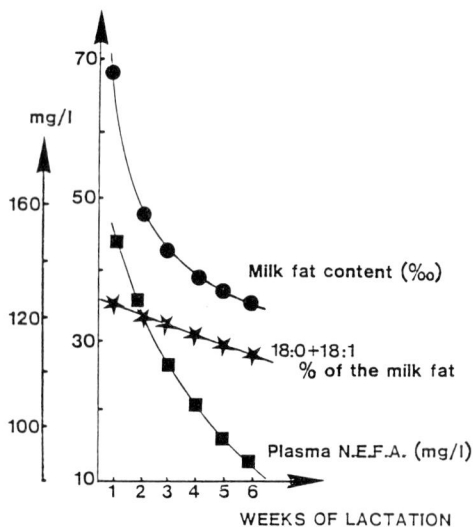

70

mg/l

160   50

Milk fat content (‰)

120   30

18:0+18:1
% of the milk fat

100

Plasma N.E.F.A. (mg/l)

10
1 2 3 4 5 6

WEEKS OF LACTATION

Mean of 124 animals.

Figure 3

LACTOSE PRODUCTION AND CONCENTRATION OF PLASMA BETA-HYDROBUTYRATE IN
GOATS DURING EARLY LACTATION

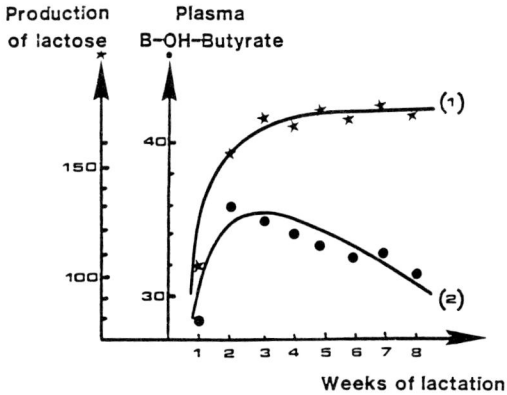

(1) Mean values of 97 animals (SAUVANT and MORAND-FEHR, 1977)

(2) Mean values of 36 animals (MORAND-FEHR, SAUVANT, BAS and ROUZEAU, 1977).

Influence of litter size

To analyse this factor, data from 157 Alpine goats during the last 6 weeks of pregnancy were studied by a within-experimental group component analysis (SAUVANT and MORAND-FEHR, 1977). The first principal component expresses the differences of voluntary dry matter or energy intake, mainly associated with live weight and lactation number (Figure 4). This first component indicates that the heavier animals, which are generally older and tend to be more prolific, do not show on excessive lipid mobilization. The second principal component indicates that when litter size increases for the same body weight, lipid mobilization shows the same tendency. These data suggest that the higher risk of toxemia in prolific goats is mainly because these animals are not able to compensate higher requirements by a higher dry matter intake. Variations in prolificacy have nutritional consequences in pregnant ewes (RUSSEL et al., 1967; TISSIER et al., 1975). For these reasons, goat pregnancy toxemia has been called "twin kid disease" or "twinning disease".

During the first week of lactation the most prolific goats tend to use more extensively their adipose reserves (SAUVANT et al., 1979), but this tendency seems to be reversed during the following weeks. However, these observations are not sufficient to determine whether more prolific goats present a higher risk of lactation ketosis.

Influence of milk yield

It was observed that for the same stage of lactation goats with higher milk yields simultanously showed a higher level of dry matter intake, increased concentrations of NEFA and beta-hydroxybutyrate in plasma and increased percentages of short ($C_{4:0}$, $C_{6:0}$) and long ($C_{18:0}$, $C_{18:1}$) chain fatty acids; these same animals had lower energy balance values and plasma glucose concentration (SAUVANT et al., 1984). Criteria of lipomobilization: NEFA, beta-hydroxybutyrate and milk long chain fatty acids increased exponentially according to milk yield whereas glycemia linearly decreased.

129

Figure 4

CORRELATIONS BETWEEN THE MAIN NUTRITIONAL PARAMETERS
AND THE PRINCIPAL COMPONENTS 1 AND 2
(SAUVANT and MORAND-FEHR, 1977)

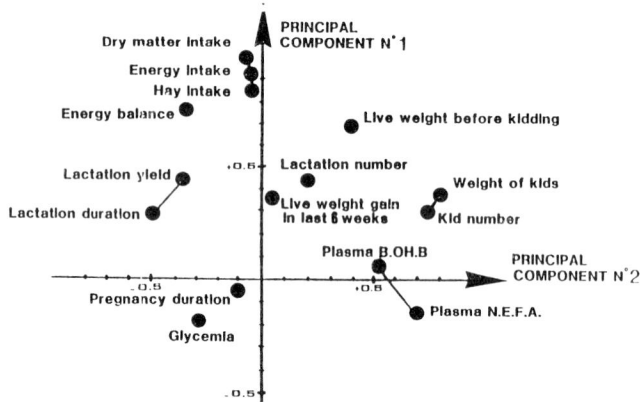

Influence of diet

Late pregnancy

An increase in energy intake reduces the risks of pregnancy toxemia by decreasing lipid mobilization. Several experiments have been conducted using two groups of pregnant goats receiving alfalfa hay ad libitum and different levels of concentrate (150 or 600 g per day) during the last six weeks of pregnancy (MORAND-FEHR and SAUVANT, 1978; SAUVANT, 1981). Data collected from 157 animals, two weeks before kidding, show a mean inter-group 14 % increase in energy intake (mean value of about 165 kcal ME/kg $W^{.75}$) and a 16 % reduction in the plasma NEFA content (mean value 65 mg/l). WENTZEL et al. (1976) conducted a similar study, measuring plasma glucose contents in 20 pregnant does on a high or a low plane of nutrition. Their conclusion was that a drop in glycemia was a primary cause of pregnancy toxemia and abortion in the Angora goat.

Several authors have obtained similar data on the influence of the level of energy intake in pregnant ewes (PATTERSON et al., 1964; RUSSEL et al., 1967; DAVIES et al., 1971; REMESY and DEMIGNE, 1976).

More information is needed on the effect of the energy balance during the period of energy accumulation, i.e. the late phase of lactation and the first phase of pregnancy. In fact, during the periods of intense anabolic activity of the adipose tissue (CHILLIARD et al., 1978, 1981), an excessive level of energy intake may lead to over-fattening of the animal. This over-conditioning induces a more intense phase of lipid mobilization, which occurs at the end of pregnancy and increases the risk of pregnancy toxemia. This tendency was observed in ewes by REMESY and DEMIGNE (1976).

The second principal component in Figure 4 indicates prolific goats produce less milk and have a shorter lactation. Moreover, we observed that prolific goats,which produce less milk recover their body reserves faster during the decreasing phase of milk production (SAUVANT, 1981). These observations emphasize the need to focus attention on the interaction of diet and prolificacy in pregnant goats. The above experimental data indicate that an increase in metabolizable energy intake in the form of increased concentrates is unable to reduce the lipid mobilization level if the mean prolificacy of the group receiving increased energy is higher (Figure 5). These data stress a relative inability of increase in energy intake to compensate for small increases in prolificacy.

Figure 5

BETWEEN-GROUP VARIATIONS IN THE PLASMA NEFA CONCENTRATION ACCORDING TO DIFFERENCES IN ME INTAKE AND LITTER SIZE

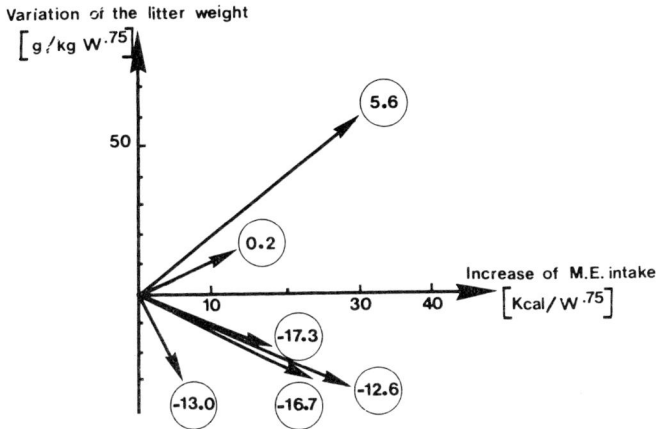

9 to 18 animals per experimental group.

Circled values indicate the between-group difference in the plasma NEFA content two weeks before parturition (mg/l).

Even if compensation for prolificacy by the diet seems to be difficult, it would be useful to have a method for determining the probable number of goat foetuses. This information may be obtained by live weight change or by hormonal control.

Present knowledge is not sufficient to indicate a precise recommendation for energy intake in late pregnancy according to live weight, prolificacy, milk production potential, and previous diet. From our data we propose a range of values for specific net energy intake (above the maintenance energy intake) from a minimum of 500 kcal, to avoid problems of pregnancy toxemia and to obtain a suitable body condition, to a maximum of 1 100 kcal milk NE (MORAND-FEHR and SAUVANT, 1978a). However, if forage quality is high, it is advisable not to use high levels of concentrates (maximum about 600 g/day) for :
- preventing problems of latent acidosis which may lead to an "off feed" period favorable to ketosis;
- maintaining a high level of forage intake which results in a high intake capacity during the post-partum period (SAUVANT et al., 1981b);
- preventing excessive plethoric body condition which may lead to dystocia;
- preventing problems of lactation ketosis (see further).

Onset of lactation

When considering the risk of lactation ketosis, it is indeed necessary to distinguish the influence of the diet ingested before and after parturition (Figure 6). A high level of energy intake before parturition induces a high level of milk production (KONDOS, 1972; MORAND-FEHR and SAUVANT, 1978b; SAUVANT et al., 1979a; SAUVANT, 1981). The energy balance after kidding is more negative (SAUVANT et al., 1979a) as a

131

consequence of an increase in lipomobilization which may enhance the risk of steatosis and ketosis as already shown in ewes (STERN et al., 1978; COWAN et al., 1980) and cows (GARDNER, 1969; FARRIES, 1975; RAYSSIGUIER et al., 1986). According to the recent work of FERNANDEZ et al. (1989) using 3 levels of energy density, milk yield and ketogenesis reached a plateau for an energy density of about 2.0 Mcal EM/ kg DM, while milk fat content and lipomobilization continuously increase (Table 2). Glucose availability causes a discrepancy between lipomobilization and ketogenesis.

The effect of the level of energy intake before parturition on the level of intake after kidding is also due to a poorer appetite in lactating goats which received more energy before kidding. Body fatness depresses the level of voluntary dry matter intake in cows (BINES et al., 1969; COWAN et al., 1980) and goats (CHILLIARD, 1985). Moreover pregnant goats which ingest more concentrates show a lower forage intake level before kidding which seems to influence intake ability in early lactation more than the level of energy intake in late pregnancy (SAUVANT et al., 1981a).

Higher levels of energy intake after parturition improve the milk production level. This improvement may be expressed in different nutritional situations. When goats are in a good body condition at parturition, the mean level of lipid mobilization is higher. In this case, an increase in concentrate and energy intake improves milk production, but reduces the level of lipid mobilization and decreases the risk of ketosis (Table 3). This trend was confirmed with goats in normal body condition and three levels of diet organic matter digestibility or energy density (SAUVANT, 1984; SAUVANT et al., 1987). For goats in a poor body condition at kidding, a higher level of concentrate intake increases the milk production with a simultaneous increase in lipid mobilization (Table 3).

The nature of the ingested energy may also influence the risk of ketosis. GUESSOUS et al. (1974) observed that an addition of sodium acetate to the diet of lactating goats increases plasma beta-hydroxybutyrate concentrations during the critical phase of lactation while a similar level of sodium propionate caused an inverse effect. Thus, concentrate feeding reduces ketosis by increasing both the level of energy intake and propionate production.

It is likely that dairy goats like dairy cows (CHILLIARD et al., 1987b), have limited protein reserves compared to energy resources. Under practical husbandry conditions, protein may be the first limiting factor of milk production in early lactation (see Chapter 17). Supplementary protein may lead to a higher level of lipid mobilization if the digestibility of the diet, and particularly the forage, is low (JOURNET and REMOND, 1981).

These data emphasize the interest in concentrate feeding for preventing ketosis in lactation. However, the rate of concentrate increase and the maximum level must not be excessive to prevent risks of "off feed" periods which may lead to ketosis. After parturition, a maximum rate of increase of 200-250 g concentrate per week can be proposed for goats which produce 4 kg milk or more at the peak of lactation.

Influence of individual variations

There are large individual variations in energy metabolism of dairy goats near parturition. Goats with clinical signs of ketosis may exhibit lower plasma ketone body contents than others without signs of disease. It is possible to observe pregnant goats with a glycemia below 0.2 g/l (normal values from 0.4 to 0.7) without signs of ketosis (BAS et al., unpubl.).

CHILLIARD et al. (1977) observed wide individual differences in the ability of goats to mobilize their adipose tissue. The capacity for mobilization was not related to the quantity of adipose tissue, which was estimated by the "weight-to DNA" ratio of omental adipose tissue, the energy balance or milk production level. From 157 pooled data in early lactating goats, statistical interpretation indicated a significant paternal influence on lipid mobilization evaluated by the NEFA plasma content and the milk stearic and oleic acid yield (SAUVANT and MORAND-FEHR, 1977). Recently, we observed that goats with higher levels of lipid mobilization had the same level of milk production and a significantly lower ratio of lactose to ash in milk (SAUVANT et al., 1981c). This can spare glucose and compensate for the higher lipid mobilization, so allowing the animals to maintain a similar plasma beta-hydroxybutyrate content.

Figure 6

INFLUENCE OF ENERGY LEVEL IN LATE PREGNANCY
ON MILK PRODUCTION DURING THE SECOND WEEK OF LACTATION
(SAUVANT et al., 1979)

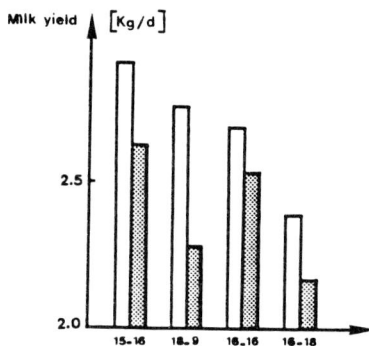

Level of energy intake during the last 6 to 8 weeks of pregnancy :
high : 166 to 183 kcal ME/kg $W^{.75}$
low : 136 to 150 kcal ME/kg $W^{.75}$
Figures in brackets are the number of goats/group.

Table 2

INFLUENCE OF THE DIET ENERGY DENSITY DURING LATE PREGNANCY (1)
ON GOAT PERFORMANCE AND BLOOD PARAMETERS
DURING THE FIRST MONTH OF LACTATION
(FERNANDEZ et al., 1989)

	Energy density (Mcal ME/ kg DM)		
	1.66	2.07	2.48
Body weight (kg)	54.1	56.3	60.0
Milk yield (kg/d)	3.08	3.73	3.38
Milk fat (g/l)	39.1	42.9	50.7
Aceto-acetate (μM)	54.1	80.6	70.4
Beta-hydroxybutyrate (μM)	770.6	1 039.4	977.0
NEFA (μEq/l)	127.4	190.3	339.6
Glucose (mg/l)	689	680	702

(1) 9 to 21 weeks, 60 multiparous Alpine goats.

Table 3

INFLUENCE OF VARIATIONS IN THE LEVEL OF NET ENERGY INTAKE BEFORE AND AFTER
PARTURITION ON BODY RESERVE MOBILIZATION AND PERFORMANCES
OF GOATS IN EARLY LACTATION
(SAUVANT et al., 1978, 1979)

	Level of net energy (milk NE) intake before parturition			
	110 kcal /kg $^{.75}$		95 kcal /kg $^{.75}$	
Level of concentrate intake after parturition (kg/goat/day)	0.94	0.58	0.93	0.56
Number of goats	8	8	8	8
Energy (milk NE) intake (kcal/day)	3 097	2 578	3 166	2 906
3.5 % fat corrected milk yield (kg/day)	5.36	4.91	4.52	3.86
Energy (milk NE) intake above maintenance per kg of 3.5 % FCM (kcal)	320	243	394	394
Energy (milk NE) balance (kcal/goat/day)	- 2 007	- 2 214	- 1 349	- 1 142
Plasma NEFA (mg/l)	140.5	163.6	121.3	89.9
$C_{18:0} + C_{18:1}$ (% of milk fat)	35.8	36.3	32.3	29.7

Influence of exercice

Exercice is known to have a favorable effect against ketosis. This is surprising because exercise increases the energy requirements. In fact, muscular contraction enhances ketone body utilization and release of lactate which is an efficient precursor of gluconeogenesis.

Influence of age

Two investigations (SAUVANT et al., 1979b; MORAND-FEHR et al., unpubl.) indicate an increase in plasma NEFA and beta-hydroxybutyrate content with the age of goats. These observations must be correlated with clinical observations which indicate that, for a similar level of milk production, ketosis occurs more frequently in the 3rd and successive lactations.

Influence of stress

Lipolysis is highly increased by epinephrine in goats. GILL (1970) and CHILLIARD et al. (1981) demonstrated that transporting the animals greatly increased the plasma non-esterified fatty acid content of adult goats, particularly if they were in a negative energy balance. Therefore, it must be kept in mind that stress might be an etiological factor of ketosis.

MINERAL METABOLISM DISORDERS; PARTURIENT PARESIS

The mechanisms of calcium homeostasis so far identified are intestinal absorption, skeletal reserves and to a less extent urinary loss. The endogenous fecal loss of calcium is not negligible (about 18 mg/kg/day in ovine) and almost constant. Consequently, the major site of calcium homeostatic regulation is the intestinal absorption which requires the intervention of a specific transport protein (calcium binding protein). Synthesis of this protein is controlled by vitamin D_3 availability.

Parturient paresis or milk fever already reviewed by BRAITHWAITE (1976) may be regarded as a failure of calcium homeostasis. This disorder may occur in dairy goats (KESSLER, 1981; PAYNE, 1983) and is characterized by a severe hypocalcemia and hypophosphatemia just after kidding. GUSS (1977) observed that parturient paresis might also occur just before parturition. Milk fever is the result of 4 events (BARLET, 1984).

1. The sudden increase in phosphorus and calcium requirements due to milk secretion. The calcium and phosphorus content of milk is maximum just after parturition. These values are reflected in the maximum ash content in milk.

2. Milk lactose concentration is low at this moment: consequently, to maintain osmotic pressure in the lumen of the secretory alveoli, mineral concentration needs to be higher.

3. The capacity for intestinal absorption of calcium increases more slowly than the requirements, as observed in cows (VAN'T KLOOSTER, 1976).

4. The thyroid secretion of calcitonin reduces osteolysis to prevent excessive skeletal demineralization (BARLET, 1974).

It is not possible to propose specific recommendations for goats but it seems necessary to take into account two main points :

1. A low calcium content in the diet leads to a higher calcium absorption (HOVE, 1984). It is possible to exploit this mechanism by giving such a diet during the last weeks of pregnancy to reduce the risk of milk fever.

2. In animals with a high risk of milk fever, it is possible to prevent the disorder by using a synthetic derivative of 1-25-dihydroxycholecalciferol, which activates the synthesis of the calcium-binding protein.

Although the dairy goat seems to be less sensitive to milk fever than the dairy cow, it is necessary to identify other nutritional factors which may be associated with milk fever. We observed that goats with a higher level of lipid mobilization, present a significant increase in the phosphorus and calcium content of milk (Figure 7). These higher phosphorus and calcium requirements may explain the inverse relation observed in cows (BRAITHWAITE, 1976) between plasma calcium and non-esterified fatty acids. The results in Figure 7 lead to the assumption that this relationship between lipid mobilization and phosphorus and calcium contents of milk is not direct, but might be a consequence of the low availability of glucose as indicated above. The relation in Figure 7 may explain why clinical observations sometimes include milk fever in the fat-cow or fat-goat syndrome.

It is also possible to have a genetic approach of the problem. In cows it is known that some breeds such as the Jersey are more sensitive to parturient paresis, and RENNER and KOSMAK (1977) demonstrated that plasma calcium content is significantly heritable within a breed. It would be interesting to determine whether such observations are valid in dairy goats and whether they may be associated with susceptibility to milk fever.

Figure 7

COMPARISON BETWEEN THE NEFA CONTENT
IN PLASMA AND MINERAL COMPOSITION OF GOAT MILK
(SAUVANT, GUEGUEN and MORAND-FEHR, 1978, unpubl.)

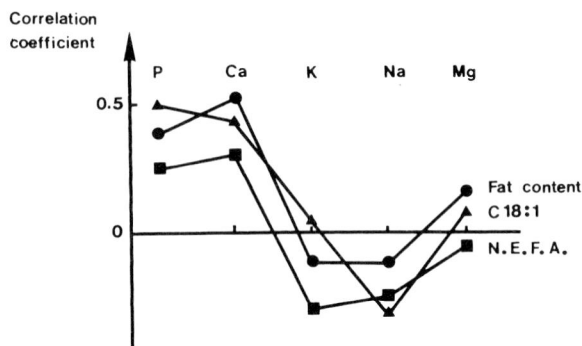

Correlations calculated on 96 data.

CONCLUSIONS

Goats are susceptible to some metabolic and nutritional disorders, in particular ketosis. These disorders are the consequence of underfeeding in extensive systems. However in intensive systems, the most frequent cause is certainly incorrect feeding (insufficient intake of forage or poor quality forages, feeds too rich in fermentable carbohydrates). In fact, in the case of ketosis, the diseases appears, frequently after a case of acidosis (MORAND-FEHR, 1989). However, unstable and unsuitable environmental conditions, in particular at the end of gestation and at the start of lactation (lack of exercise by the goats, too high a stocking rate) favour the triggering off of these problems, especially if other predisposing factors are present. It is then recommended to correct the ration and improve environmental conditions. Unfortunately it is often too late and the goat performance is likely to be largely affected. For this reason it is advisable to take preventive measured by maintaining a satisfactory balance of nutrients in the ration favoring the ingestion of forage and especially ensuring that the goats are neither too fat nor too lean before and after kidding. Recently, several simple methods for the estimation of body condition of goats have been developed and now help the farmer in this area (see Chapter 20).

Problems due to mineral nutrition seem in general less frequent. However, these disorders may also be due to insufficient or unbalanced supplies of calcium, phosphorus or magnesium but also to an unsatisfactory system of feed distribution which leads to large differences in the mineral intake of goats within the herd.

It is quite evident that information on this subject is still very limited in goats. However, in order to progress in this field it seems imperative to simultaneously integrate clinical and nutritional research work. In this way research could be used to produce reliable parameters for estimation of the risk of disorders, linked to a particular diet, or the amplitude of the metabolic imbalance in an animal with clinical symptoms.

SUMMARY

This chapter deals with the etiological factors of nutritional and metabolic disorders of goats. Ketosis which is the main kind of energy and metabolic disorders in goats, is due to the accumulation of ketone bodies which become toxic over a threshold of concentration in the body fluids. Generally it is the consequence of a discrepancy between glucose availability, glucose demand and lipid mobilization. In late gestation, an increase in lipomobilization due to a decrease in feed intake and an enhanced fetal demand for energy result in intense ketogenesis and can induce pregnancy toxemia. In early lactation, ketosis is induced by the increased level of lactose secretion and an intense lipomobilization. Frequently ketosis appears when goats exhibit latent acidosis due to an high proportion of concentrates in the diet or a too limited forage intake or a poor forage quality. Unsuitable environmental conditions (in particular lack of exercise or a too high stocking rate favour the triggering of lactosis).

Parturient paresis may occur in dairy goats but are generally less frequent than ketosis. This disorder is characterized by a severe hypocalcemia and hypophosphatemia just after kidding.

To limit the risk of nutritional and metabolic disorders in goats, it is advisable to adopt a feeding programme reducing overfattening and preventing underfeeding before kidding to favour a high level of forage intake and to avoid suddent changes in diets and other stress.

Keywords: Dairy goat, Pregnancy toxemia, Ketosis, Parturient paresis, Lipomobilization, Gluconeogenesis, Hypocalcemia, Forage intake.

RÉSUMÉ

Ce chapitre traite des facteurs étiologiques des maladies nutritionnelles et métaboliques chez la chèvre. La principale maladie liée au métabolisme énergétique, la cétose, est due à l'accumulation de corps cétoniques qui deviennent toxiques à un certain seuil de concentration dans l'organisme. Généralement, c'est la conséquence d'un déséquilibre entre la disponibilité et la demande en glucose, et la mobilisation des lipides. En fin de gestation, un accroissement de la lipomobilisation lié d'une part, à une baisse du niveau d'ingestion et d'autre part, à une demande énergétique du fœtus augmentée conduit à une cétogenèse intense et peut induire une toxémie de gestation. En début de lactation, la cétose est induite par l'augmentation de la sécrétion de lactose et par une intense lipomobilisation. Souvent, elle apparaît à la suite d'une acidose due à une proportion élevée de concentrés dans le régime, d'un faible niveau d'ingestion de fourrages ou d'une mauvaise qualité des fourrages. Des conditions d'environnement inadaptées (manque d'exercice, densité d'animaux trop élevée) favorise les déclenchements de cétose.

La fièvre de lait peut arriver chez la chèvre laitière mais de façon moins fréquente. Cette maladie se caractérise par des hypocalcémies et des hypophorphotéines importantes juste après le chevrottage.

Pour limiter le risque d'apparition de ces maladies nutritionnelles et métaboliques chez la chèvre, il est conseillé d'adapter un programme alimentaire qui n'aboutisse pas à un engraissement excessif ou à une maigreur importante avec la mise-bas, de favoriser un haut niveau d'ingestion de fourrages et à éviter un brutal changement de régime ou tout autre stress.

REFERENCES

BARLET J.P., 1974. Rôle physiologique de la calcitonine chez la chèvre gestante ou allaitante (Role of calcitonin in pregnant or lactating goats). Ann. Biol. Anim. Bioch. Biophys., 14: 447-457.

BARLET J.P., 1984. Calcium, phosphore, troubles du métabolisme phosphocalcique (Calcium, phosphorus and Ca and P metabolism diseases), p. 393-398. In : YVORE P. and PERRIN G. (Eds.) : Goat diseases, Oct. 9-11, 1984, Niort (FRANCE). INRA Publ. (Les Colloques de l'INRA n° 28), Paris (FRANCE).

BAS P., ROUZEAU A. and MORAND-FEHR, 1980. Variations diurnes et d'un jour à l'autre de la concentration de plusieurs métabolites sanguins chez la chèvre en lactation (Diurnal and daily variations of blood parameter concentration in lactating goats). Ann. Rech. Vet., 11: 409-420.

BAS P., 1980. Facteurs de variation analytiques des métabolites sanguins (Causes of variation of blood parameters concentration), p. 11-70. In : CAAA, Les profils métaboliques en alimentation animale. ADEPRINA, Paris (FRANCE).

BINES J.A., SUZUKI S. and BALCH C.C., 1969. The quantitative significance of long term regulation of food intake in the cow. Brit. J. Nutr., 23: 695-704.

BRAITHWAITE G.D., 1976. Calcium and phosphorus metabolism in ruminants with special reference to parturient paresis. J. Dairy Res., 43: 501-520.

CHILLIARD Y., 1985. Métabolisme du tissu adipeux, lipogenèse mammaire et activités lipoprotéine-lipasiques chez la chèvre au cours du cycle gestation-lactation (Metabolism of adipose tissues, mammary lipogenesis and lipoproteine-lipase activities in goats during gestation and lactation). Thèse Doct. Etat ès-Sci., Univ. Paris 6 (FRANCE).

CHILLIARD Y., SAUVANT D., HERVIEU J., DORLEANS M. and MORAND- FEHR P., 1977. Lipoprotein lipase activity and composition of omental adipose tissue as related to lipid metabolism of the goat in late pregnancy and early lactation. Ann. Biol. Anim. Bioch. Biophys., 17: 1021-1033.

CHILLIARD Y., DURAND G., SAUVANT D. and MORAND-FEHR P., 1978. Activité métabolique du tissu adipeux de la chèvre au cours de la gestation et en début de lactation (Metabolic activity of adipose tissues of goats during gestation and early lactation). C.R. Acad. Sci., Paris, 287 : D, 1131-1134.

CHILLIARD Y., SAUVANT D. and MORAND-FEHR P., 1979a. Goat mammary, adipose and milk lipoprotein lipases. Ann. Rech. Vet., 10: 401-403.

CHILLIARD Y., MORAND-FEHR P., DURAND G. and SAUVANT D., 1979b. Évolution de l'activité métabolique du tissu adipeux chez la chèvre au cours des deux premiers mois de lactation. Relation avec la sécrétion lactée (Evolution of metabolic activity in goat adipose tissues during the first two months of lactation ; relationship with milk secretion). Bull. Acad. Vét. France, 52: 417-422.

CHILLIARD Y., SAUVANT D., BAS P., PASCAL G., and MORAND-FEHR P., 1981. Importance relative et activités métaboliques des différents tissus adipeux de la chèvre laitière (Importance and metabolic activities of various adipose tissues of dairy goats), Vol. 1, p. 80-89. In : MORAND-FEHR P., BOURBOUZE A. and DE SIMIANE M. (Eds.) : Nutrition and systems of goat feeding. Symposium International, Tours (FRANCE), May 12-15, 1981. INRA-ITOVIC, Paris (FRANCE).

CHILLIARD Y., RÉMOND B., AGABRIEL J., ROBELIN J. and VÉRITÉ R., 1987a. Variations du contenu digestif et des réserves corporelles au cours du cycle gestation-lactation (Variations of digesta mass and body reserves during gestation and lactation). Bull. Tech. CRZV Theix, INRA, (70), 117-131.

CHILLIARD Y., SAUVANT D., MORAND-FEHR P. and DELOUIS C., 1987b. Relations entre le bilan énergétique et l'activité métabolique du tissu adipeux de la chèvre au cours de la première moitié de la lactation (Relationship between energy balance and metabolic activities of goat adipose tissues during the first half of lactation). Reprod. Nutr. Dévelop., 27: 307-308.

COULON J.B., 1981. Capacité d'ingestion des vaches laitières en début de lactation (Food intake capacity in early lactating dairy cows). Thèse de Docteur Ingénieur, ENSA, Montpellier (FRANCE).

COWAN R.T., ROBINSON J.J., Mc DONALD I. and SMART R., 1980. Effects of body fatness at lambing and diet in lactation on body tissue loss, feed intake and milk yield of ewes in early lactation. J. Agric. Sci. (Camb.), 95: 497-514.

DAVIES P.J., JOHNSTON R.G., ROSS D.B., 1971. The influence of energy intake on plasma level of glucose, non esterified fatty acids and acetone in the pregnant ewe. J. Agric. Sci. (Camb.), 77: 261-265.

DEMIGNÉ C., YACOUB C., MORAND C. and REMESY C., 1988. Les orientations du métabolisme intermédiaire chez les ruminants (Orientation of intermediate metabolism in ruminants). Reprod. Nutr. Dévelop., 28: 1-17.

FARRIES E., 1975. Untersuchungen zur Futterungsniveau für trockenstehende Kühe. Der Tierzuchter, 11: 476-480.

FERNANDEZ J.A., SAHLU T., LU C.D. and POTCHOIBA M.J., 1989. Effect of energy intake during late pregnancy on subsequent milk yield and metabolic status of Alpine does. J. Dairy Sci., 72, suppl. 1 : 511 (Abst.).

FINGERLING G., 1911. Beiträge zur ververtung von Kalk-und Phosphorsäure Verbindungen durch den tierischen Organismus, I, Landw. Versuchßtat, 75: 1.

GARDNER R.W., 1969. Interaction of energy levels offered to Holstein cows, prepartum and postpartum. J. Dairy Sci., 52 : 1973-1984.

GIGER S. and SAUVANT S., 1982. Utilisation des acides gras non estérifiés plasmatiques pour la prévision des bilans et des besoins énergétiques de la chèvre en lactation (Use of plasma NEFA to predict energy balances and requirements in lactating goats). Ann. Zootech. 31: 149-160.

GILL R.A., 1970. Studies on lipid metabolism in ruminants. Thesis ph D, Univ. Wisconsin (US).

GUESSOUS F., MORAND-FEHR P. and DELAGE J., 1977. Influence de l'acétate et du propionate de Na ajoutés au régime avant et après mise bas sur le métabolisme, la production et la composition lipidique du lait de chèvre (Effect of Na acetate and propionate added to diets before and after parturition on metabolism, production and composition of goat milk fat). Ann. Biol. Anim. Biophys. Biochim., 14: 251-269.

GUSS S.B., 1977. Management and diseases of dairy goats. Dairy goat Journal Publ. Corp. Scottsdale, Arizona (USA).

HAENLEIN G.F.W., 1980a. Mineral nutrition of goats. J. Dairy Sci., 63: 1729-1748.

HAENLEIN G.F.W., 1980b. Nutrient requirements of dairy goats, Past an Present. Intern. Goat and Sheep Res., 1: 79-95.

HARDWICK D.C., LINZELL J.L. and PRICE S.M., 1961. The effect of glucose and acetate on milk secretion by the perfused goat udder. Biochem. J., 80: 37-45.

HARMEYER J. and MARTENS H., 1980. Aspects of urea metabolism in ruminants with reference to the goat. J. Dairy Sci., 63: 1707-1728.

HOVE K., 1984. Intestinal radio-calcium absorption in the goat : measurement by a double isotope technique. Br. J. Nutr., 51: 145-156.

JOURNET M. and RÉMOND B., 1981. Response of dairy cows to protein level in early lactation. Livest. Prod. Sci., 8: 21-35.

KESSLER J., 1981. Éléments minéraux majeurs chez la chèvre. Données de base et apports recommandés (Major elements in goats : basic data and allowances). Vol. 1, p. 197-209. In : MORAND-FEHR P., BOURBOUZE A. and DE SIMIANE M. (Eds.) : Nutrition and systems of goat feeding. Symposium International, Tours (FRANCE), May 12-15, 1981, INRA-ITOVIC, Paris (FRANCE).

KING N.B., 1972. The recognition and treatment of pregnancy toxemia in goats, p. 30-31. In : Proc. National Goat Breeders Conference, Paddington (AUSTRALIA).

KONDOS A.C., 1972. Pre- and post-partum nutrition of the goat and its effect on milk production, p. 7-12. In : Proc. National Goat Breeders Conference, Paddington (AUSTRALIA).

MELBY, 1987. Disease among dairy goats in 27 Norwegian herds, Vol. 2, p. 1351-1352. 4th Intern. Conf. on goats. March 8-13, 1987. Brasilia (BRAZIL).

MENAHAN L.A., 1966. Interrelationships between ketone body production, carbohydrate utilization and fat mobilization in the ruminant. Ph D thesis, Univ. of Wisconsin, Madison (USA).

MORAND-FEHR P., SAUVANT D., BAS P. and ROUZEAU A., 1977. Paramètres caractérisant l'état nutritionnel de la chèvre (Parameters relative to nutritional status of goats), p. 195-203. Symp. on goat breeding in mediterranean countries, Oct. 3-7, 1977, Malaga-Granada-Murcia (SPAIN).

MORAND-FEHR P. and SAUVANT D., 1978a. Alimentation des caprins (Goat Nutrition), Chap. 15, p. 449-467. In: JARRIGE R. (Ed.): L'alimentation des ruminants, INRA publ., Versailles (FRANCE).

MORAND-FEHR P. and SAUVANT D., 1978b. Nutrition and optimum performances of dairy goats. Livest. Prod. Sci., 5: 203-213.

MORAND-FEHR P., HERVIEU J. and SANTUCCI P., 1989. Notation de l'état corporel des chèvres (Body condition scoring of goats). La Chèvre (175), p. 39-42.

NCR, 1981. Nutrient requirement of goats. National Academy Press, Washington D.C. (USA).

PATTERSON D.S., BURNS K.N., CUNNIGHAM N.F., HEBERT C.N. and SABA B., 1964. Plasma concentration of glucose and non esterified fatty acids in the pregnant and lactating ewe and the effect of dietary restriction J. Agric. Sci. (Camb.), 62: 253-262.

PAYNE J.M., 1966, cited by LINDAHL D.H. Nutrition and Feeding of goats, p. 372-388. In : CHURCH D.C. (Ed.) : Digestive physiology and nutrition of ruminants, O & B Books Inc., Corvallis, Oregon (USA).

RAYSSIGUIER Y., MAZUR A., REMOND B., CHILLIARD Y. and GUEUX E., 1986. Influence de l'état corporel au vêlage et du niveau d'alimentation en début de lactation sur la stratose hépatique chez la vache laitière (Effect of body condition at calving and feeding level in early lactation on liver steatosis in dairy cows). Reprod. Nutr. Develop., 26: 359-360.

REMESY C. and DEMIGNE C., 1976. Genetische aspeckte zur Gehalt und Mineralstoffen in der Milch. Züchtungskunde, 49: 99-109.

RUSSEL A.J.F., DONEY J.M. and REID R.L., 1967. The use of biochemical parameters in controlling nutritional state in pregnant ewes, and the effect of undernourishment during pregnancy on lamb birth weight. J. Agric. Sci (Camb.), 68: 351-358.

RUSSEL A.J.F., 1979. The nutrition of the pregnant ewe, p. 221-241. In : British Council (Ed.) : Management and diseases of sheep. Commonwealth Agricultural Bureaux, Slough (UK).

SAUVANT D. and MORAND-FEHR P., 1977. Facteurs de variation du risque de toxémie de gestation et de cétose chez la chèvre laitière (Effects of various factors on the risk of pregnancy toxaemia and cetosis in dairy goats), p. 150-171. 3e Journées de la Recherche ovine et caprine, INRA-ITOVIC, Paris (FRANCE).

SAUVANT D., 1978. Les profils biochimiques chez la chèvre laitière, intérêt et limites (Interest and limits of biochemical profiles in dairy goats), p. 84-108. In: CAAA: L'alimentation de la chèvre laitière. ADEPRINA, Paris (FRANCE).

SAUVANT D., 1981. Alimentation énergétique des caprins (Energy nutrition in goats), Vol. 1, p. 55-79. In: MORAND-FEHR P., BOURBOUZE A. and DE SIMIANE M. (Eds.): Nutrition and systems of goat feeding. Symposium International, Tours (FRANCE), May 12-15, 1981, INRA-ITOVIC, Paris (FRANCE).

SAUVANT D., CHILLIARD Y. and MORAND-FEHR P., 1979. La mobilisation des réserves énergétiques par la chèvre laitière (Body energy mobilization in dairy goats). Les Dossiers de l'Elevage, 3: 49-58.

SAUVANT D., CHILLIARD Y., BAS P. and MORAND-FEHR P., 1979a. Goat adipose tissue mobilization and milk production level. Ann. Rech. Vet., 10: 404-407.

SAUVANT D., CHILLIARD Y., HERVIEU J., MORAND-FEHR P. and DELAGE J., 1979b. Influence de l'âge sur la mobilisation des réserves adipeuses de la chèvre en première semaine de lactation (Effect of age on body fat mobilization in goats during the first week of lactation). Journées de l'Association Française de Nutrition, Nancy (FRANCE).

SAUVANT D., GIGER S. and MORAND-FEHR P., 1981a. Influence du régime alimentaire de la fin de gestation sur la capacité d'ingestion de la chèvre au démarrage de la lactation (Effect of diets during late gestation on the food intake capacity in goats at the beginning of lactation). 32th Annual Meeting EAAP, Zagreb (YUGOSLAVIA).

SAUVANT D., GIGER S., CHILLIARD Y. and MORAND-FEHR P., 1981b. Étude de la capacité d'ingestion de la chèvre en début de lactation (Food intake capacity in goats during early lactation), Vol. 2, p. 572-581. In : MORAND-FEHR P., BOURBOUZE A. and DE SIMIANE M. (Eds.), Nutrition and systems of goat feeding. Symposium International, Tours (FRANCE), May 12-15, 1981, INRA-ITOVIC, Paris (FRANCE).

SAUVANT D., CHILLIARD Y., BAS P. and MORAND-FEHR P., 1981c. Interactions alimentation x individus en production laitière caprine (Interactions between feeding and individuals in goat milk production), p. 311-331. 6e Journées de la Recherche ovine et caprine. INRA-ITOVIC, Paris (FRANCE).

SAUVANT D., MORAND-FEHR P. and BAS P., 1984. Facteurs favorisant l'état de cétose chez la chèvre (Factors increasing ketosis risks in goats), p. 369-378. In YVORE P. and PERRIN G. (Eds.), Goat diseases. Colloque International, Oct. 9-11, 1984, Niort (FRANCE), INRA Publ. (Les Colloques de l'INRA n° 28), Paris (FRANCE).

SAUVANT D., HERVIEU J., GIGER S., TERNOIS F., MANDRAN N. and MORAND- FEHR P., 1987. Influence of dietary organic matter digestibility on goat nutrition and production at the onset of lactation. Ann. Zootech., 36 : 335-336 (Abst.).

SCHWALM J.W. and SCHULTZ L.H., 1975. Blood and liver metabolites in fed and fasted diabetic goats. J. Dairy Sci., 59 : 262-269.

SENGAR O.P.S., 1980. Indian research on protein and energy requirements of goats. J. Dairy Sci., 63 : 1655-1670.

SKJEVDAL T., 1979. Effect of feeding various levels of concentrate during the dry period in dairy goats, Report 194, Dep. Anim. Nutr. Agric. Univ., As (NORWAY).

STERN D., ADLER J.H., TAGARI H. and EYAL E., 1978. Responses of dairy ewes before and after parturition to different nutritional regimes during pregnancy. 2 - Energy intake, body weight changes during lactation and milk production. Ann. Zootech., 27 : 335-346.

TISSIER M., THERIEZ M. and MOLENAT G., 1975. Evolution des quantités d'aliment ingérées par la brebis à la fin de la gestation et au début de la lactation. Incidences sur leurs performances. Etudes de deux rations à base de foin de qualité différentes (Level of feed intake in ewes during late gestation and early lactation ; Effect on performance by using 2 diets with various quality hays). Ann. Zootech., 24 : 711-727.

VAN'T KLOOSTER A.T., 1976. Adaptation of calcium absorption from the small intestine of dairy cows to changes in the dietary calcium intake and at the onset of lactation. Zeitschr. für Tierphysiol., Tierernaehrg., Futtermittkde., 37 : 169-182.

WENTZEL D., ROUX M. and BOTHA L.J.J., 1976. Effect of the level of nutrition on blood glucose concentration and reproductive performance of pregnant angora goats. Agro-animalia, 8 : 59-61.

Part 2

Evaluation and utilization of feeds

Chapter 13

FORAGE UTILIZATION IN GOATS

C. MASSON, R. RUBINO and V. FEDELE

INTRODUCTION

Like all ruminants, the goat is primarily a forage consumer. According to technical and economic analysis, the use and efficiency of forage from areas grazed by goats are often insatisfactory. Hence, advances can be obtained even if they are to be modulated with the different production systems. Goat feeding systems have changed since 1981. In temperate areas, grass and corn silages now often replace green forages in zero grazing while hay is still used. Grazing of permanent and sown pastures can now be used in some regions and complete rations (silage + hay + concentrate) have recently also appeared in goat farming.

This is why we are updating the report of DE SIMIANE et al. (1981) using in particular results on the nutritive value of dry forages (hay, straw) and green forages grazed by goats. It was not deemed necessary to discuss silages since DE SIMIANE et al. (1981) have already treated this subject thoroughly and no further information has arisen.

DRY FORAGES

Feeding behaviour

The original feeding behaviour of goats is undoubtedly one of the main characteristics which differentiates them from other ruminants. Selective behaviour is very pronounced for legume forages (MORAND-FEHR et al., 1987 ; GIGER et al., 1987) and is also seen in the case of pasture hay and even straw (MASSON et al., 1989). The fractions eaten are generally those with the lowest cellulose and the highest protein content. Thus, as a result of selection by goats, the nutritive value of the hay actually ingested may be substantially different from that of the hay offered (MORAND-FEHR et al., 1980 ; see Chapter 1).

In addition, the feeding behaviour characteristics of goats and sheep have been studied in order to determine whether they could explain the differences in the levels of feed intake (see Chapter 1). Results obtained from studies in France have shown that the differences between the two species are variable.

Level of dry forage intake

Table 1 and Table 2 show the results of the dry matter intake of hays of variable quality given alone to sheep and goats fed at a maintenance level and to lactating goats. Even though the differences between sheep and goats in the levels of intake were sometimes significant, they were not systematically lower in one species than in the other. Nevertheless, goats in comparison to sheep prefered legumes rather than grasses. In addition, the mean dry matter intake was higher (+ 53 %) in lactating goats than in dry animals: 91.3 vs 59.8 g DM/kg $W^{.75}$, i.e. about 2.0 and 1.3 kg DM per animal, respectively (GIGER et al., 1987). Furthermore, untreated or NaOH-treated straw was eaten better by goats than by sheep (ALRAHMOUN et al., 1985, 1986). BROWN and

JOHNSON (1981) fed young growing females on rations containing 35-50 and 65 % wheat straw. They observed that the total dry matter intake was higher in goats than in sheep with the 65 % straw ration and lower with the 35 % straw ration.

Table 1
DRY MATTER INTAKE OF SHEEP AND GOATS AT THE MAINTENANCE

Forages (hays)	Chemical composition (% DM*)			Dry matter intake (g/kg $W^{.75}$)		References
	OM*	CP*	CF*	Sheep	Goats	
Italian Rye Grass Hay	92.7	8.3	33.1	61.9	55.9[+]	ALRAHMOUN et al. (1985)
Natural Meadow Hay	92.6	9.2	35.6	51.7	58.9[+]	-
Timothy Hay	95.2	9.7	37.3	57.5	51.4[+]	DE SIMIANE et al. (1981)
Lucerne Hay	90.5	15.2	34.4	61.0	87.2[+]	BLANCHART et al. (1980)
Lucerne Hay	-	-	-	75.9	83.7[+]	OWEN and WAHED (1985)
NaOH treated Barley Straw	90.7	3.7	45.9	35.0	53.7	
Untreated Barley Straw (+ 120 g Soya oilmeal)	90.4	7.5	45.9	54.5	56.9	ALRAHMOUN et al. (1986)
NaOH treated Barley Straw + Urea (12 g/kg)	92.6	8.3	47.9	65.6	59.9	-
NH_3 treated Barley Straw	-	-	-	45.3	57.7	OWEN and WAHED (1985)

(*) DM: dry matter ; OM: organic matter ; CP: crude protein ; CF: crude fibre
(+) significant: P≤ 0.05

Table 2
DRY MATTER INTAKE OF LUCERNE HAY IN ALPINE AND SAANEN GOATS
(GIGER et al., 1987)

	Chemical composition of hay ingested (% DM*)		Dry matter intake (g/kg $W^{.75}$)
Dry and pregnant goats	23.0	31.1	55.0
	14.2	37.7	61.9
	18.4	35.2	50.4
	17.3	31.1	64.2
Lactating goats	19.8	34.2	101.9
	20.2	32.6	82.2
	20.5	35.5	86.8

(*) OM, CP, CF (see Table 1)

These results are to be compared with those of DEVENDRA (1978, 1981), who showed that goats tended to eat larger amounts of dry forages of low nutritive value than did sheep. The nitrogen supplementation of these poor forages, however, was more effective in sheep than in goats (HIGGINS, 1981 ; ALRAHMOUN et al., 1986).

MASSON et al. (1989) reported the results of a study of the feeding behaviour of goats fed with barley straw. In comparison to sheep, the rates of ingestion chewing and rumination were higher in goats, whereas the unitary time spent eating, chewing and ruminating (mn/g DM/kg $W^{.75}$) was lower.

These authors put forward the hypothesis that the adaptation mechanism of feeding behaviour in goats is more efficient than that of sheep, and all the more marked as the quality of the forage is low. Thus, the goat can apparently adapt to the nature and quality of the forage and hence, can survive well in a variety of regions, ranging from temperate to arid areas (see Chapters 2 and 19).

Apparent digestibility

Observations on the apparent digestibility are primarily derived from metabolism crate experiments in which a single forage with a highly variable nutritive value (from lucerne to cereal straw), with or without supplementation, is given to goats and sheep. These experiments were carried out on castrated bucks and rams in maintenance conditions, dry goats and more rarely lactating goats (see Chapter 5).

The digestibilities of hay of high or medium nutritive value are almost similar in goats and sheep (Table 3), confirming prior conclusions of DEHORITY and GRUBB (1977) and of MORAND-FEHR (1981).

GIGER et al. (1987) reported that the mean digestibility of seven good quality lucerne hays measured in lactating goats (Table 4) was 61.5 %. The authors also reported that the selective behaviour of goats improved the nutritive value of forage by 0.60 points of organic matter digestibility, i.e. 0.01 UFL/kg DM.

However in the case of tropical forages of poor quality and straws, the digestibility of dry matter and crude fibre and to a lesser extent of crude protein, was significantly higher in goats than in sheep (Table 5). Nevertheless, when concentrate was added to these forages, the differences obtained with the forage alone were reduced.

The differences in digestibility between breeds and the adaptability of local breeds were well shown by SILANIKOVE (1986). Thus, wheat straw digestibility in Bedouin goats was higher than in Saanen goats (Table 6).

Goats thus tend to digest better forages of poor nutritive value, i.e. rich in cell wall carbohydrates and poor in crude protein, as suggested by DEVENDRA (1978, 1981) based on the study of tropical forages from Southeast Asia. MASSON et al. (1984), ALRAHMOUN et al. (1986) and TISSERAND et al. (1986) state a better recycling of urea and a cellulolytic flora more resistant to a nitrogen deficiency as currently the being most probable arguments for explaining why goats can better eat low quality forages (see Chapter 5). Studies on the recycling of nitrogen in goats, carried out in particular by BRUN-BELLUT et al. (1987) should provide more details in this field (see Chapter 8).

Utilization for milk production

As a result of their high feed requirements, lactating goats are seldom fed exclusively on hay. GIGER et al. (1987), however, offered only lucerne hay to lactating goats (Table 2), which on average satisfied the needs for production of 1 to 2.5 kg of milk. This is somewhat lower than the amounts obtained with green forage (2 to 3 kg) (DE SIMIANE, 1981). Hay from grasses with the same ingestibility led to a higher production, but generally these kinds of hays had a lower ingestibility in goats than legume hays.

Table 3

DRY FORAGE DIGESTIBILITY IN GOATS AND SHEEP

	Digestibility % OM*		CP*		CF*		References
	Sheep	Goats	Sheep	Goats	Sheep	Goats	
Italian Rye Grass Hay	63.2	64.3	52.3	53.7	58.6	55.6	MASSON et al. (1986)
Timothy Hay	55.6	57.8	52.0	49.5	64.1	64.5	DE SIMIANE et al. (1981)
NaOH treated Barley Straw	41.8	63.6	-	-	56.9	79.1	MASSON et al. (1984)
3% NH3 treated Rice Straw	-	-	-	37.0	-	67.9	GIHAD (1988) (personnal communication)
5% Urea treated Rice Straw	-	-	-	31.7	-	63.6	

(*)   OM, CP, CF (see table 1)

Table 4

DIGESTIBILITY OF SEVEN LUCERNE HAYS IN GOATS

(GIGER et al., 1987)

Digestibility (in %)	Hay 1	2	3	4	5	6	7
Organic matter (OM*)	61.0 ±1.3	62.3 ±1.7	62.5 ±1.8	63.4 ±2.6	57.9 ±2.0	60.0 ±1.9	63.8 ±1.9
Crude protein (CP*)	71.3 ±1.9	64.3 ±1.8	74.1 ±1.7	75.7 ±1.1	65.2 ±3.5	73.0 ±1.9	74.4 ±2.4
Crude fibre (CF*)	40.3 ±2.0	48.1 ±2.8	45.3 ±4.2	40.0 ±2.9	36.7 ±3.0	35.3 ±4.9	44.5 ±2.9

(*)   OM, CP, CF (see Table 1).

As a result of this, a diet composed exclusively of hay requires supplementation with concentrates in order to meet the requirements for high milk production. Thus, freely available hay supplemented with 600 to 800 g of concentrate can support a milk production of 3 to 4.5 kg. Straw is very rarely included in the ration of dairy goats in Europe, but can account for a large proportion of the diet in semi-arid and arid areas. Depending on the quality of straw, its real intake by mid-lactation goats generally ranges between 300 and 800 g dry matter/day, meeting at most 20 to 50 % of the energy requirements for maintenance. This insufficient intake requires supplementation with concentrates, which can become excessive and dangerous for milk yields exceeding 3 kg per day. However, treating straw with NaOH or ammonia can be of value, since it increases the level of intake by 15 to 30 %.

Table 5

COMPARISON OF THE DIGESTIBILITY IN GOATS AND IN SHEEP
FROM 35 BIBLIOGRAPHIC DATA
(MORAND-FEHR, 1989)

Digestibility	Dry matter			Crude protein			Crude fiber		
	+(1)	0(2)	- (3)	+	0	-	+	0	-
Berseem or alfalfa	0	4	0	0	4	0	0	4	0
Maize Silage	0	1	0	1	0	0	0	1	0
Tropical forages of poor quality	5	4	0	3	6	0	7	2	0
Straw (barley, wheat or rice)	3	1	0	2	2	0	3	1	0
Total :									
- Forage of good quality	0	5	0	1	4	0	0	5	0
- Forage of poor quality	8	5	0	5	8	0	10	3	0

(1) + digestibility significatively higher in goats
(2) 0 digestibility non significatively different between goats and sheep
(3) - digestibility significatively less high in goats.

The number indicate the assays where the goat digestibility was more (+) or less (-) high than the sheep digestibility or not different (0).

Table 6

VOLUNTARY INTAKE AND DIGESTIBILITY OF DIFFERENT FORAGES
IN BEDOUIN AND SAANEN GOATS
(SILANIKOVE, 1986)

Forage	Breed	Weight (kg)	Intake (kg DM/g $W^{.75}$)	DM digestibility (%)
Lucerne hay	B*	17.6	63.9	71.6
	S*	28.3	95.0	66.8
Rhodes grass	B	19.5	72.7	68.1
	S	49.1	72.5	61.0
Wheat straw	B	26.1	31.6	53.5
	S	28.7	29.6	38.9

(*) B: Bedouin ; S: Saanen

GREEN FORAGES

Sown pastures used with rationed grazing

Measuring the level of feed intake

Only a few studies have been devoted to measuring levels of feed intake by goats on sown pastures. The methodology of MASSON et al. (1983) using the differences between amounts offered and refused, is summarized.

The grazing was determined daily and a motorized scythe was passed along its length. From a practical standpoint, this was done to enable the installation of an electric fence and also to calculate the grass yield. It was also possible to determine the quality of material available to the animals. The motorized scythe was used daily to cut the refusals of the field which has been available to the goats for 48 hours. These refusals were weighed and their dry matter content determined. The difference between amounts of dry matter offered and refused accounted for the levels of dry matter intake.

These experiments were conducted with lactating goats (100-250 days of lactation) in Burgundy region (France) from April to September.

Level of feed intake

By pooling the results from all forages, for all grazing cycles and for all the years used for experiments, the mean daily quantity of dry matter ingested was 1.42 kg DM (75.7 DM/kgW$^{.75}$). As seen in Table 7, the best ingested forages in the first cycle were red clover, lucerne-orchard grass and fescue. These were followed by Italian ryegrass and Lucifer and Prairial orchard grasses. There was the same hierarchy in the second cycle, but the differences in the levels of intake were very low. These results are daily means, however, and there are large variations (coefficient of variation between 30 and 35). The variation factors which can affect this level of intake are the forage composition, grazing management, concentrate supplementation and climatic factors.

Table 7

DRY MATTER INTAKE OF GREEN FORAGE IN ALPINE GOATS
IN MID-LACTATION

Forages	Cycle	kg DM/day**
Italian Ray Grass (Tiara)*	1	1.20
	2	0.95
Orchard-Grass (Prairial)	1	0.64
	2	0.61
Orchard-Grass (Lucifer)	1	0.90
	2	0.88
Fescue (Clarine)	1	1.49
	2	0.90
Red Clover (Alpille)	1	1.95
	2	1.45
Lucerne-Orchard Grass	1	1.83
	2	1.28

(*)   In brackets: this name of varieties used.
(**) kg of green forage DM/day/goat.

Forage composition

There was no correlation between level of intake of green forage and its chemical composition. These observations on pasture differ from those of DE SIMIANE et al. (1981), who noted a close relationship in zero grazing between the level of intake and the dry matter content, up to 16-17 %. When the spring was rainy, however, levels of intake during grazing closely followed the variations in dry matter content.

Grazing management

The correlations between the level of green forage intake and the parameters characterizing grassland exploitation show the effects of the amount of dry matter offered and the amount of refusals. The positive correlation between the level of intake and the quantity of dry matter offered was high ($r = +0.7$). The quantity of dry matter offered explained 32 % of the variation in the level of intake, when only available dry matter amounts higher than 2 kg per day and per goat were considered. In addition, even though there was a negative correlation between the level of intake of green forage and the amount of refusals, there was no correlation between these amounts and the quantity of dry matter offered. This suggests that the part of variation in the level of intake which is not linked to the quantity of dry matter offered, is partially explained by the variation in the amount of refusals.

Proportion of refusals

The proportion of refusals was high, varying from 8.5 to 85 % with a mean of 44 %. The chemical composition of refusals differed from that of the grass offered as a result of selection by goats. This selective behaviour increased by 8 and 20 %, respectively the energy and protein values of grass ingested relative to grass offered.

Concentrate supplementation

Under our experimental conditions, no statistically significant difference in the green forage intake was seen between the two concentrate levels examined: 900 and 700 g/day on the one hand, and 600 and 400 g/day on the other hand. The very similar levels of concentrate made it difficult to estimate the effect of substitution.

Climate factors

Regardless of the analysis, there was no significant correlation between the level of green forage intake and the climatic variables recorded (temperature, rainfall, humidity, sunshine). It should nevertheless be noted that Alpine goats cannot stand high temperatures and intermittent rainfall. As has been reported for grazing cows, it is difficult to distinguish between direct effects (on the physiology of the animals) and indirect effects on the level of feed intake and thus on milk production.

Milk production

The average milk yield 2.20 kg/day and per goat, was satisfactory in animals whose potential was close to 600 kg of milk per year and in the experimental period corresponding to mid-lactation. During the year, however, one group of dairy goats was maintained at 3 kg of milk/day in these grazing conditions.

Conclusion

Based on our observations, the most important factors affecting the level of grass intake in conditions of restricted grazing were the forage species and the quantity of dry matter offered. The other factors had no effect or their effect was too masked to appear clearly. In order for the quantity of dry matter offered not to be a limiting factor, it was necessary to offer forage quantities corresponding to at least 2 kg of dry matter. Climatic factors in our French conditions are not very important, except perhaps in extreme cases. The absence of any variation in the grass intake during grazing related to the level of concentrate supply and to milk production could not be considered. Additional experiments are required to explore these points.

Natural pastures

Feeding behaviour

The feeding behaviour of naturally grazed goats varied throughout the year and was closely related to the daily time spent grazing, the choice of plant species and the selection of the different morphological parts of the plant.

The daily time spent grazing generally varied according to day length, temperature and the moisture content of the plants.

In agreement with observations by MORAND-FEHR (1981), FEDELE (unpubl.) observed that during the hottest days in southern Italy goats grazed early in the morning or late in the evening after sunset. During the rest of the day, the animals remained in the shade offered by any trees or shrubs present. In cooler or colder seasons, when dew formed, goats grazed during the warmer hours, when the dew had disappeared.

Without taking the time spent grazing into account, the degree of selection by goats depended on the plant species offered and their growth stage. In the presence of a large number of palatable species, goats ate a large part of this vegetation. In natural grazing conditions, however, the flora is very often heterogeneous and the goats will select according to a given palatability gradient. Thus, at the end of a grazing cycle, it was easy to recognize refusals, which were areas containing the highest grass. In addition, species consumed well were not always used in the same way. As an example, in the case of grazing in the south of Italy (Basilicate), goats consumed *Borrago officinalis,* entirely, but ate only the foliar apex of the leaves and the flowers of *Cichorium spp.* and *Ranunculus bulbosus.* Finally, they consume only the upper half of the leaves and spikes of *Festuca arundinacea, Brachypodium ramosum* and *Lolium perenne* (FEDELE et al., 1988).

Kind of plants ingested as a function of the season

Numerous studies have dealt with the composition of the ration of goats as a function of the season, especially on rangelands (MALACHEK and PROVENZA, 1981). These authors reported that the composition of the ration in these extensive systems underwent considerable modifications depending on the climate and the year. RUBINO et al. (1988b) observed the same in natural pastures in southern Italy.

Under grazing conditions in southern Italy, FEDELE et al. (1988b) and PIZZILLO et al. (1988) observed the composition of ingesta in goats fitted with an œsophageal fistula. In the springtime, the forage species present in the œsophageal contents were primarily grasses (65-75 %), followed by other non-grass and non-legume families (22-25 %) and legume families (3-10 %). The percentage of grasses diminished in summer (55-60 %), the other families increased (40-45 %) and legumes tended to disappear. The grasses held a very important place in autumn (80 %), more than in spring, the other families decreased (20 %) and only traces of legume species were present.

The presence of different forage species in the ration varied throughout the year. Certain, such as *Lolium perenne* among the grasses, *Daucus carota* and *Cichorium spp.* were present in high proportions in the ration throughout the whole grazing season. Others, on the contrary, were present only in the spring: *Ranunculus bulbosus, Vicia sativa,* others only in summer (*Avena barbata, Convolvulus arvensis*) or in the autumn (*Festuca arundinacea, Dactylis glomerata*).

Among the 50 species present in the pasture, only seven were well ingested throughout the year, 14 were not consumed and 29 were used only during certain periods of the year (Table 8).

Table 8

PREFERENCE INDEX* OF THE PRINCIPAL SPECIES OF A NATURAL PASTURE**
(FEDELE et al., 1988a)

Species	Spring	Summer	Autumn	Species	Spring	Summer	Autumn
Grasses				**Other families**			
Agrostis sp.	1	2	-	*Asperula odorosa*	5	3	1
Antoxantum odoratum	0	0	-	*Borrago officinalis*	5	5	-
Avena barbata	5	5	-	*Carex* sp.	4	2	5
Brachypodium ramosum	5	4	5	*Cerastium* sp.	0	0	-
Bromus mollis	2	0	0	*Cichorium* spp.	4	5	1
Bromus sterilis	0	0	0	*Convolvulus arvensis*	1	5	1
Cynodon dactylon	0	3	0	*Crepis* sp.	0	1	0
Dactylis glomerata	4	3	5	*Daucus carota*	3	4	0
Festuca arundinacea	4	3	5	*Euphorbia* sp.	0	0	-
Holcus lanatus	1	3	2	*Fœniculum* sp.	1	5	5
Lolium perenne	5	5	5	*Galium verum*	5	4	1
Phalaris cærulescens	4	4	-	*Geranium molle*	0	0	0
Phleum pratensis	3	1	1	*Malva silvestris*	0	0	-
Poa pratensis	3	1	1	*Mentha* spp.	0	1	0
Trisetum sp.	0	1	-	*Muscari racemosum*	0	0	-
				Plantago spp.	0	1	0
Legumes				*Papaver roheas*	0	2	-
				Picris echioides	0	0	0
Dorycnium pentaphyllum	0	2	1	*Poligonum aviculare*	0	2	0
Lotus corniculatus	0	2	0	*Potentilla* sp.	0	0	-
Medicago polimorfa	0	0	0	*Ranuculus bulbosus*	3	1	0
Melilotus sulcata	0	1	0	*Rumex* sp.	0	0	-
Ononis spinosa	5	1	1	*Sanguisorba minor*	1	2	-
Trifolium pratense	1	1	0	*Scabiosa* sp.	4	4	-
Trifolium repens	0	0	0	*Soncus arvensis*	2	5	0
Trifolium subterraneum	1	1	-	*Veronica* sp.	0	0	-
Vicia sativa	2	4	-				

(*) The preference index (score between 0 and 5) indicates the level of plant utilization (in %):
0: 0-1 % - 1: 1-20 % - 2: 21-40 % - 3: 41-60 % - 4: 61-80 % - 5: 81-100 %.

(**) Natural pasture in South Italy (Basilicate).

Level of feed intake

A large number of results published on the level of feed intake in goats have been obtained primarily at the trough, on rangelands or on artificial pastures. RANJAHN (1980) showed that the level of intake varied from 1.47 to 3.65 % of live weight. MORAND-FEHR (1981) reported that the mean level of intake in a lactating Alpine goat was 3.5 % of live weight. In natural pastures, the level of intake in goats varied with the concentrate level and the grazing system, but also with the breed and production level.

FEDELE et al. (1988b) observed that the level of intake in Maltese and Syrian goats varied from 1.9 % of live weight in spring to 3 % in summer. RUBINO et al. (1988b) reported the effect of pasture management with Maltese goats : in the pasture system grazing + hay making in the same pastures, level of intake was demonstrated to be higher than in the continuous or rotational grazing system (Table 9).

FEDELE et al. (1988a) studied the effect of 150 and 550 g/day/goat of concentrates in Maltese and Syrian goats. Statistically significant differences in the level of grass intake (Table 10) were seen as a function of the level of concentrate. In addition, the highest concentrate level was accompanied by the lowest level of grass intake.

Digestibility

Digestibility was highly influenced by climate and the physiological stage of the plant. VAN SOEST (1987) reported that the digestibility of young leaves was 60-70 % and that of old leaves 52 to 62 %. Similarly, the digestibility of the apical part of the stems was 46 to 62 % and that of the basal part between 41 and 45 %.

In Mediterranean areas, it has been shown (FEDELE et al., 1988b ; PIZZILLO et al., 1988 ; RUBINO et al., 1988a) that *in vitro* digestibility of grass in the ration of the goat and that in the pasture differed, with a high variability throughout the year (see Figure 1).

Table 9

LEVEL OF HERBAGE DRY MATTER INTAKE IN GOATS DURING 3 GRAZING PERIODS
(April, June, September) (RUBINO et al., 1988b)

Period of lactation	Level of dry matter intake (g $DM/day/kg.W^{.75}$)*					
	Hay maked from the same pasture and continuous grazing		Rotational grazing		Continuous grazing	
Early lactation (April)	34.1a	(8.4)**	34.3a	(7.8)**	26.1b	(6.9)**
Mid lactation (June)	37.2a	(6.8)	33.2ab	(7.1)	28.3b)	(7.0)
Late lactation (September)	39.0a	(9.2)	32.'b	(7.0)	31.2b	(6.8)

(*)   g herbage dry matter per day and per kg of metabolic weight.
(**)   In brackets: Standard deviation.
        In the same line, two figures which have a same letter are not significantly different (P ≤ 0.05).

Table 10

DRY MATTER INTAKE OF MALTESE AND SYRIAN GOATS FED TWO LEVELS OF
CONCENTRATE DURING LACTATION
(FEDELE et al., 1988a)

Period of lactation	Level of concentrate							
	150 g/goat/day				550 g/goat/day			
	Maltese		Syrian		Maltese		Syrian	
Early lactation	52.4a	(8.7)*	37.3b	(10.0)*	18.1d	(6.2)*	31.2c	(9.1)*
Mid lactation	52.2a	(10.9)	44.6b	(9.7)	24.3d	(13.4)	34.2c	(10.9)
Late lactation	44.1a	(9.4)	31.2b	(7.6)	27.2b	(13.7)	27.1b	(6.7)

(*)   see Table 9.

Figure 1

SEASONAL VARIATION OF *IN VITRO* ORGANIC MATTER DIGESTIBILITY
OF PASTURE GRASS AND INGESTED GRASS.

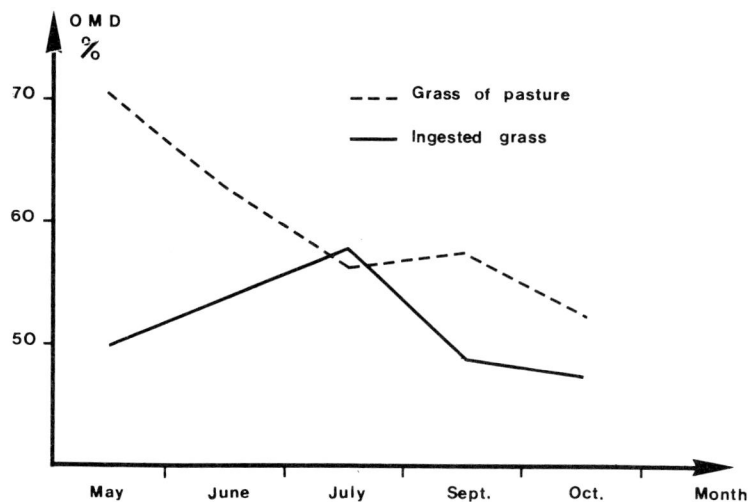

Feeding value

In natural pastures, the chemical composition of the ration taken by goats varied during the year as a function of selection by the animal (FEDELE et al., 1988a ; PIZZILLO et al., 1988). RUBINO et al. (1988b) analysed variations in the protein content of ingested forage. In the spring, the protein content of ingested grass was 13.5 %, while it was only 12.8 % for grass offered at grazing. In summer, these figures were 6.3 and 4.5 %. FEDELE et al. (1988a) observed the same for the crude fibre content and the energy value (UFL). Thus, the crude fibre content of ingested feed was 30, 24 and 28 % in the spring, summer and fall, respectively, while it was 22, 30 and 26 % in the feed offered. The energy values (UFL) in the ingesta during the three periods of the year were 0.55, 0.66 and 0.50 and were 0.84, 0.63 and 0.57 for the feed offered.

These results were obtained in the south of Italy, a region characterized by a Mediterranean climate. They showed that the selective behaviour of goats buffered variations in the chemical composition of the natural pastures.

Concentrate supplementation

In the south of Italy, as in many other regions, the nutritive value of rangelands is sufficient in the spring, but decreases considerably during the dry summer with a variable improvement in autumn (RUBINO et al., 1986). High yielding goats thus cannot meet their dietary needs with pasture grass alone and it is necessary to offer concentrate, but the quantity and quality of the concentrate should be ajusted for maximum utilization of the pasture. The strategy of concentrate supplementation of grazing goats throughout the year is thus delicate. It is nevertheless more interesting to offer a fixed quantity during the entire grazing season. PILLA et al. (1986) reported that goats given a constant quantity of concentrates produced almost the same quantity of milk (- 6 %) as those receiving a quantity of concentrate proportional to their needs and they also consumed less concentrate (- 30 %) and more grass (+ 40 %).

After comparing two levels of concentrate (150 and 500 g/day/goat), RUBINO et al. (1988a) observed that milk production was relatively unchanged, but that grass consumption while grazing, was higher in the first group (+ 32 %). In addition, when given equal quantities of concentrate, goats receiving concentrate containing 11.5 % total crude protein (CP) produced more milk (+ 10 %) than goats receiving 16.5 % CP concentrate (RUBINO, unpublished).

Strategy for using natural pastures during the year

In southern Italy, goat farming is expanding and the production systems are based on the optimal use of natural pastures, i.e. favouring the intake of grass without compromising milk production. The utilization of pastures must be continuous, but it is important to reserve a part of the grazing area (40 to 60 %) for the production of good quality hay in the spring; the close connection ratio between animals and the vegetation is thus satisfactory. During the grazing season, the concentrate should be given in a constant amount; work by the goat farmer is then simplified and the goat appears to be able to adapt. With natural pastures, however, it is recommended to offer the following to dairy goats: in the spring about 500 to 600 g of 11-12 % CP concentrate and during the other seasons 200 to 300 g of 16-18 % CP concentrate.

Finally, it is clear that the characteristic selective behaviour of goats in natural grazing conditions leads to a decrease in the efficiency of utilization of the pasture, but that it is possible to master this by well adapted grazing management system.

CONCLUSION

At the current state of knowledge, it is reasonable to apply to goats the values of ingestibility and digestibility obtained with sheep for medium and good quality hay. In the case of low quality dry forage (e.g. straw), however, it is possible in some cases that ingestibility and digestibility, especially of wall carbohydrate is higher in goats.

Goats and in particular dairy goats can adapt to intensive grazing of artificial pastures provided they are offered a minimum of 2 kg DM per day per animal (MASSON et al., 1983).

The selecting behavior of dairy goats in natural pastures enables them to adapt very well to the feeding value of the pasture which changes throughout the year. Supplementation with concentrate is nevertheless necessary, but in order to be effective, care must be taken to distribute the correct quality and quantity of concentrate as a function of the nature and value of the natural pasture.

SUMMARY

Results concerning the nutritive value of dry and green forages (used in grazing) obtained since the 1981 International Symposium in Tours are reviewed. Data obtained with goats and sheep fed in maintenance conditions and with lactating goats show that the level of voluntary intake of moderate or high quality hay offered alone is not systematically higher in one species than in the other. Cereal straw, either untreated or treated with NaOH, is better consumed by goats than by sheep. This is to be compared to former results which showed that dry forages of poor nutritive value were better ingested by goats. The digestibilities of hay of high or medium nutritive value are similar in goats and sheep. On the other hand, with certain tropical forages of moderate or poor quality and in the case of straws, the dry matter and crude fiber digestibility seems to be higher in goats.

Artificial pastures can be used for restricted grazing of dairy goats. The quantity of dry matter offered and the forage species are the main factors influencing the level of grass intake.

Studies performed in southern Italy using natural pastures show that the selective behaviour of dairy goats enables them to adapt to the nutritive value of the pasture which changes during the year, but concentrate supplies remain necessary.

Keywords: Goat, Level of forage, Digestibility, Hay, Grazing, Straw, Green forage, Pasture, Concentrate supply.

RÉSUMÉ

La valeur alimentaire des fourrages secs et des fourrages verts (utilisés en pâturage) fait l'objet d'une mise au point à partir des résultats obtenus depuis le symposium international de Tours de 1981. A partir des données obtenues sur des caprins et des ovins alimentés à l'entretien et sur des chèvres en lactation, il apparaît que le niveau d'ingestion de foin de moyenne ou bonne qualité distribué seul n'est pas systématiquement plus élevé dans une espèce que dans l'autre. En revanche, la paille de céréales traitée ou non à la soude est mieux consommée par les caprins que par les ovins, ce qui est à rapprocher de résultats antérieurs qui signalaient que les fourrages secs de médiocre valeur nutritive étaient mieux ingérés par les caprins. Les digestibilités de foin de bonne ou moyenne valeur nutritive sont assez proches chez les caprins et les ovins. En revanche, dans le cas de certains fourrages tropicaux de moyenne ou de mauvaise qualité et dans le cas des pailles, la digestibilité de la matière sèche et de la cellulose brute semble supérieure chez les caprins.

Les prairies artificielles peuvent être utilisées en pâturages rationnés par les chèvres laitières. La quantité de matière sèche offerte et l'espèce fourragère sont les principaux facteurs influençant le niveau d'ingestion d'herbe.

Sur prairies naturelles, des travaux menés en Italie du Sud montrent que la chèvre laitière, par son comportement de tri, s'adapte bien à la valeur nutritionnelle à la valeur du pâturage qui évolue au cours de l'année mais la complémentation en concentré s'avère nécessaire.

REFERENCES

ALRAHMOUN W., MASSON C. and TISSERAND J.L., 1985. Étude comparée de l'activité microbienne dans le rumen chez les caprins et les ovins. I. Effet de la nature du régime (Comparison of rumen microbial activity in goats and sheep: Effect of diets). Ann. Zootech., 34: 417-428.

ALRAHMOUN W., MASSON C. and TISSERAND J.L., 1986. Étude comparée de l'activité microbienne dans le rumen chez les caprins et les ovins. II. Effet du niveau azoté et de la nature de la source azotée (Comparison of rumen microbial activity in goats and sheep. Effect of protein level and source). Ann. Zootech., 35: 109-120.

BLANCHART G., BRUN-BELLUT J. and VIGNON B., 1980. Comparaison des caprins aux ovins quant à l'ingestion, la digestibilité et la valeur alimentaire de diverses rations (Comparison of the level of intake, digestibility, nutritive value of various rations in sheep and goats). Reprod. Nutr. Dévelop., 20: 1731-1737.

BROWN L.E. and JOHNSON W.L., 1981. Intake and digestibility of wheat straw rations fed to goats and sheep, p. 385. 73rd Ann. Meeting Amer. Soc. Anim. Sci., Raleigh, North California (USA).

BRUN-BELLUT J., BLANCHART G., LAURENT F. and VIGNON B. , 1987. Nitrogen requirement for goats, Vol. 2, p. 1205-1223. 4th Intern. Conf. on Goats, March 8-13, 1987, Brasilia (BRAZIL).

DEHORITY B.A. and GRUGL J.A., 1977. Characterization of predominant bacteria occuring in the rumen of goats (Capra hircus). Appl. Environ. Microbiol., 33: 1030-1036.

DE SIMIANE M., GIGER S. and BLANCHART G., 1981. Valeur nutritionnelle et utilisation des fourrages cultivés intensivement (Nutritive value and utilization of intensively cultived roughages), Vol. I, p. 274-299. In: MORAND- FEHR P., BOURBOUZE A. and DE SIMIANE M. (Eds.): Nutrition and systems of goat feeding. Symposium International, Tours (FRANCE), May 12-15, 1981, INRA-ITOVIC, Paris (FRANCE).

DEVENDRA C., 1978. The digestive efficiency of goats. World Rev. Anim. Prod., 14: 9-22.

DEVENDRA C., 1981. The utilization of forages from Cassava figea pea, laucaena and groundnut by goats and sheep in Malaysia. Vol. 1, p. 338-345. In: MORAND-FEHR P., BOURBOUZE A. and DE SIMIANE M. (Eds.): Nutrition and systems of goat feeding. Symposium International, Tours (FRANCE), May 12-15, 1981, INRA-ITOVIC, Paris (FRANCE).

FEDELE V., PIZZILLO M., RUBINO R. and MORAND-FEHR P., 1988a. The effect on the supplement feeding level on pasture utilization, behaviour and level of intake in two goat breeds. Seminar of FAO Subnetwork on Goat Nutrition and Feeding, Oct. 3-5, 1988, Potenza (ITALY).

FEDELE V., ZARRIELLO G. and DI TRANA A., 1988b. Osservazioni preliminari sul pascolamento di capre di ceppo locale (Preliminary observations on pasture grazing of local goats). Agricoltura ricerca, 81: 63-65.

GIGER S., SAUVANT D., HERVIEU J. and DORLEANS M., 1987. Valeur alimentaire du foin de luzerne pour la chèvre (Nutritive value of lucerne hay for goats). Ann. Zootech., 36: 139-152.

HIGGINS A.J., 1981. The effect of sodium hydroxide spray treatment on digestibility of barley straw in sheep and goats. Agricultural wastes, 3: 145.

MALECHEK J.C. and LEINWEBER C.L., 1972. Forage selectivity by goats on lightly and heavily grazed ranges. J. Range Manag., 25: 105-111.

MALECHEK J.C. and PROVENZA F.D., 1981. Feeding behaviour and nutrition of goats on rangelands. Vol. 1, p. 411-428. In: MORAND-FEHR P., BOURBOUZE A. and DE SIMIANE M. (Eds.): Nutrition and systems of goat feeding. Tours (FRANCE), May 12-15, 1981, INRA-ITOVIC, Paris (FRANCE).

MASSON C., HACALA S. and KERBOEUF D., 1983. Utilisation du pâturage rationné par la chèvre laitière (Utilization of rationed grazing for dairy goats), p. 294-324. 8e journée de la recherche ovine et caprine, INRA-ITOVIC, Paris (FRANCE).

MASSON C., ALRAHMOUN W. and TISSERAND J.L., 1984. Étude comparée de l'activité microbienne dans le rumen chez les caprins et les ovins (Comparison of rumen microbial activity in goats and sheep). Seminar of FAO Subnetwork on Goat Nutrition, Oct. 16-18, 1984, Grangeneuve (SWITZERLAND).

MASSON C., KIRILOV D., FAURIE F. and TISSERAND J.L., 1989. Comparaison des activités alimentaires et méryciques d'ovins et de caprins recevant de la paille d'orge traitée ou non à la soude (Feeding behaviour and rumination in sheep and goats fed with hydroxyde treated or untreated barley straw). Ann. Zootech., 38: 73-82.

MORAND-FEHR P., 1981. Caracteristique du comportement alimentaire et de la digestion des caprins (Characteristics of feeding behaviour and digestion in goats). In: MORAND-FEHR P., BOURBOUZE A. and DE SIMIANE M. (Eds.): Nutrition and systems of goat feeding, Symposium International, Tours (FRANCE), May 12-15, 1981, INRA-ITOVIC, Paris (FRANCE).

MORAND-FEHR P., HERVIEU J. and SAUVANT D., 1980. Contribution à la description de la prise alimentaire de la chèvre (Contribution to the description of feeding pattern). Reprod. Nutr. Dévelop., 20: 1641-1644.

MORAND-FEHR P., SAUVANT D. and DE SIMIANE M., 1981. L'alimentation de la chèvre (Goat Feeding). World Rev. Anim. Prod., 1: 45-74.

158

MORAND-FEHR P., GIGER S., SAUVANT D., BROQUA B. and DE SIMIANE M., 1987. Utilisation des fourrages secs par les caprins (Utilization of dry dorages in goats), p. 391-422. In: DEMARQUILLY C. (Ed.), Les fourrages secs, récolte, traitement, utilisation. INRA Publ., Paris (FRANCE).

MORAND-FEHR P., 1989. Goat nutrition and its particularities in dry tropics, p. 215-229. In: GALAL E.S.E., ABOUL-ELA M.B. and SHAFIE M.M. (Eds.): Ruminant production in dry subtropics: contraints and potential. Intern. Symp., Cairo (EGYPT), Nov. 5-7, 1988, EAAP, Publ. No 38, Pudoc, Wageningen (THE NETHERLANDS).

PILLA A.M., SCARDELLA P., RUBINO R., PIZZILLO M. and DI TRANA A., 1986. Bilancio alimentare di capre Maltesi alimentate al pascolo senza e con integrazione di concentrati (Feeding balance of Maltese breed goats fed on pasture with and without concentrate supplements). Ann. Ist. Sper. Zootec., 2: 97-108.

PIZZILLO M., DI TRANA A. and FEDELE V., 1988. Comportamento alimentare al pascolo di capre Maltesi et locali (Grazing feeding behaviour of Maltese breed and local goat). Agricoltura Ricerca, (81), 69-78.

RANJAHN S.K., 1980. Animal nutrition in the tropics. Bikas publication of house, Dist. Ghabidad (INDIA).

RUBINO R., PILLA A.M., PIZZILLO M., SCARDELLA P. and DI TRANA A., 1986. Effetto d'll'integrazione del pascolo con fave sull'efficienza produttiva di capre Maltesi (Effect of oats supplementation to pasture on the productive efficiency of Maltese breed goats). Ann. Ist. Sper. Zootech., 2: 85-92.

RUBINO R., DAMIANO R., CIAFARONE N. and MORAND-FEHR P., 1988a. Effect on the supplement feeding level on the milk yield and composition on certain blood constituents in two goats breeds. Seminar of FAO Subnetwork on Goat Nutrition and Feeding, Oct. 3-5, 1988, Potenza (ITALY).

RUBINO R., PIZZILLO M. and SCARDELLA P., 1988b. Confronto fra diversi sistemi di pascolamento in un modello intensivo con capre Maltesi (Comparison between different grazing systems in an intensive breeding model of Maltese goats). Agricoltura Ricerca, (81), 79-86.

SILANIKOVE N., 1986. Interrelationships between feed quality, digestibility feed consumption and energy requirements in desert (Bedouin) and temperate Saanen goats. J. Dairy Sci., 69: 2157-2162.

TISSERAND J.L., BELLET B. and MASSON C., 1986. Effet du traitement des fourrages par la soude sur la composition de l'écosystème microbien du rumen des ovins et des caprins (Effect of NaOH treated roughages on composition and microbial ecosystem of goat and sheep rumen). Reprod. Nutr. Dévelop., 26: 313-314.

Chapter 14

EVALUATION AND UTILIZATION OF RANGELAND FEEDS BY GOATS

M. MEURET, J. BOZA, H. NARJISSE and A. NASTIS

INTRODUCTION

Rangeland is a very heterogeneous pasture with multistratified distribution of forage resources, subject to important quantitative and qualitative variations depending on the season. Livestock owners in many goat-raising systems count on these cheap forage resources (see Chapter 19). The latter are used through a management scheme which is a compromise between production aims and a management plan ensuring self-renewal of the forage resource.

For nutritionists, the importance of factors related to feeding behaviour (see Chapters 1 and 2) and to grazing management practices, as well as those related to forage composition, renders partly unapplicable the nutritional references acquired in more intensive feeding situations. For a goat farmer, banking on a resource means knowing its "value" for the animals and determining whether there is room for improvement by changing the feeding management.

Few references are available on the use and conversion of rangeland resources by lactating goats. Even more than in the case of cultivated forage, understanding of rangeland vegetation nutritional utilization should consider the feeding behaviour, which conditions the dynamic relation between the animal and the vegetation.

Goats frequently express their capacity to better ingest and digest coarse rangeland forage than other livestock (see Chapters 3 and 13). This ability results from a significant selection of the plants eaten, from the development of a nitrogen recycling process and from an occasionally increased tolerance to antinutritional metabolites (see Chapter 2). This is observed particularly when conditions of animal management have favored a habituation to this type of situation (MALECHEK and PROVENZA, 1981; MERRILL and TAYLOR, 1981; SIDAHMED et al., 1981; PFISTER and MALECHEK, 1987).

This has led us to suggest and put forward arguments in favor of the development of a new methodology in the study of the feeding value of rangeland vegetation. This methodology will favor the study of animal-plant relations in a real-life situation, and will attempt ultimately to reproduce the context of this relation in controlled experimental conditions, e.g. in the digestibility crate. For the analysis of the chemical composition of very diverse rangeland forages, this method will rely on the grading of ingested species and on assessment of the change in their composition in space and time.

NUTRITIVE VALUE OF RANGELAND FORAGE FOR A GOAT

Relative palatability and influence of grazing on the "value" of the species

The concept of "nutritive value" should be considered relatively, with reference to the choices the animal will be led to make in a given situation. Placed in a situation of free choice on the rangeland, goats will select among the species present and will only very rarely consume one species to the exclusion of all others. Authors have already suggested that the "palatability" of rangeland plants is related to the diversity and abundance of available species, and that conclusions on the "feeding value" of one species (nutritive value time the quantity of its dry matter intake) should be related to the conditions of herd and grazing management in which they were formulated (HARRINGTON and WILSON, 1980).

For example, it cannot be concluded that a relatively rare shrub selected on a given rangeland will necessarily be willingly consumed if it is artificially multiplied on the same rangeland. Rare species are selected for reasons often little related to their actual palatability (VAN DYNE et al., 1980).

In order to estimate the feeding value of species on a rangeland, it is necessary to determine their representation in the diets selected, as well as their "consumption level" (ratio between the quantities eaten and the quantities available) during different grazing seasons. Important variations in these levels are sometimes reported within the same rangeland species. Thus, LIACOS and MOULOPOULOS (1967) distinguished five types of Kermes oak (*Quercus coccifera*) in Greece which had a varying degree of palatability for local goats. In this case, it appeared that the animals were influenced preferentially by the morphological differences of the foliage. In addition, in all cases the Kermes oak is less consumed when mixed with other shrubs. Besides, for a given species, the abundance of consumable foliage and its chemical composition are influenced by browsing. Work by TSIOUVARAS (1987) using clipping to simulate increasing consumption intensities of annual Kermes oak production (a shrub with abundant root system), showed that optimum annual leaf production at the end of spring did not concord with the optimum chemical composition (total crude protein, TCP) and *in vitro* digestibility (IVD). Although light browsing significantly improves the chemical composition (TCP + 2.5 % and IVD + 13 % when browse intake increases from 0 to 20 %), the best productivity / quality compromise is around a 60 % consumption.

Another factor in the variation of rangeland plant chemical composition concerns their ecological growth conditions. A great number of studies do not mention with sufficient accuracy the characteristics of the site in which a sample was collected, the phenological state of this sample and any treatments it may have undergone. This preclude adequate comparison of results obtained in different studies. LACHAUX et al. (1987) showed that in the case of *Quercus pubescens* leaves, there could be a difference of more than 5 % TCP during the same day along an altitude gradient between 600 and 1100 m on the same mountain slope.

In one grazing day, goats are capable of travelling more than 4 km and climbing more than 800 m in altitude. They thus encounter a heterogeneous vegetation, with variations occurring even within the same species. By selecting their diet, goats affect the growth dynamics and the subsequent value of the forage species.

THE CHEMICAL COMPOSITION OF PLANTS SHOULD BE ESTIMATED IN RELATIVE TERMS

We have discussed several sources of variation in the chemical composition of plants ingested on rangelands. Assessing the heterogeneity of rangelands by conventional chemical methods is time-consuming and very costly. In addition, the results are unsatisfactory due to the particular composition of woody forages by comparison to cultivated forages for which standard analytical methods were developed (URNESS et al., 1977; REED, 1986).

Problems encountered in conventional chemical analyses

Fiber assay is disturbed by the possible insolubilization of a part of the proteins, caused by their interaction with phenolic compounds. Proanthocyans (or condensed tannins) can react with proteins to form an insoluble complex which will have the same harmful effects as lignin on ruminal digestion of sugars and will be confused with it during analysis (BARRY and MANLEY, 1986). These compounds occur particularly when tree leaves are being digested. When the crude lignin content of woody fodders and of fecal matter are being determined, it is necessary to correct the error related to residual nitrogen (JARRIGE, 1980; SILVA-COLOMER, 1987). In general, we can question the relevance of the VAN SOEST fibre fractioning method for tree and shrub foliage, since certain stages in the analysis can lead to the formation of insoluble complexes in the 103°C oven. This causes us to raise the following question when variations in the chemical composition of rangeland forages are being characterized: could a technique qualifying species in terms of relative values be sufficient to predict the dietary choices and effectiveness of the selection made by the animals?

A proven method for predicting the quality of varied forages from diverse origins is Near Infrared Reflectance Spectroscopy (NIRS) (SHENK et al., 1979; BROWN and MOORE, 1987). NIRS will not solve the problems of usual chemical analyses, but, 1. its efficiency and low cost in sorting very large numbers of samples for the determination of a sub-sample representative of all variations are interesting when large rangeland surveys are to be carried out; 2. this analysis is very rapid and is nondestructive, enabling the "laboratory effect" to be tested by exchanging samples; 3. it enables the detection of samples having undergone poor conditioning. We recently applied this technique to plants and goat feces on rangelands with satisfying results (WAELPUT et al., 1989).

THE GOAT SHOULD BE ALLOWED TO PARTLY CHOOSE ITS DIET IN DIGESTIBILITY CRATE MEASUREMENTS

If we admit that an optimal value can be obtained from a heterogeneous range by letting the animal select its diet (i.e. choose between the plants as well as between the organs of the same plant), this selection must also be made possible in digestibility crate. This remark has been made earlier for the utilization of tropical grasses by grazing animals. Only a non fistulated goat with certain habits and within a certain mode of feeding management will be more or less capable of profiting from a plant.

In addition, the goat farmer will be more interested to know to what extent he can affect the "value" of the diet eaten on the rangeland (for example by making it more ingestible for a lactating animal by feeding a supplement) rather than precisely determining its "standard" value for an animal consuming it to 75 % of its maintenance needs.

Table 1 summarizes the most recent data on the ingestion and digestibility of tree and shrub foliage by goats, alone, supplemented or fed as mixed diet (at more than 50 % of total dry matter). They include 34 digestibility tests with forages generally having an ADL (VAN SOEST, 1963) content equal to or greater than 10 % DM. These studies are partially comparable since the animals all belonged to local breeds fed regularly on rangelands and were on average close to the 30 kg adult liveweight (except for trials 29 and 30). On the other hand, almost all the tests were carried out with dry females or castrated males, only five involving females in active milk production.

As these experiments differed primarily by the physical state and moisture content of the forage offered to the animals, we have classed the data in Table 1 into three groups (Figure 1).

Researchers have always sought an optimum way to condition tree foliage in order to measure its digestibility. Studies with wild animals (in particular fallow deer) are highly suggestive in this context (ULLREY et al., 1964; HJELFORD et al., 1982). These studies have shown that :

1. a preselection of plant organs offered which allows the animal to feed on a forage with practically constant chemical composition and produces a quality of intake that rarely concords with observations on voluntary intake on the rangeland (generally, the proportion of twigs distributed is too high);

2. dessication of forages, particularly at temperatures above 60°C, can transform their internal structure and alters their palatability in all cases (decreasing it in most cases);

3. grinding, even when coarse, affects the conversion of forage to a great extent, by increasing the rate of intestinal transit, by decreasing rumination time and the salivation level (a very important factor in goats).

Figure 1 shows the results concerning:
- tests with conditioned forage (white symbols in cloud "A");
- tests with fresh precut forage or with piles of branches (black symbols in cloud "B") and
- tests with fresh forage on whole branches distributed individually and renewed constantly (black symbols in cloud "C").

The intake value of these foliages (g DM/kg $W^{.75}$) is the factor which most differentiates the results, the majority of diets being close to a 50-55 % organic matter (OM) digestibility. The means and standard errors in each group are:
- 40 ± 20 g DM/kg $W^{.75}$ for conditioned (A);
- 65 ± 10 g DM/kg $W^{.75}$ for fresh precut or in batches of branches (B) and
- 100 ± 10 g DM/kg $W^{.75}$ for fresh on individual branches (C).

A pure diet of *Quercus ilex* is eaten by dry goats at 48 g DM/kg $W^{.75}$ when it is heated and compacted to 77 % DM (GUERRERO and BOZA, 1983) and at 90 g when it is distributed as fresh individual branches at 58 % DM (MEURET, 1986). Its intake can increase to 100 g when supplemented (at 14 % DM of total diet) with a molassea-urea mixture, thereby constituting a production diet for lactating goats (MEURET, 1988). Although both measurements lead to the same OM digestibility value of *Quercus ilex* unsupplemented in winter (49 % OM), the same is not true for crude protein (CP) digestibility (36 % CP for conditioned *Quercus ilex* and 53 % for fresh), since conditioning certainly decreases the solubility of protein. The two estimates also differ in terms of the feeding value of this oak foliage. These results confirm that the goat can profit well from tree foliage if it is able to ingest substantial quantities of this relatively mediocre forage (55 % OM digestibility on average).

Table 1

THE MAIN RECENT RESULTS ON INTAKE AND DIGESTIBILITY OF TREE AND SHRUB FOLIAGE BY GOATS

REFERENCE	TYPE OF GOAT	L.N. (kg)	PHYSIOL. STAGE	DIET
WILSON (1977)	Local feral	30	Mainten.	Acacia pendula
				Casuarina cristala
				Heterodendrum oleifolium
				Geijera parviflora
Mc CAMMON-FELDMAN et al. (1980)	Nicaraguayan wether	35	M.	Cordia dentata
				Pithecolobium dulce
MASSON and DECAEN (1980)	Alpine	45	M.	Populus sp. summer
				Fraxinus sp. summer
SIDAHMED et al. (1981)	Spanish	20	M.	Quercus dumosa (40%)
				Adenostoma fasciculatum (35%)
				Arctostaphylos glandulosa (25%)
NASTIS and MALECHEK	Spanish	36	M.	Quercus gambelii June (80%) + Alf.
		34	M.	Quercus gambelii Aug. (80%) + Alf.
DEVENDRA (1981)	Local Malaysian	22	M.	Leucaena leucocephala
GUERRERO and BOZA (1981)	Granada	35	M.	Quercus ilex
				Olea europea
NASTIS (1982)	Local Greek	20	M.	Quercus coccifera spring
				Quercus coccifera summer
				Quercus coccifera fall
				Quercus coccifera winter
				Q.C. winter + Sugarbeet pulp (21%)
				Q.C. winter + soybean meal (26%)
ANTONIOU and HADJIPANAYIOTOU (1985)	Damascus vasectomized	68	M.	Acacia cyanophylla
SILVA-COLOMER et al. (1986)	Granada	39	M.	Atriplex nummularia (50%)
				+ (Alfalfa/Barley)
				Acacia salicina (50%) + Alf./Barley)
NARJISSE and BEN OMAR (1986)	Local Moroccan	-	M.	Quercus ilex winter (80%)
				+ oat straw
MEURET (1986)	Rove	25	M.	Quercus ilex winter
MEURET (1988)	Rove	35	Lactation	Quercus ilex winter
				+ Urea-molasses (14%)
				Quercus ilex winter
				+ Soybean-meal (7%)
MEURET (unpubl.)	Rove	36	Lactation	Quercus ilex winter
				+ Urea-molasses (14%)
		38		Quercus ilex winter
				+ Beet-pulp urea (13%)
MEURET (unpubl.)	Rove	38	M.	Quercus pubescens summer
			Lactation	Quercus pubescens summer
				+ Urea-molasses (11%)
DICK (1988)	Spanish wether	44	M.	Quercus gambelii June (95%)
				+ chopped Alfalfa hay
		37	M.	Quercus gambelii August (80%)

FORAGE STAGE	C.P. IN DM(%)	ADL IN DM(%)	CONDITIONS AD LIBITUM	TOTAL INTAKE to body size (g DM/kg $W^{.75}$)	TREE FOLIAGE DIGESTIBILITY	
					DMD(%)	OMD (%)
dried 50 + milled	16.9	13.6	-	57	-	46.7
+ winnoved	9.4	14.3	-	48	-	35.0
	12.5	10.6	-	38	-	43.1
	15.0	11.7	-	33	-	60.2
Fresh	11.4	15.8	+	67	-	52.9
	20.4	12.3	+	ɔ2	-	58.1
Fresh (29% DM)	11.5	8.5	+	60	-	65.7
Fresh (39% DM)	16.3	9.8	+	76	-	55.7
Air-dried + ground + chopped (97% DM)	6.5	17.2	+	19	-	46.9
Pelleted	16.3	9.8	+	87	54.0	-
Pelleted	15.6	10.5	+	102	48.9	-
Fresh	20.8	-	+	61	56.4	57.0
Dried + chopped (77%DM)	7.7	-	+	48	45.0	49.3
Fresh (69%DM)	7.7	-	+	80	57.0	60.4
Dried (92%DM)	7.0	-	+	71	52.2	55.2
Fresh	7.8	13.8	+	Oak:79 Tot:79	70	-
	6.5	15.1	+	oak:51 Tot:51	53	-
	7.2	16.1	+	oak:59 Tot:59	55	-
	7.1	16.2	+	oak:58 Tot:58	55	-
Fresh	-	-	+	oak:52 Tot:66	54	-
	-	-	+	oak:72 Tot:97	55	-
Chopped	14.4	15.2	-	oak:32	47.4	50.6
Dried 70° (89% DM)	17.0	9.8	-	-	57.4	50.4
	17.3	11.6	-	-	46.8	48.0
Air-dried (% DM) + ground (89% DM)	6.5	17.0	+	-	-	47.7
Fresh (56%DM)	8.3	17.5	+	oak:90 Tot:90	47.9	49.0
Fresh (58% DM)	8.2	18.0	+	oak:101 Tot:119	53.3	53.7
	-	-	+	oak:97 Tot:104	51.7	52.7
Fresh (58% DM)	8.3	19.1	+	oak 101 Tot:119	50.1	51.2
	-	-	+	oak:96 Tot:110	44.9	47.2
Fresh (45% DM)	10.8	14.0	+	oak:104 Tot:104	50.9	53.1
	-	-	+	oak:123 Tot:138	53.5	55.1
Fresh (35% DM)	13.1	7.8	+	oak:58 Tot:61	66	-
Fresh (45% DM)	11.9	11.8	+	oak:74 Tot:93	57	-

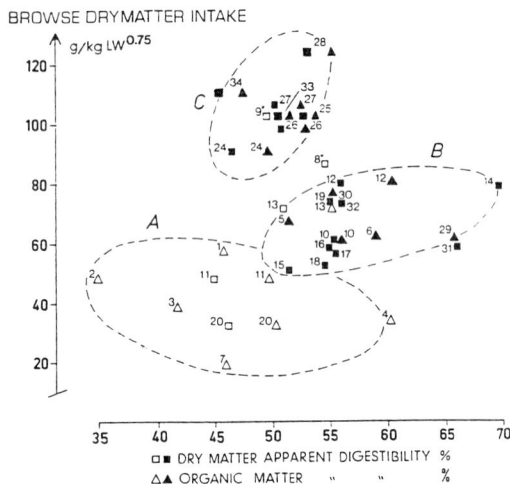

Figure 1

COMPARATIVE RESULTS OF MEASUREMENTS OF INTAKE AND APPARENT DIGESTION OF DRY MATTER (SQUARES) AND ORGANIC MATTER (TRIANGLES) IN GOAT DIETS BASED ON BROWSE

White symbols: conditioned forage; black symbols: fresh forage (numbers correspond to those of the last column in table 1).

A: conditioned forage.
B: fresh precut forage or piles of branches.
C: fresh forage with whole branches constantly renewed.

HIGH INTAKE OF BROWSE CAN COVER 3/4 OF THE ENERGY NEEDS OF A DAIRY GOAT FED ON A RANGELAND

Three Alpine oats in their 7th month of lactation were browsing in *Quercus pubescens* coppices (MEURET, 1989a). The animals were forced to eat only that species (each animal was on a leash held by an experimenter). The animals received daily a small supplement (17 g DM/kg $W^{.75}$) composed of a mixture of barley-alfalfa-urea (40 g urea/100 kg LW. day). The quantity and quality of *Quercus pubescens* intake were estimated by simultaneously using the external marker chromium oxide and direct observation of the feeding behavior, with the method described by MEURET et al. (1985), improved by the stratification of bites into distinct units of weight and structural composition (MEURET, 1989b).

The results showed that these animals, raised on rangeland since birth, could regularly ingest an average 120 g DM of oak/kg $W^{.75}$ (Figure 2). The difference between the two measurements were small (less than 5 %) for estimating intake, provided that diet digestibility is known before hand. In addition, estimations were very close to those carried out earlier on the same terrain with 5 goats (MEURET et al., op. cit.).

These Alpine goats can maintain their weight (63 kg LW on average) while producing 86 ± 3 g corrected milk/kg $W^{.75}$. day (total annual production: 650 kg over 280 days). Using the energy balance method, we were able to assess that oak on the range ingested covered the theoretical maintenance requirements, those related to travel on the range (48 % of the theoretical maintenance requirements) and up to 50 % of milk production requirements. This confirms the possibility of obtaining a satisfying production from woody forage, even with a high yielding animal, providing the management conditions permit and even stimulate substantial voluntary intake on the range.

Figure 2

ESTIMATES OF INTAKE BY TWO ALPINE GOATS FORCED TO BROWSE EXCLUSIVELY
QUERCUS PUBESCENS FOLIAGE ON RANGELAND
(MEURET, 1989b) by direct observation of feeding behavior (black symbols) and by using the chromium oxide marker (white symbols).

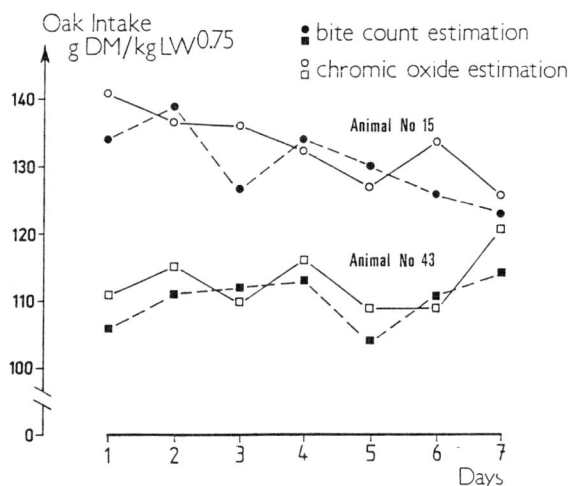

SUPPLEMENTING THE DIET EATEN ON THE RANGELAND CAN BE A MANAGEMENT TOOL FOR THE IMPACT OF THE ANIMAL ON THE VEGETATION

Under appropriate management, we have seen that high-yielding Alpine goats can draw benefits from coarse rangeland vegetation in southern France. There is a simultaneous attempt in Greece to define the conditions of a better compromise between optimizing animal productivity and the control of the woody vegetation on *Quercus coccifera* and grassy rangelands by using various techniques including goat browsing. In contrast to *Quercus pubescens* cited above, *Q. coccifera* is characterized by very prickly foliage growing on short and very hard twigs. This physical characteristic makes it difficult for the goat to obtain a rate of dry matter intake (g DMI/min. of grazing), motivating it sufficiently to consume large quantities of this foliage.

NASTIS (1982) showed (Figure 3) that although the foliar density of *Quercus coccifera* is sufficient in all seasons, the intake level of non-conditioned foliage is sufficient to cover the maintenance requirements of local goats (20 kg LW) only when these consume immature springtime foliage. This ingestion can become almost sufficient when the oak is supplemented with quality proteins. Since it is not economically reasonable to propose massive imports of soybean into the mountains of Greece, studies are examining the best techniques for treating rangeland vegetation (mechanical grinding, controlled winter burning, oversowing of perennial legumes, etc.) in order to offer the goat an environment containing both shrubs and grasses, where selected forages spontaneously complement each other. A diversification of the forage on offer can lead the animal to select a sufficiently stimulating diet which will lead it to voluntarily increase its intake. This substantial ingestion can satisfy the objective of controlling woody vegetation (MERRILL and TAYLOR, 1981).

Figure 3

INGESTED DIGESTIBLE ENERGY AND LEVEL OF DIGESTIBLE ENERGY OF FORAGE IN THE CASE OF LOCAL GREEK GOATS INGESTING FRESH FOLIAGE OF *QUERCUS COCCIFERA* IN FOUR DISTINCT PHENOLOGICAL STATES, SUPPLEMENTED IN WINTER WITH 100 g BEET PULP OR 200 g SOYBEAN OILCAKE. COMPARISON WITH INTAKE OF ALFALFA HAY BY THE SAME ANIMALS (NASTIS, 1982)

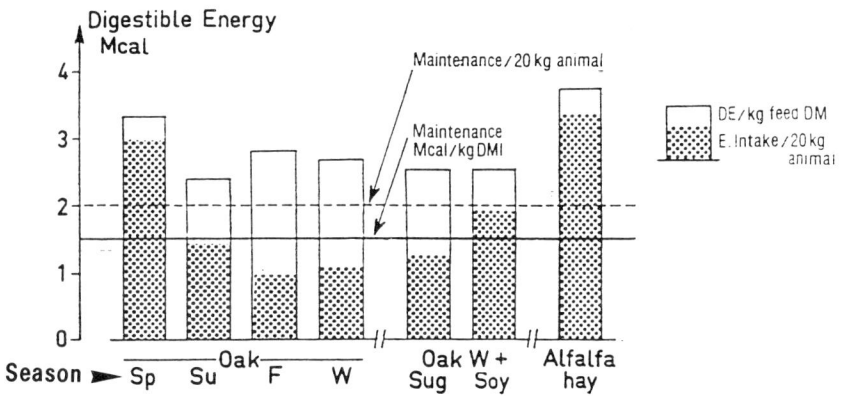

CONCLUSION

Forage plants making up the diets selected on rangelands are characterized by their high content of poorly digestible fiber, of occasionally repulsive secondary constituents and by their particular distribution in the pastoral space (three dimensions). Investigating how the goat extracts maximal value in this context is possible in studies with caged animals only by keeping very close to the conditions encountered on the rangeland. For example, *Quercus coccifera* in the above example is no longer prickly when ground, and it is only in a grazing situation that the animal's intake will be limited, when confronted with whole branches.

A given available forage on the rangeland is not of value in itself; its value is partly conditioned by the form of herd management chosen by the goat keeper. A goat may be stimulated to consume large quantities of coarse rangeland forage by feeding it an appropriate supplement, or by offering it a sufficiently diversified vegetation. The domestic goat can be influenced, including in its dietary choices. It is a highly adaptable herbivore when faced with contrasted situations.

Rather than refine the analysis of the "value" of rangeland plants in comparison to cultivated forage species, we should examine to what degree it is possible, with a reasonable input of investments, to incite a goat to mobilize this *a priori* mediocre resource.

SUMMARY

Rangelands consist of heterogeneous forage resources that may be abundant but are nutritionally unbalanced. Forage availability on the rangeland is not of value in itself. Instead, its value results from the selection which the goat is free or not to make and from the animal's motivation to ingest coarse forage. The plant species included in the diet have a relative palatability. In addition, the grazing impact may induce their chemical and physical composition to change. A lactating goat which is free to select its food and is stimulated by a nitrogen-rich supplement can browse fresh branches of Mediterranean oaks in a digestibility crate, at levels similar to those recorded for diets with high nutritional value (+ 120 g DM/kg liveweight $W^{.75}$). As a result of high intake level, a diet eaten on a wooded rangeland in summer can cover almost 3/4 of the total energy requirements of an Alpine goat regularly producing 80 g corrected (3,5 % fat) milk/kg $W^{.75}$/day.

Keywords: Goat, Rangeland, Fodder tree, Digestibility, Level of intake, Supplementation.

RÉSUMÉ

Un parcours comporte des fourrages hétérogènes, parfois abondants, mais déséquilibrés au niveau nutritionnel. Une certaine disponibilité fourragère sur parcours n'a pas une valeur en soi, cette valeur est le résultat de la sélection que la chèvre est libre ou non d'y opérer, et de sa motivation à ingérer des fourrages grossiers. Les végétaux constitutifs des rations ont une appétibilité relative, et de plus, l'impact du pâturage peut modifier leurs compositions physique et chimique. Libre de réaliser des choix, et stimulée par une complémentation riche en azote, une chèvre en lactation peut brouter des branches fraîches de chênes méditerranéens, en cage à digestibilité, à des niveaux comparables à ceux mesurés avec des rations de haute valeur nutritive (± 120 g MS/kg $P^{0.75}$). Grâce à cette ingestion élevée, une ration prélevée sur des parcours boisés en été peut couvrir près des 3/4 des besoins énergétiques totaux d'une chèvre Alpine produisant régulièrement 80 g lait corrigé/kg $P^{0.75}$ par jour.

REFERENCES

ANTONIOU T. and HADJIPANAYTIOTOU M., 1985. The digestibility by sheep and goats of five roughages offered alone or with concentrates. J. Agric. Sci. (Camb.), 105 : 663-671.

BARRY T.N. and MANLEY T.R., 1986. Interrelationships between the concentrations of total condensed tannin, free condensed tannin an lignin in Lotus sp. and their possible consequences in ruminant nutrition. J. Sci. Food Agri., 37 : 248-254.

BROWN W.F. and MOORE J.E., 1987. Analysis of forage research samples utilizing a combination of wet chemistry and near-infrared reflectance spectroscopy. J. Anim. Sci., 64 : 271-282.

DEVENDRA C., 1981. Feeding systems for goats in the humid and sub-humid tropics, Vol. 1, p. 394-410. In: MORAND-FEHR P., BOURBOUZE A.and DE SIMIANE M. (Eds.): Nutrition and systems of goat feeding. Symposium International, Tours (FRANCE), May 12-15, 1981, INRA-ITOVIC, Paris (FRANCE).

DICK B.L., 1988. Gambel oak for spanish goats: A digestion-balance evaluation of nutrient availability. M. S. Thesis Utah State Univ..

GUERRERO J. and BOZA J.E., 1981. Valeur alimentaire de quelques sous-produits agricoles pour la chèvre, Vol. 2, p. 635-642. In: MORAND-FEHR P., BOURBOUZE A., DE SIMIANE M. (Eds.): Nutrition and systems of goat feeding. Symposium International, Tours (FRANCE), May 12-15, 1981, INRA-ITOVIC, Paris (FRANCE).

GUERRERO J. and BOZA J.E., 1983. Holm-oak twigs in the feeding of goats. Avances en Alimentacion y Mejora Animal. 24 : 287-289.

HARRINGTON G.N. and WILSON A.D., 1980. Methods of measuring secondary production from browse. pp. 255-259. In: LE HOUEROU H.N. (Ed.) : Browse in Africa: the current state of knowledge. Intern. Symp., April 08-12, 1980, Addis Abeba (ETHIOPIA).

HJELFORD O., SUNDSTOL F. and HAAGENRUD H., 1982. The nutritional value of browse to moose. J. Wild. Manag., 46 : 333-343.

JARRIGE R., 1980. Chemical methods for predicting the energy and protein value of forages. Ann. Zootech., 29 : 299-323.

LACHAUX M., MEURET M. and DE SIMIANE M., 1987. Composition chimique des végétaux ligneux pâturés en région méditerranéenne française : problèmes posés par l'interprétation des analyses (Chemical composition of tree plants pastured in French mediterranean areas: interpretation of chemical analysis), p. 231-267 in: HUBERT B. (Ed.), La forêt et l'élevage en région méditerranéenne française, Fourrages, n° hors série.

LIACOS L.G. and MOULOPOULOS C., 1967. Contribution to the identification of some range types of *Quercus coccifera* L. North Greece Forest Res. Center Res. Bull. 16.

MALECHEK J.C. and PROVENZA F.D., 1981. Feeding behaviour and nutrition of goats on rangelands, Vol. 1, p. 411-428. In: MORAND-FEHR P., BOURBOUZE A. and DE SIMIANE M. (Eds.), Nutrition and systems of goat feeding. Symposium International, Tours (FRANCE), May 12-15, 1981, INRA-ITOVIC, Paris (FRANCE).

MASSON C. and DECAEN C., 1980. Composition chimique et valeur alimentaire des jeunes pousses de peuplier (*Populus*) et de frêne (*Fraxinus*) (Chemical composition and nutritive value of young sprouts of poplars (*Populus*) and ashs (*Fraxinus*)). Ann. Zootech., 29 : 195-200.

McCAMMON-FELDMAN B., GARRIGUS U.S. and VAN SOEST P.J., 1980. Differences in digestive response to grass and browse species by goats. J. Anim. Sci., 51 : 242 (Abst.).

MERRILL L.B. and TAYLOR C.A., 1981. Diet selection, gazing habits and the place of goats in range management, p. 233-252. In: GALL C. (Ed.) Goat Production, Academic Press, London (UK).

MEURET M., BARTIAUX-THIL N. and BOURBOUZE A., 1985. Estimation de la consommation d'un troupeau de chèvres sur parcours forestier. - Méthode d'observation directe des coups de dents. - Méthode du marqueur oxyde de chrome (Estimation of intake of a goat herd using a wooded rangeland. Method of direct observations of bites. Method of chromium oxide marker). Ann. Zootech., 34 : 159-180.

MEURET M., 1986. Digestibilité du feuillage de chêne vert (*Quercus ilex*) distribué frais à des caprins entraînés au pâturage sur parcours (Digestibility of fresh foliage of holm oak (*Quercus ilex*) in goats using range pasture). Reprod. Nutr. Dévelop. 28 : 91-92.

MEURET M., 1988. Feasibility of *in vivo* digestibility trials with lactating goats browsing fresh leafy branches. Small Rum. Res., 1 : 273-290.

MEURET M., 1989a. Utilization of native mediterranean fodder trees by dairy goats. 16th Intern. Grassld. Cong., Oct. 4-11, 1989, Nice (FRANCE).

MEURET M., 1989b. Feuillages, fromages et flux ingérés (Foliages, cheeses and rate of intake). Thesis in Agronomic Sciences, Gembloux (Belgium), INRA-SAD, Avignon (FRANCE).

NARJISSE H. and BEN OMAR S., 1986. Valeur nutritive des feuilles de chêne vert et influence de leur taux d'incorporation sur la valeur alimentaire de la ration (Nutritive value of live oak leaves and effect of their proportion on the nutritive value of diets). In: Les aliments pour ruminants. April 10-11, 1986. IAV Hassan II, Rabat (MAROC).

NASTIS A.S. and MALECHEK J.C., 1981. Digestion and utilization of nutrients in oak browse by goats. J. Anim. Sci., 53 : 283-290.

NASTIS A.S., 1982. Nutritive value of oak browse (*Quercus coccifera* L.) foliage for goats at various phenological stages. Dissert. Dept. of Forest., University of Thessaloniki (GREECE).

PFISTER J.A. and MALECHEK J.C., 1987. Dietary selection by goats and sheep in a deciduous woodland of Northeastern Brazil. J. Range Manag., 39 : 24-28.

REED J.D., 1986. Relationships among soluble phenolics, insoluble proanthocyanidins and fiber in East African browse species. J. Range Manag., 39 : 5-7.

SHENK J.S., WESTERHAUS M.O. and HOOVER M.R., 1979. Analysis of forage by infrared reflectance. J. Dairy Sci., 62 : 807.

SIDAHMED A.E., MORRIS KOONG L.J. and RADOSEVITCH S.R., 1981. Contribution of mixtures of three chaparral shrubs to the protein and energy requirements of spanish goats. J. Anim. Sci., 53 : 1391-1400.

SILVA-COLOMER J., FONOLLA J., RAGGI L.A. and BOZA J., 1986. Valoracion nutritiva des *Atriplex nummularia* en ganado caprino (Nutritive value of *Atriplex nummularia* in goats). Rev. Arg. Prod. Anim., 6 : 661-665.

SILVA-COLOMER J., 1987. Evaluacion de los recursos alimenticios de la zona arida del ambito del proyecto LUCDEME en ganado caprino. Doctoral thesis. ETSIA, University of Cordoba (SPAIN).

TSIOUVARAS C.N., 1987. Ecology and management of kermes oak (*Quercus coccifera* L.) shrublands in Greece: A review. J. Range Manag., 40 : 542-546.

ULLREY D.E., YOUATT W.G., JOHNSON H.E., KU P.K. and FAY L.D., 1964. Digestibility of cedar and aspen browse for the white-tailed deer. J. Wild. Manag., 28 : 791-797.

URNESS P.J., SMITH A.D. and WATKINS R.K., 1977. Predicting dry matter digestibility of mule deer forages. J. Range Manag., 30 : 119-121.

WAELPUT J.J., BISTON R. and MEURET M., 1989. Near-Infrared reflectance spectroscopy for analysis of fodder trees browsed by dairy goats in digestibility cages and on rangeland. 6e Journées Rech. Alim. Nutr. Herbivores, March 16-17, 1989, Paris (FRANCE).

WILSON A.D., 1977. The digestibility and voluntary intake of the leaves of trees and shrubs by sheep and goats. Aust. J. Agric. Res., 28 : 501-508.

VAN DYNE G.M., BROCKINGTON N.R., SZOCS Z., DUEK J. and RIBIC C.A., 1980. Large herbivore subsystem, p. 269-537. In: BREYMEYER A.I. and VAN DYNE G.M. (Eds.): Grasslands, systems analysis and man, Cambridge Univ. Press (UK).

VAN SOEST P.J., 1963. Use of detergents in the analysis of fibrous feeds II. A rapide method for the determination of fiber and lignin. J. Ass. Off. Anal. Chem., 46 : 829-835.

Chapter 15

EVALUATION AND UTILIZATION OF CONCENTRATES IN GOATS

Sylvie GIGER-REVERDIN and D. SAUVANT

INTRODUCTION

The use of concentrate feeds in goat diets depends essentially on the production system and objectives (e.g. milk, meat, hair), the availability of forages or other feeds and the goat product market.

In temperate climates, concentrates are currently used for milk or meat production, but to various extents according to the breed and physiological status of the animals and to the level of production.

Forages are often available in large quantities at relatively low costs. In dairy goats, only maintenance requirements and 0 to 3 kg milk production can be met by the forages depending on their quality. Higher production requirements can be satisfied by concentrate supplies because it is the best way of adjusting energy, protein, mineral and vitamin supplies both quantitatively and qualitatively (see Chapter 17).

For growing weaned kids, the percentage of concentrate in the diet varies from 0 to 90 % according to the feeding system (see Chapter 23). Proportions between forages and concentrates are calculated to reach the energy and protein concentrations required by the weight gain objectives. In intensive production systems, high levels of production can be reached with large proportions of concentrates.

In other cases, and especially in tropical and sub-tropical climates, the availability of forages is often reduced. Two cases may then occur. In the first one, concentrates are given in large proportions to reach a sufficient level of production. In the second one, like in arid countries where the agro-climatic and socio-economic constraints are very heavy, concentrate feeds are scarce and expensive. Therefore, concentrate supplies are not economically justified. Moreover, the level of intake is limited by roughage availabilities.

In all cases, the performances obtained with given feed allowances should be accurately determined through a better knowledge of the nutritive value of the concentrate feeds. It is also of importance to compare the efficiency and costs of the different diets. This chapter deals only with the technical point.

Results on the evaluation and utilization of compound concentrate feeds for goats are relatively scarce. It is of interest to notice that no chapter on this specific topic was included in the Proceedings of the Tours Symposium in 1981. Only a few paragraphs in different chapters concerned this matter (DE SIMIANE et al., 1981; SKJEVDAL, 1981). In the present book, the use of concentrates for milk and meat production is dealt with in Chapters 17 and 23, respectively. The present chapter analyses the palatability of concentrates in goats, the level of intake and the substitution rate between forages and concentrates, their digestion in goat rumen, their digestibility and their "milk value". Some proposals are made for their use in goat husbandry.

Concentrate feeds are used by farmers as a tool to influence goat performance. Moreover, it must be kept in mind that the type of forage is generally imposed because it depends on agro-climatic conditions. We will not discuss here the use and interest of each concentrate feed (see Chapters 17 and 23), but we would like to present the information required to define the general strategy for concentrate use in goat production, so as to prevent errors in feed formulation for goats. For this purpose, we report results from experiments using compound feed concentrates, the composition of which can be easily modified.

INTAKE AND SUBSTITUTION

In goats given forages *ad libitum* the main factor influencing the performances is the level of dry matter intake (MORAND-FEHR and SAUVANT, 1978). Concentrates may improve the level of forage intake by supplying fermentescible carbohydrates or proteins. Moreover, goats can ingest large quantities of concentrates to reach high performances (MORAND-FEHR and SAUVANT, 1987). With concentrate feeds offered *ad libitum*, their level of intake depends on different factors: their physical form, their palatability, the type of forages and the fact that concentrates are in mixture or not. Nevertheless, it is well known that with an *ad libitum* concentrate allowance, the risks for development of digestive and metabolic disorders increase (see Chapter 12), and that it is frequently unsuitable without including a minimum amount of fibre.

Palatability of concentrate feeds

Chapter 1 indicates that goats are very sensitive to the palatability of feeds, and generally more than sheep or cattle. Indeed, goats are able to detect the presence of about 1 % low quality meat meal in compound feeds in the tests described in Chapter 1. Refusal of concentrates is more frequent in late gestation and early lactation. Moreover, goats are very sensitive to badly stored feeds and particularly to moldy or oxidized ones.

Figure 1 presents a feed palatability scale set up on the basis of all the palatability tests published between 1984 and 1989 with more or less palatable concentrate feeds (MORAND-FEHR, HERVIEU and CORNIAUX, 1989, unpubl.). Some feeds, for example fats or meat meals, can strongly reduce the palatability of compound feeds. In fact, the palatability of one ingredient depends on its level of incorporation into compound feeds.

Figure 2 shows that palatability decreases according to an S shaped curve when the level of incorporation increases. However, the palatability of a feed depends also on its origin, on its botanical variety and on the technological treatments used to obtain it. A lower palatability is observed with unprotected rather than protected fat, with 50 % rather than 60 % fat meat meal (Figure 1), with variety 0 rather than variety 00 of rapeseed meal (Figure 2).

Composition of compound concentrates

The influence of compound concentrate composition on the level of intake has mainly been studied in dairy goats. It was possible to formulate three couples of compound feeds with the same organic matter digestibility (OMD) and energy value measured *in vivo*, but with very different fibre contents (GIGER et al., 1987a). Each couple consisted of a "low-fibre" mixture containing mainly cereal grains and their by-products and rich in starch ("S" compound), and of a "high-fibre" mixture mainly made of sugar beet pulp or soybean hulls, and rich in fibre of high digestibility ("FH" compound). These three couples of compound feeds were studied in several experiments using different kinds of forages and animals at different physiological states. The ingredient composition, chemical composition as well as the digestibilities of these compounds when offered with lucerne hay, are given Table 1.

Figure 1

SCALE OF FEED PALATABILITY IN GOATS
(MORAND-FEHR, HERVIEU and CORNIAUX, 1989, unpubl.)

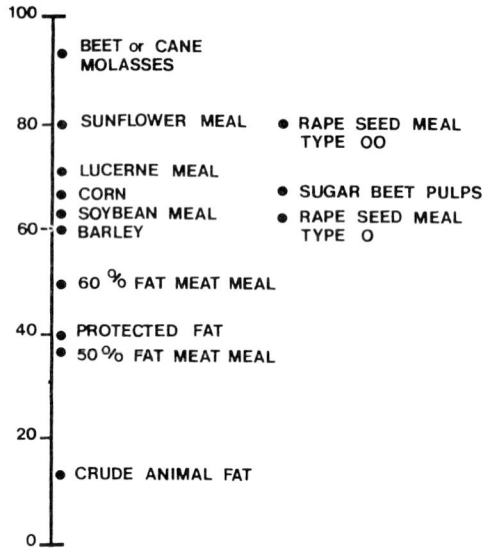

Figure 2

EFFECT OF RAPESEED VARIETY AND PERCENTAGE IN CONCENTRATE
ON COMPOUND FEED PALATABILITY INDEX
(MORAND-FEHR and HERVIEU, 1988)

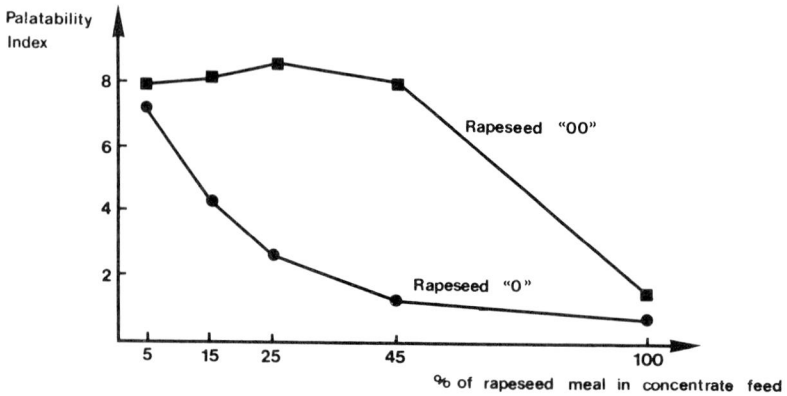

The influence of the composition of these compound feeds on the level of intake and milk production was tested in goats between the 7th and the 20th week of lactation. They were given either lucerne hay or maize silage or a mixture of lucerne hay and maize silage (GIGER et al., 1987b). Forages were offered *ad libitum* with a level of refusals between 10 and 35 % of offered dry matter. Concentrates were offered according to the level of production at a rate of 350 g dry matter/kg of milk produced above the requirements satisfied by the forages. Results concerning level of intake and milk production are given Table 2. When all the variates were in g. per kg $W^{.75}$, dry matter intake (DMI) was highly correlated with milk yield (MY) and concentrate intake (CONC) :

$$DMI = 50.5 + 0.330 \ MY + 0.329 \ CONC$$

$$(r = 0.86, \ n = 84, \ RSD = 11)$$

Table 1

COMPOSITION AND NUTRITIVE VALUE OF COMPOUND FEEDS
(GIGER, 1987; GIGER et al., 1987a)

Compound feeds	S 1	FH1	S 2	FH2	S 3	FH3	FL3
Composition of ingredients (% Dry Matter)							
Cereals and by-products	68		80		75		30
Soybean hulls		6		20		20	
Beet or citrus pulps		47		40		40	10
Leguminous by-products				22		10	30
Oilmeals	22	37	12	10	15	20	20
Molasses and minerals	10	10	8	8	10	10	10
Chemical composition (% Dry Matter)							
NDF	19.4	48.3	20.4	25.6	15.1	33.8	34.1
ADF	8.9	25.3	6.8	16.4	6.4	18.6	20.2
ADL	1.2	3.8	1.4	1.3	1.2	1.4	6.7
CP	19.5	16.3	17.1	21.7	18.8	19.6	22.0
NDF *in sacco* digestibilities (%)							
6h incubation	21.4	36.1	22.5	12.4	14.9	22.1	11.3
12h incubation	23.5	56.6	38.7	51.2	44.9	47.0	29.3
24h incubation	33.2	75.9	46.8	79.2	50.0	74.1	39.9
48h incubation	49.2	84.6	53.1	86.0	54.2	82.1	46.2
***In vivo* digestibilities measured with lucerne hay as roughage (%)**							
Organic matter	78.5	80.5	81.4	85.1	83.7	83.1	65.8
NDF	25.9	77.3	35.9	63.4	39.3	71.2	31.5
ADF	24.0	68.6	36.4	81.1	26.9	67.4	25.2
CP	78.5	64.1	88.5	83.0	85.4	88.3	73.8

Compounds : S : rich in starch; FH : rich in fibre of high digestibility; FL : rich in fibre of low digestibility.

When taking into account variations in the DMI explained by milk production and concentrate intake, it was possible to study the residual of this equation. There was no effect of the kind of forage on DMI and no interaction between forage and concentrate types in this residual. DMI was influenced only by the type of compound feeds, when given in large proportions (40 % dry matter or more). The goats which received a "FH" compound ate a little more than the others (Table 2). The difference was 6.26 g DM/kgW$^{.75}$/day or 135 g DM/day for a 60 kg goat. It represented about 10 % more forage dry matter intake. This was in agreement with the results of MEIJS (1986) in dairy cows fed *ad libitum* with perennial rye-grass silage and either a starchy or a fibrous concentrate. They observed higher levels of intake (2.8 kg DM/day/cow) with the fibrous compound feed compared to the starchy one. COULON et al. (1989) observed also a higher level of forage intake with the fibrous one when the level of concentrate intake was the same for these two kinds of concentrates.

These results showed that all the concentrates did not, in fact, have the same fill value, contrary to the principle of the French Fill Unit system. When given with lucerne hay (GIGER, 1987) or a mixture of lucerne hay and maize silage (SAUVANT et al., 1980), an "S" concentrate had a higher fill value and a higher rate of substitution, than a FH concentrate. It corresponded to a decrease in the quantity of ingested forage, expressed in kg, when an extra 1 kg of concentrate was eaten. These results were also in agreement with those of BERGE and DULPHY (1985) who observed that in sheep, barley and wheat bran had a higher fill value than sugar beet pulps. The opposite was observed when maize silage was used in goat diets (GIGER, 1987).

Table 2

INFLUENCES OF CONCENTRATE INTAKE AND MILK PRODUCTION
ON DRY MATTER INTAKE
(GIGER et al., 1987b)

Forage	**LH***		**LH + MS***		**LH**		**LH + MS**		**LH**		**MS**	
Concentrate	S1	FH1	S1	FH1	S2	FH2	S2	FH2	S3	FH3	S3	FH3
Number of observations	11	9	8	8	4	4	4	4	9	9	9	5
Dry Matter Intake (g/kg W$^{.75}$)	112	122	108	118	118	127	113	117	111	122	103	96
Concentrate intake (g/kg W$^{.75}$)	73	65	29	28	51	56	49	45	58	56	42	37
Milk production (g/kg W$^{.75}$)	124	144	158	169	146	152	172	168	132	147	122	91

(*) LH : lucerne hay; MS : maize silage

These results were confirmed by studies of degradabilities in nylon bags, especially after a 12 h degradation (HAKIM et al., 1988), which gives a good estimate of the level of intake (DEMARQUILLY et al., 1981). Indeed, the "low-fibre" feed was more degraded than the "high-fibre" feed when the forage offered was lucerne hay, but the results were reversed with maize silage. The decrease in cellulolysis in the rumen observed with a maize silage diet when compared to a hay diet, was linked to a low level of fermentescible nitrogen and to a decrease in pH values in the rumen.

Therefore, the composition of a compound concentrate seems to be less important than its level of incorporation into the diet. The rate of substitution is close to 1 when the level of concentrate intake is high (more than 800 g per day for a 60 kg lactating goat). This rate of substitution is less than 1 when the level of concentrate is below 500 g concentrate/day in mid-lactation (OPSTVEDT, 1967; GARMO, 1986; MORAND-FEHR and SAUVANT, 1987).

Even if these results have been obtained on just a few types of compound feeds, they show that goats do not differ from other ruminants like cows or sheep.

Physical form and presentation of concentrates

Concentrate feeds are available as whole grains or ground grains or as meals or pellets. The particle size of meals must not be too small, otherwise the particles can enter the pulmonary tract of goats, and the loss in a dust form can be high. To be suitable for goats, grinding of concentrate feeds must be rough enough.

Diets containing high proportions of concentrates (more than 75 % of the total dry matter ingested in dairy goats or more than 85 % in growing goats) can cause acidosis or acido-cetosis because of the large decrease in rumination (see Chapter 12). Cereals are as acidogenic as industrial by-products rich in highly digestible fibre, such as sugar beet or citrus pulps and soybean hulls. Finely ground meal or a large quantity of concentrates offered in a single meal per day increase the risk of acidosis. Mixing a concentrate with a chopped forage, as maize silage, in a complete diet should reduce this risk. However, the use of complete diets is more difficult in goats than in other ruminants, since goats try always to select some specific components of the complete diet.

RUMEN DIGESTION

Data about digestion in the rumen are of two kinds : some of them concern the rumen parameters measured *in vivo* in goats receiving a mixed diet containing concentrates and forages, and others have been obtained by the nylon bag method.

Carbohydrate composition of concentrates can influence ruminal digestion parameters, as shown in a study including three concentrates associated with lucerne hay (GIGER et al., 1988). Two of them were the S3 and FH3 described before. The third one, called FL3, was formulated to get the same fibre content thant the FH3 (NDF : 34.1 vs 33.8 %), but a low (65.8 vs 83.1 %) organic matter digestibility (Table 1). The influence of the kind of concentrate on rumen pH was very marked for concentrates rich in highly degradable cell-wall (FH3) or cereal constituents (S3) and less noticeable for poorly digestible fibre (FL3) (see Figure 3). The diet also influenced the molar proportion of volatile fatty acids. Compared to a diet containing only hay, a diet containing the same hay and the "low-fibre" concentrate rich in cereals (S3) did not modify the percentage of propionic acid, increased significantly that of butyric acid and decreased that of acetic acid. "High-fibre" concentrates did not modify acetic acid proportion, but increased propionic acid percentage when the concentrate had a good organic matter digestibility (FH3) and increased butyric acid percentage when the compound concentrate had a poor organic matter digestibility and contained a lot of cereal by-products, like the FL_3 (GIGER et al., 1988).

These results were similar to those obtained in sheep, when comparing, on the one hand, the compound cereal rich concentrate (S3) and barley, and on the other hand, the highly digestible fibrous compound (FH3) and sugar beet pulp (BERGE and DULPHY, 1985). Decrease in pH values and intensive fermentation activities were observed with the FH3 concentrate and degradation essentially concerned cell contents. This result can be explained by the use of ingredients rich in pectins like sugar beet or citrus pulps. It should be noticed that the FH3 concentrate of which the organic matter is degraded in the hours following a meal at a lower rate than that of the S3 concentrate, may result in a larger pH decrease than S3. This observation is of importance when considering rumen cellulolytic activity or even risks of acidosis.

Figure 3

RUMINAL PH VARIATION DURING THE FIRST FOUR POST-PRANDIAL HOURS
(GIGER et al., 1988)

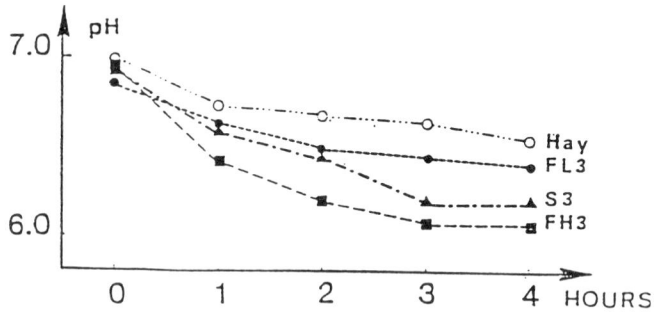

In *in sacco* experiments (GIGER et al., 1987a), variabilities in cell wall degradation were large between the three types of compound feeds (Table 1). The differences were enhanced with the duration of incubation. Concentrates containing high percentages of cereals and their by-products had a poor neutral detergent fiber (NDF) degradation after 48 h of incubation, independently of the level of organic matter digestibility (65 or about 80 %).

These results were in agreement with those observed on the ingredients alone (SAUVANT et al., 1985; SAUVANT et al., 1986). In agreement with NOCEK (1988) working on dairy cows, it seems important to balance goat diets with adequate amounts and types of carbohydrates and to supply nitrogen and non structural carbohydrate sources (starch, pectins) together rather than separately. Microbes are then allowed to find simultaneously nitrogen and carbon skeletons for their protein synthesis. Complete diets are an interesting feeding strategy, because mixing all the ration components insures that goats receive the proper proportions of carbohydrates and nitrogen at each meal. Blending all diet components is probably more important and more useful for a good utilization of feeds by animals than the kind of concentrate when it has a good nutritive value. In fact, fibrous concentrates are often made of pulps rich in highly digestible fibre, but also in pectins which have a bad effect on pH.

Thus, it is quite important to know the kinetic of energy and protein degradation of the different types of compound concentrates in order to reach the best complementarity between a concentrate and a forage. Results obtained with the *in sacco* method will help to achieve this objective and will have to be taken into account in a near future for the determination of all the feed unit systems for ruminants, including those concerning goats.

DIGESTIBILITY

About twenty different compound concentrates were tested in dairy goats kept in metabolic crates (GIGER, 1987 and unpubl.). They were of one of the three types described earlier : S, FH or FL. Most of them were given to lactating goats fed according to their milk production, i.e. 350 g DM of compound feed/kg milk above the requirements met with the roughage. The digestibilities of compound feeds were calculated according to the method proposed and discussed by GIGER and SAUVANT (1983).

When the composition and percentages of ingredients in the different compound concentrates were known, it was possible to calculate the organic matter digestibility values using literature data, generally obtained in sheep fed at a maintenance level (DEMARQUILLY et al., 1978). Indeed, when the ingested forage was lucerne hay, the digestibilities observed *in vivo* in goats were not significantly different from the calculated ones. However, some negative digestive interactions appeared when maize silage replaced lucerne hay (GIGER et al., 1987d). The observed values were lower when the same concentrate was associated with maize silage than with lucerne hay (76.0 vs 83.1 for FH3, and 77.4 vs 83.7 for S3). The different components of the organic matter were not equally affected (GIGER, 1987). When maize silage replaced lucerne hay, crude protein and cell wall digestibilities decreased more than that of the cell content.

The level of production did not seem to have any significant influence on the organic matter digestibility. Thus, the organic matter digestibility (OMD) of a whole diet can be predicted with a high accuracy only from its chemical composition. Indeed, the residual standard deviation (RSD) of the OMD was 2.3 when using neutral detergent fibre (NDF), acid detergent fibre (ADF), acid detergent lignin (ADL) and crude protein (CP) percentages of the organic matter (GIGER et al., 1986) :

$$OMD = 75.9 - 0.294 \text{ NDF} + 0.563 \text{ ADF} - 2.477 \text{ ADL} + 0.288 \text{ CP}$$

$$(r = 0.80, n = 224, RSD = 2.3)$$

The precision could be improved to 2.0 of RSD if the botanical composition (% of maize silage or MS, % of cereals or Ce) was also included in the model and it was then of the same order than the uncertainty of measurement (GIGER et al., 1986).

$$OMD = 89.9 - 0.45 \text{ ADF} - 1.41 \text{ ADL} - 0.012 \text{ CP} - 0.079 \text{ MS} - 0.129 \text{ Ce}$$

$$(r = 0.86, n = 224, RSD = 2.0)$$

The factors of variation of OMD were studied on a set of 83 compound concentrates of various origins and types. Results from eighteen compound concentrates were obtained in goats and the rest in sheep (GIGER-REVERDIN et al., unpublished data). Chemical composition and/or enzymatic digestibilities explained the OMD with a high precision, since the residual standard deviation was about 2.4. In the residuals of the equations, results obtained in goats did not significantly differ from those obtained in sheep. This allows goat farmers to use data given in tables issued from experiments in sheep, for goat diet formulation. When taking into account results obtained on diets with a poor digestibility (Chapter 5), it appears that the superiority of goats on other ruminants decreases when the level of energy value increases and even disappears with diets of good nutritive value.

Our study showed huge variations in the digestibility of the different cell wall components measured as crude fibre (CF : 22.8 to 71.0 %), neutral detergent fibre (NDF : 25.9 to 78.7 %), acid detergent fibre (ADF : 24.0 to 70.5 %) and small variations for the total cell content (83.8 to 93.5 %) (GIGER, 1987). The *in vivo* digestibilities were highly correlated with the *in sacco* digestibilities (GIGER et al., 1987a and Table 1), and the differences between the two methods were constant when considering the 24 and 48 h incubations. This is of practical importance, since a determination is easier to perform *in sacco* than *in vivo*. The differences between a 6 h *in sacco* degradation and an *in vivo* digestibility were larger when concentrates had a high cell wall digestibility rather than a low one.

MILK VALUE

When goats received either lucerne hay (LH) or maize silage (MS) offered with a compound concentrate with low or high-fibre values, lucerne hay gave better milk values than maize silage (GIGER and SAUVANT, 1983). Goats receiving lucerne hay had a higher milk persistence than the others, but a lower live-weight gain. The concentrate with the low-fibre value gave a better milk value than the other one, even if they had the same organic matter digestibility.

However, from results pooling different couples of compounds having the same OMD, ("S" and "FH" types) and given with different roughages, the carbohydrate nature of the concentrate appeared to have no influence on milk composition (fat and protein contents, fatty acid composition) and quantities of fat production (GIGER et al., 1987c). These results were also close to those observed in cows by STEG et al. (1985).

It must be kept in mind that concentrate composition had no or a small effect on milk production unless unfavourable digestive interactions and acidosis in the reticulo-rumen appeared. They led to limit the dietary supply of starch and soluble sugars to a maximal value of about 40 % of the dry matter (see Chapter 12).

In conclusion, as indicated by MORAND-FEHR and SAUVANT (1987), goat performance results with concentrates are very similar to those obtained in cattle or sheep.

CONCLUSION

Compound feeds supplies allow to balance energy, nitrogen and minerals brought by forages and consequently to improve the forage utilization. Moreover, they can express the production potential of goats and permit to keep good nutritional status and sanitary conditions.

These results point out that when kept in the same conditions, goats digest concentrate feeds of a high nutritive value to the same extent than other ruminants. Nevertheless, it is important not to consider a compound concentrate separately, but to take into account the type and the availability of the forage to which it is added, in order to meet the requirements of the animal.

Concentrate feeding can be simplified without reducing performances of dairy or growing goats (see Chapter 17). However, an error concerning the quantity or the composition of the concentrate given can induce metabolic disorders or a decrease in performances (see Chapter 12). The decision of goat farmers must take into account the technical and economical efficiencies of concentrate supplies.

SUMMARY

The use of concentrate feeds in goat diets depends on the production system, the availability of forages or other feeds and the goat products market. The evaluation and utilization of compound feeds depend on their composition, which may vary a lot and thus, influence the different components of their nutritive value.

The types of ingredients in compound feeds influence the level of forage intake offered *ad libitum,* and then the substitution rate between forages and concentrates. Modifications also appear for rumen digestion estimated *in vivo* or through the *in sacco* method when comparing different feeds. But, it seems that the level of incorporation of compound feeds, and especially of highly degradable carbohydrates is of greater importance than the type of concentrate itself. Cell wall *in sacco* degradability or *in vivo* digestibility may vary a lot. Moreover, level of production does not influence organic matter digestibility. Milk value and milk persistence are influenced by the type of forage, but not by the type of concentrate. It must also be stressed that, when kept in the same conditions, goats act as other ruminants for digestion and efficiency.

Keywords : Goat, Concentrate feeds, Level of intake, Digestion, Digestibility, Milk value.

RÉSUMÉ

L'utilisation des aliments concentrés dans l'alimentation de la chèvre dépend du système de production et de ses objectifs, de la disponibilité du fourrage et des autres aliments, ainsi que du marché des produits caprins. L'évaluation et l'utilisation des aliments composés est liée à leur composition, qui est très variable, et qui, de ce fait, peut influencer les différentes composantes de leur valeur nutritive.

Les qualités des matières premières entrant dans la composition des aliments composés influence le niveau du fourrage distribué *ad libitum*, et par conséquent, le taux de substitution entre le fourrage et le concentré. Le remplacement d'un aliment par un autre peut entraîner des modifications au niveau de la digestion ruminale estimée *in vivo* ou par la méthode des sachets de nylon. Cependant, le niveau d'incorporation des aliments concentrés, et plus particulièrement la teneur en glucides hautement dégradables est plus importante que le type de concentré lui-même.

La dégradabilité *in sacco* des parois végétales ou leur digestibilité *in vivo* peuvent être très variables. Cependant, il est à noter que le niveau de production ne semble pas modifier la digestibilité de la matière organique. Les performances laitières sont influencées par le type de fourrage, mais non par celui du concentré.

Il ressort également de ces résultats que lorsque les chèvres sont placées dans les mêmes conditions que les autres ruminants, leurs réponses sont équivalentes pour la digestion et l'efficacité de transformation.

REFERENCES

BERGE Ph. and DULPHY J.P., 1985. Étude des interactions entre fourrage et aliment concentré chez le mouton. I. Facteurs de variation du taux de substitution (Interactions between forages and concentrates in sheep. I. Variations of substitution rate). Ann. Zootech., 34: 313-334.

COULON J.B., FAVERDIN P., LAURENT F. and COTTO G., 1989. Influence de la nature de l'aliment concentré sur les performances des vaches laitières (Effect of kind of concentrates on dairy cows performances). INRA Prod. Anim., 2: 47-53.

DE SIMIANE M., GIGER S., BLANCHART G. and HUGUET L., 1981. Valeur nutritionnelle et utilisation des fourrages cultivés intensément (Nutritive value and utilization of intensively cultivated forages). Vol I, p. 274-299. In: MORAND-FEHR P., BOURBOUZE A. and DE SIMIANE M. (Eds.), Nutrition and systems of goat feeding. Symposium International, Tours (FRANCE), May 12-15, 1981, INRA-ITOVIC, Paris (FRANCE).

DEMARQUILLY C., ANDRIEU J. and SAUVANT D., 1978. Tableaux de la valeur nutritive des aliments (Tables of feeds nutritive value), p. 519-555. In: JARRIGE R. (Ed.): Alimentation des ruminants, INRA Publ., Versailles (FRANCE).

DEMARQUILLY C., ANDRIEU J. and WEISS Ph., 1981. L'ingestibilité des fourrages verts et des foins et sa prévision (Ingestibility of green forages and hays and its prevision), p. 155-167. In: DEMARQUILLY C. (Ed.): Prévision de la valeur nutritive des aliments des ruminants, INRA Publ., Versailles (FRANCE).

GARMO T.H., 1986. Dairy goats grazing mountain pastures. 1. Effect of supplementary feeding. Meld. Nor. Landbrukshoegsk, 241: 1-19.

GIGER S., 1987. Influence de la composition de l'aliment concentré sur la valeur alimentaire des rations destinées au ruminant laitier (Effect of concentrates composition on the nutritive value of dairy ruminant diets). Thèse de Docteur Ingénieur, INA-PG, Paris (FRANCE).

GIGER S. and SAUVANT D., 1983. Données sur la valeur laitière des concentrés. Observations sur chèvres (Data on milk value of concentrate feeds from experiments on goats), p.I1-I14. In: CAAA. Quels aliments pour les vaches fortes productrices de lait ?, ADEPRINA, Paris (FRANCE).

GIGER S., SAUVANT D., HERVIEU J. and DORLEANS M., 1986. Étude de la prévision de la digestibilité des rations mixtes distribuées à des chèvres laitières par ses caractéristiques analytiques (Predicting ability of different analytical criteria for digestibility of mixed diets given to dairy goats). Ann. Zootech., 35: 137-160.

GIGER S., SAUVANT D. and HAKIM S., 1987a. Degradation of carbohydrates in compound feeds for ruminants and its impact on feeding value, p. 48-52. Cornell Nutr. Conf. for Feed Manuf., Oct. 26-28, 1987, Syracuse, Ithaca (USA).

GIGER S., SAUVANT D. and HERVIEU J., 1987b. Influence de la nature de l'aliment concentré complémentaire sur l'encombrement des rations par la chèvre en lactation (Fill effect of concentrate type on voluntary intake in lactating goats). Reprod. Nutr. Dévelop., 27: 201-202.

GIGER S., SAUVANT D. and HERVIEU J., 1987c. Influence of the kind of compound on the goat milk production and composition. Ann. Zootech., 36: 334-335.

GIGER S., SAUVANT D. and HERVIEU J., 1987d. Digestive interactions between forages and concentrates in lactating goat diets. Vol. 2, p. 1421-1422. 4th Intern. Conf. on Goats, March 8-13, 1987, Brasilia (BRAZIL).

GIGER S., SAUVANT D., DURAND M. and HERVIEU J., 1988. Influence de la nature de l'aliment concentré sur quelques paramètres de la digestion dans le rumen (Influence of the concentrate source on some rumen digestion parameters). Reprod. Nutr. Dévelop., 28: 117-118.

HAKIM S., SAUVANT D., GIGER S. and CHAPOUTOT P., 1988. Influences comparées du foin de luzerne et de l'ensilage de maïs sur la dégradabilité in sacco d'aliments concentrés riches en amidon ou en paroi végétale dégradable (Comparative effects of lucerne hay and maize silage on in sacco degradability of concentrates based on cereal or by-products rich in digestible cell walls). Reprod. Nutr. Dévelop., 28: 173-174.

MEIJS J.A.C., 1986. Comparison of starchy and fibrous concentrates for grazing dairy cows. Occasional Symp. Brit. Grassland Soc., 19: 129-137.

MORAND-FEHR P. and SAUVANT D., 1978. Nutrition and optimum performances in dairy goats. Livest. Prod. Sci., 5: 203-213.

MORAND-FEHR P. and SAUVANT D., 1987. Feeding strategies in goats. Vol. 2, p. 1275-1303. 4th Intern. Conf. on Goats., March 8-13, 1987, Brasilia (BRAZIL).

MORAND-FEHR P. and HERVIEU J., 1988. Acceptabilité des aliments composés contenant des tourteaux de colza par des tests de préférence sur chèvres (Acceptability of concentrate feeds containing rapeseed oilmeals by cafetaria tests on goats). Reprod. Nutr. Dévelop., 28: 101-102.

NOCEK Jr E., 1988. Feeding management to enhance carbohydrate metabolism in ruminants, p. 27-34. Cornell Nutr. Conf. Feed Manuf., Oct. 25-27, 1988, Syracuse, Ithaca (USA).

OPSTVEDT J., 1967. Feeding experiments with goats. 1. Studies on the effect of energy level and feed combination feed intake, milk yield, composition and organoleptic properties. Techn. Bull. Inst. Nutr. Agric. Coll. Norway (134), 1-114.

SAUVANT D., BERTRAND D. and GIGER S., 1985. Variations and prevision of the in sacco dry matter digestion of concentrates and by-products. Anim. Feed Sci. Technol., 13: 7-23.

SAUVANT D., GIGER S., CHAPOUTOT P. and DORLEANS M., 1986. Variations of the feed cell wall degradation kinetics in the rumen. Cell walls 86, Sept. 10-12, 1986, Paris (FRANCE).

SAUVANT D., GIGER S. and MORAND-FEHR P., 1980. Influence de la composition glucidique de l'aliment concentré sur sa valeur d'encombrement (Influence of the carbohydrate composition on the bulkiness of feed compounds). 31st EAAP Annual Meeting, Sept. 1-4, 1980, Munich (GERMANY).

SKJEVDAL T., 1981. Effect on goat performances of given quantities of feedstuffs, and their planned distribution during the cycle of reproduction. Vol. 1, p. 300-318. In: MORAND-FEHR P., BOURBOUZE A. and DE SIMIANE M. (Eds.): Nutrition and systems of goat feeding, Symposium International, Tours (FRANCE), May 12-15, 1981, INRA-ITOVIC, Paris (FRANCE).

STEG A., VAN DER HONING Y. and DE VISSER H., 1985. Effect of fibre in compound feeds on the performance of ruminants, p. 113-119. In: HARESIGN W. and COLE D.J.A. (Eds.): Recent advances in animal production, Butterworths, London (UK).

Chapter 16

GROWTH PROMOTERS FOR FATTENING KIDS

P. SCHMIDELY and M. HADJIPANAYIOTOU

INTRODUCTION

There is a considerable interest for the use of growth promoters to improve animal production efficency which can be achieved in two main ways. Firstly, altering ruminal digestion with ionophore antibiotics (monensin, lasalocid, narasin, salinomycin) or non ionophore antibiotics (avoparcin) may favourably modify the quantity and quality of nutrients entering the body. Secondly, it is possible to manipulate metabolism processes and utilisation of absorbed nutrients with hormonal anabolic agents. However, the use of these products in goat kids is very limited. The present paper will outline research work on the use of growth promoters in fattening kids conducted by the institutes participating in the FAO research sub-network on goats and compare them with those obtained in other ruminants.

ANTIBIOTICS

Ionophores are molecules with an antimicrobial activity mainly on rumen protozoa and Gram (+) bacteria. Their mode of action is related to their ability to modify ion flux and protonmotive force across lipid bilayer membranes of these microorganisms, by increasing the intracellular Na^+ concentration and concomitantly decreasing the K^+ concentration (RUSSEL, 1987). The most widely studied ionophores are monensin and lasalocid. A new non-ionophore antibiotic, avoparcin, seems to produce the same effects as ionophores on Gram (+) bacteria, ruminal digestion and growth performance in cattle (see below). In the EEC, recommended doses for fattening cattle range between 15-45 mg/kg feed (optimum : 20-30) for avoparcin and between 20-60 mg/kg feed (optimum 30) for monensin.

Influence of ionophores on kid performance

Monensin and lasalocid sodium have been used in trials with kids conducted in Cyprus (HADJIPANAYIOTOU et al., 1988) and in Egypt (MEHREZ et al., 1982). In these studies, addition of lasalocid sodium (37 mg/kg concentrate) improved the average daily gain (ADG) of kids fed on high concentrate diets (Table 1), but feed intake of lasalocid-fed kids was variable among experiments ; in all cases, the feed conversion efficency (ADG / ingested dry matter) was improved. In the literature, differences in feed intake have been related to concentrate/roughage ratio of the diet: it has been shown in cattle (review of GOODRICH et al., 1984) that on high concentrate diets, ionophore fed animals did not improve their ADG, but reduced their level of intake. With high forage diets, ADG was increased with no change of intake. These differences probably are a result of modifications of energy and protein digestion in the rumen after treatment (see below).

Table 1

EFFECT OF LASALOCID SODIUM AND TYPE OF FEEDING TROUGH ON GROWTH AND
INCIDENCE OF DIARRHOEA IN GROWING DAMASCUS KIDS
(from HADJIPANAYIOTOU et al., 1988)

| Feeding trough protection | Concentrate mixture | | | | SE and level of significance | |
| | Without Lasalocid | | Lasalocid (37 ppm) | | | |
	Yes	No	Yes	No	(1)	(2)
Number of kids	12	12	12	11		
Initial weight (kg)	14.3	14.4	14.4	14.5	1.92	NS
Final weight (134 d) (kg)	33.1	36.0	37.5	38.5	2.65	**
Daily weight gain (g/day)	241	277	263	293	37	**
Total feed intake (kg)						
Concentrate	77	85	83	85		
Hay	7.5	7.5	7.5	7.5	Group feeding	
Feed/gain (kg DM/kg)	3.9	3.7	3.5	3.4		
Diarrhoa incidence						
All cases	4	0	0	1		
Coccidiosis	3	0	0	0		

(1) Effect of lasalocid : NS : non significant, ** $P < 0.01$

(2) No effect of feeding trough protection.

Ionophores and diet digestibility

Adding lasalocid sodium (37 mg/kg concentrate, as fed basis) to concentrate mixtures fed *ad libitum* to Damascus kids did not significantly alter dry matter (DM) and crude protein (CP) digestion coefficients or CP absorption and retention (Table 2). The literature is not conclusive on the effect of ionophores on DM or CP digestibility of the diet. Significantly higher digestion coefficients for lambs fed monensin have been reported by JOYNER et al. (1979) and for steers by BEEDE et al. (1986). On the other hand, apparent DM digestibility was not affected in monensin fed cattle (THOMPSON and RILEY, 1980), goats (BEEDE et al., 1985, 1986) or sheep (POOS et al., 1979). This can be attributed to a modification in the site of digestion and absorption as reported for avoparcin. For CP, these differences may also be explained by the extent of the decrease in ruminal nitrogen degradation and microbial protein synthesis are affected by ionophores.

Effects of ionophores on rumen fermentation

A very well known effect of ionophores on rumen fermentation is the increase in propionic acid and decrease in acetic acid proportion leading to an increase in propionic/ acetic acid ratio (SCHELLING, 1984), which favourably alters metabolism of meat producing animals. This is due to an increase in the level of insulin (which is an anabolic hormone) and because propionate is a precursor of fat via gluconeogenesis. In our studies however (Table 3), addition of lasalocid sodium resulted in an increase in propionate and decrease in n-butyrate proportions without any significant change in acetate which can be attributed to acetate-butyrate interconversions. Addition of lasalocid sodium did not affect ruminal pH or total VFA concentration. Our data are in accordance with those on cattle and sheep reviewed by BERGEN and BATES (1984) and BEEDE et al. (1985) with growing goats. The lack of difference between the control diet and that supplemented with lasalocid sodium is in agreement with the data of NEUENDORF et al. (1985) but at variance with HORTON et al. (1980), who reported lower total ruminal VFA concentration in diets containing monensin.

Table 2

Table 2

EFFECT OF LASALOCID ON DIET DIGESTIBILITY AND CRUDE PROTEIN (CP) BALANCE IN
GROWING DAMASCUS KIDS
(from HADJIPANAYIOTOU et al., 1988)

	Concentrate mixture		
	Without Lasalocid	Lasalocid (37 ppm)	SE (1)
Dry matter intake (g/day)	814	813	
Digestibility of DM	0.76	0.78	0.01
Digestibility of CP	0.69	0.74	0.02
CP intake (g/day)	149	146	2.32
CP absorbed (g/day)	103	107	4.15
CP retained (g/day)	67	65	6.42

(1) No significant effect of lasalocid sodium.

Table 3

RUMEN METABOLITES IN GROWING DAMASCUS KIDS ON CONCENTRATE MIXTURE WITH
OR WITHOUT LASALOCID
(from HADJIPANAYIOTOU et al., 1988)

	Without Lasalocid	With Lasalocid	SE and level of significance (1)	
Number of kids	10	10		
Total VFA concentration (mmol/1)	74.9	69.5	11.4	NS
Rumen NH_3 - N (mg/l)	212	229	53	NS
Rumen pH	6.1	6.0	0.2	NS
Molar proportion of :				
- Acetate	47.4	43.3	2.3	NS
- Propionate	39.2	48.6	2.8	*
- Isobutyrate	1.02	1.15	0.6	NS
- n-butyrate	12.4	6.9	2.0	*

(1) lasalocid effect : NS : non significant; * : $P < 0.05$

As a result of propionic acid pattern of fermentation, methane production decreased by 10 % *in vivo* (THORNTON and OWENS, 1981) and up to 30 % *in vitro* (JOYNER et al., 1979) resulting in an overall better utilisation of dietary gross energy.

Influence of antibiotics on nitrogen utilisation

Ruminal ammonia concentration was not affected by lasalocid sodium feeding to kids (Table 3). However, a well documented effect of ionophore and non ionophore antibiotics (avoparcin) in cattle and sheep is the decrease of NH_3 production (HANSON and KLOPFENSTEIN, 1979) even when the diet contains a high proportion of urea (POOS et al., 1979), reflecting a decrease of ruminal protein hydrolysis, the extent of which depends on protein degradability (JOUANY and THIVEND, 1983). Decrease of NH_3 production is one of the reasons used to explain the decrease in microbial protein synthesis often observed in vitro (VAN NEVEL and DEMEYER, 1979) and in vivo (POOS et al., 1979). It may also be attributed to uncoupling fermentation as shown by a modification of the propionate/ acetate ratio. In the studies of BEEDE et al. (1985) with growing goats, the response to monensin supplementation was twice as high at low as compared to high CP levels, suggesting a possible sparing effect due to monensin feeding.

Effect of ionophores on animal health

Coccidiosis in kids over six weeks of age is considered as the main cause of diarrhoea, anorexia and weight losses (AUMONT et al., 1982). Feeding lasalocid (37 mg/kg concentrate) or monensin (33 mg/kg concentrate) reduced coccidiosis in kids in the studies of HADJIPANAYIOTOU et al. (1988) and ANTONIOU and CHRISTOFIDES (unpubl.). Monensin has also been reported to be effective in suppressing clinical infections of coccidia in cattle (GOODRICH et al., 1984), lambs (VAN VUUREN and NEL, 1983) and kids (HINKLE and GRAIG, 1980).

Ionophore toxicity

No toxicity due to ionophore feeding has been noted in the studies conducted in Cyprus with kids (ANTONIOU and CHRISTOFIDES, unpublished data; HADJIPANAYIOTOU et al., 1988) or lambs (ECONOMIDES et al., 1988). It should be underlined, however, that farmers must be aware of possible hazzards associated with inaccurate on-farm mixing. An outbreak of monensin toxicity in cattle has been reported by WARDROPE et al. (1983) and was associated with overdosage. In fact, the most important problem is the ionotropic effect of ionophores (especially those carrying bivalent cations as lasalocid) on heart muscle function.

ANABOLIC AGENTS

Nature and practical use of anabolic agents

Anabolic agents are chemical substances that enhance N retention in the body. The major anabolic agents used are : testosterone (TEST), trenbolone acetate (TBA) and oestradiol-17B (E-17B). TEST is a potent androgen mainly produced in the testes, and improves the growth rate. TBA is a synthetic androgen with similar properties as TEST in promoting growth rate but with no side effect (aggressive behaviour) as with TEST. E-17B is a natural oestrogen produced by ovaries and testes. Another known-non steroid-anabolic agent with oestrogen-like activity produced by *Giberella zeae* is zeranol (ZER). Androgens are mainly used in females and wethers andoestrogens in males. In accordance with studies in steers and bulls, combination of oestrogens and androgens in male and female kids resulted in higher ADG than when a single agent was used (Table 4). In the EEC, these products are forbidden since January 1989, because of health problems they might cause. Most common doses were 20 mg E-17B + 140 mg TBA in veal calves, steers and bulls, and 180 or 300 mg TBA in heifers.

Effects of anabolic agents on growth

The use of anabolic agents in bovine and ovine species has been studied extensively (VAN DER WAL and BERENDE, 1984), but studies on goats are limited. ADG increase in goats ranged from 4 % (single implantation of E-17B + TBA) to 84 % (repeated ZER implants). In some cases, implantation failed to induce a significant extra growth response, probably because levels of endogenous hormones were very high and the growth rate had already reached its maximum. Mean increase in ADG as a result of implants (Table 4) was 13 % ± 5 % and it was close to values obtained in sheep (ROCHE and QUIRKE, 1986).

Feed conversion effiency was associated with either decreased (3 to 5 %) DM intake in milk fed kids treated with E-17B + TBA (SCHMIDELY et al., 1987) or increased roughage and concentrate intake in kids (MEHREZ et al., 1982) and lambs (COELHO et al., 1981).

The effectiveness of anabolic agents to enhance ADG and the amplitude of response depends on the nature of the molecule, dose used, sex and age of kids and nature of diet. SCHMIDELY et al. (1987) reported higher growth rates with kids implanted with E-17B + TBA, when fed with lactose-rich rather than lipid-rich milk. These differences were amplified when the dose used was 5 mg E-17B + 35 mg TBA rather than 5 mg E-17B + 17 mg TBA.

Anabolic agents induce modifications in growth rate as revealed by changes in body weight, carcass weight and characteristics when animals are compared at constant age. In line with data on lambs by COELHO et al. (1981), SCHMIDELY et al. (1987) found a carcass weight increase (+ 2 %) and modifications of carcass characteristics after treatment ; neck and ribs increased (+ 4 %) and long legs decreased (- 2 %). These changes were correlated with better conformation and a reduction of the fattening score (Table 5).

Table 4

RESPONSE OF MALE AND FEMALE GOATS TO DIFFERENT ANABOLIC AGENTS

Anabolic agent	Sex	Age of implantation (days)	Final liveweight (1)	LS (2)	Reference
TBA (140 mg)	Female	240	100		RANAWEERA and
	Male	240	100		THANGARAJAH (1982)
ZER (3 x 12 mg)	Male	14	163	*	
	Castrated	14	135	*	
	Male	56	184	*	SILVA and
	Castrated	56	153	*	BERENGUER (1984)
	Male	112	128	*	
	Castrated	112	111	NS	
ZER (6 x 12 mg)	Castrated	-	103	NS	
					SNOWDER et al. (1982)
TEST prop. (47 mg)	Castrated	-	106	NS	
TBA (30 mg)	Male		108	NS	MEHREZ et al. (1982)
ZER (2 x 12 mg)	Male	-	118	*	LANDAU (1987)
TBA + E-17B (17.5 mg + 5 mg)	Male	7	104	*	SCHMIDELY et al. (1987)
TBA + E-17B (35 mg + 5 mg)	Male	21	106	*	SCHMIDELY et al. (unpubl.)

(1) Values are expressed as a percentage of the control.
(2) LS : Level of significance ; NS : non significant ; * : P < 0.05.

Effect of anabolic agents on body composition

Treatment of bulls, steers and sheep with anabolic agents showed variable results (PATTERSON and SALTER, 1985). There are no studies on the effect of anabolic agents on composition of whole goat carcasses. Some parameters, however, indicate that there is a trend towards a decrease in body fat content in milk fed kids (Table 5).

Table 5

BODY COMPOSITION OF KIDS TREATED WITH ANABOLIC AGENTS
(from SCHMIDELY et al., 1987, and SCHMIDELY et al., unpubl.)

Treatment	Fattening score (1)	Intermuscular + subcutaneous fat in right long leg (g/kg)	Total body fat dissected at slaughter (% of empty body)
Control (2)	3.5 (0.25)a(3)	2.77 (0.20)a(3)	3.75 (0.5)a(3)
TBA + E-17B (17. 5 mg + 5 mg) (2)	3.0 (0.20)b	2.30 (0.30)b	3.10 (0.20)b
Control (2)	3.3 (0.30)c		4.70 (0.20)c
TBA + E-17B (35 mg + 5 mg) (2)	3.3 (0.30)c		3.82 (0.15)a
TBA + E-17B (70 mg + 10 mg) (2)	2.5 (0.30)d		3.42 (0.17)b

(1) Score from 1 (lean) to 5 (fat).
(2) Twelve kids in each group.
(3) Values in parentheses indicate the standard deviation. Values in the same column with different subscripts differ significantly (P < 0.05).

Results showed a positive relation between the dose administered and the reduction in body fat content. These data partially contradict those obtained in bulls treated with TBA + E-17B where the fat content increased after treatment. Differences might be due either to age at treatment or puberty differences. Since there was no increase of energy deposition in liveweight gain (VAN DER WAL and BERENDE, 1984), it can be assumed that the decrease in fat deposition was partially - but not in the same proportions - compensated by the increase in protein deposition, as indicated by a higher percentage of muscle in the right long leg (SCHMIDELY et al., 1987). This might explain why differences in final liveweight were not significant.

Mode of action of anabolic agents

Anabolic effects are primarily revealed by increasing nitrogen (N) retention in the body and particularly in the muscle. Results obtained in kids treated with TBA + E-17B (SCHMIDELY et al., 1988) showed no differences with controls for N-digestibility, but a significant decrease in blood and urinary urea and urinary N (Table 6), leading to a higher N-retention. This effect, however, seems to be reduced with time after treatment because hormone concentrations decrease in the blood (GALBRAITH, 1982). LANDAU (1987) failed to show decrease in blood urea and a relationship between ADG and blood urea in kids treated with ZER. This can be due to different modes of action of oestrogens and androgens in increasing N-retention and ADG.

Table 6

EFFECT OF TBA (35 mg) COMBINED WITH E-17B (5 mg) ON NITROGEN RETENTION IN KIDS
(from SCHMIDELY et al., 1988) (1)

Days after treatment	7	(2)	21	(2)
N digestibility	0.99	NS	0.96	NS
Urinary urea	0.73	**	0.79	*
Urinary nitrogen	0.64	**	0.71	*
N retention	1.30	**	1.21	*

(1) Values are expressed as percentage of controls.
(2) Implant effect : NS : non significant, * : P < 0.05, ** : P < 0.01.

The mode of action of anabolic agents in sheep and cattle has been extensively studied. Work on goats is limited, but results obtained so far did not show any difference between goats and sheep or cattle. The mechanism of action of anabolic agents is not fully understood. Androgens and oestrogens may act in different ways: firstly, androgens (and oestrogens) act directly at the level of the muscle cell, by binding to a cytoplasmic receptor. The hormone-receptor interacts with nuclear receptors located in the chromatin and enhances protein synthesis (and probably protein degradation). Although this mode of action is clearly demonstrated for TEST, it is not valid for TBA which has been shown to reduce protein synthesis and, to a greater extent protein deposition (BUTTERY and VERNON, 1983). This can be achieved either by reducing the adrenal function and secretion of glucocorticoids (which are catabolic hormones in the muscle) or the affinity of glucocorticoids for their receptors in the muscle, or more probably by reducing the number of glucocorticoid receptors (SHARPE et al., 1986). An indirect action has also been involved in N retention increase with anabolic agents. After treatment with œstrogens (E-17B or ZER) but also with TEST, the growth hormone and insulin probably increase while free thyroxin decreases. Androgen-oestrogen combination decrease only the thyroxin index. This can reduce energy and protein requirement for maintenance (HEITZMAN, 1981).

CONCLUSION

The use of anabolic agents and ionophores in goats gives similar results to those obtained in sheep and cattle. Ionophores improve feed/gain ratio and suppress coccidiosis which is a serious problem in growing kids. Use of anabolic agents, if permitted, will improve kid fattening economics and result in the production of leaner carcasses. However, this may not be useful since more fat is often needed for commercial purposes.

SUMMARY

Ionophores improve average daily gain (ADG) and feed conversion efficiency of growing kids by 10 percentage units. As in cattle, ionophore and non ionophore antibiotics induced modifications of rumen fermentation, increasing propionate/acetate ratio. In contrast to results obtained in cattle and sheep, no apparent modifications of nitrogen metabolism in the rumen were observed in our experiments with kids. Controlling partition of nutrients absorbed with anabolic steroids and consequences on growth performance are reviewed in goat kids: average increase in ADG was 13 % but results are very heterogenous. In milk fed kids, early utilization of anabolics agents reduced the fat/muscle ratio and generally improved the nitrogen retention, though to a lesser extent with time after treatment. Practical use and legislation of these coumpounds within EEC are discussed.

Keywords : Growing goat, Growth promoters, Anabolic agents, Fattening, Body composition, Diet digestibility.

RÉSUMÉ

Les substances ionophores améliorent le gain moyen quotidien (GMQ) et l'efficacité alimentaire des chevreaux en croissance de 10 pour cent. Comme chez les bovins, les antibiotiques ionophores et non ionophores modifient les fermentations du rumen et accroissent le rapport Propionate / Acétate. Dans nos expériences, aucune modification apparente du métabolisme azoté du rumen n'a été observée à l'inverse des résultats obtenus sur bovins et ovins, ce qui peut refléter des différences inter-espèces. Le contrôle de la répartition des nutriments absorbés en présence de traitement aux stéroïdes anabolisants et les conséquences sur les performances de croissance ont été passés en revue chez les chevreaux. L'augmentation moyenne du GMQ est de 13 % mais les résultats sont très hétérogènes. Chez les chevreaux de lait, l'implantation précoce d'anabolisants réduit le rapport Tissus adipeux / Muscles de la carcasse et améliore généralement la rétention azotée quoique ces résultats semblent s'estomper avec le temps après traitement. L'utilisation pratique et la législation en vigueur dans la CEE relative à ces composés sont discutées.

REFERENCES

AUMONT G., YVORE P. and ASNAULT E., 1982. Coccidiosis in young goats, p. 506, (Abst.). 3rd Intern. Conf. on Goat Production and Disease, Jan 20-21, 1982, Tucson, Arizona (USA).

BEEDE D.K., SCHELLING G.T., MITCHELL G.E. JR. and TUCKER R.E., 1985. Utilization by growing goats of diets that contain monensin and low or excess crude protein: comparative slaughter experiments. J. Anim. Sci., 61: 1230-1242.

BEEDE D.K., SCHELLING G.T., MITCHELL G.E.JR., TUCKER R.E., GILL W.W., KOENIG S.E. and LINDSEY T.O., 1986. Nitrogen utilization and digestibility by growing steers and goats of diets that contain monensin and low crude protein. J. Anim. Sci., 62: 857-863.

BERGEN W.G. and BATES D.B., 1984. Ionophores: their effect on production efficiency and mode of action. J. Anim. Sci., 58: 1465-1483.

BUTTERY P.J. and VERNON B.G., 1983. Protein metabolism in animals treated with anabolic agents. Vet. Res. Communications, 7: 11-22.

COELHO J.F.S., GALBRAITH H. and TOPPS J.H., 1981. The effect of combination of trenbolone acetate and estradiol-17B on growth performance, blood, carcass and body characteristics of wether lambs. Anim. Prod., 32: 261-268.

ECONOMIDES S., GEORGHIADES E., PARACHRISTOFOROU C. and HADJIPANAYIOTOU M., 1988. The effect of feed antibiotics and of vitamin A on the performance of Chios lambs and Friesian calves. Techn. Bull., 101, Agric. Res. Inst., Nicosia (CYPRUS).

GALBRAITH H., 1982. Growth, hormonal and metabolic response of a post-pubertal entire male cattle to trenbolone acetate and hexoestrol. Anim. Prod., 35: 269-276.

GOODRICH R.D., GARRET J.T., CAST D.R., KIRICK M.A., LARSON D.A. and MEISKE J.C., 1984. Influence of monensin on performance of cattle. J. Anim. Sci., 58: 1484-1497.

HADJIPANAYIOTOU M., PAPACHRISTOFOROU C. and ECONOMIDES S., 1988. Effect of lasalocid on growth, nutrient digestibility and rumen characteristics in Chios lambs and Damascus kids. Small Rum. Res., 3: 217-227.

HANSON T.L. and KLOPFENSTEIN T.J., 1979. Monensin, protein source and protein levels for growing steers. J. Anim. Sci., 48: 474-479.

HEITZMAN R.J., 1981. Mode of action of anabolic agents, p. 129-139. In: "Hormones and metabolism in Ruminants". Workshop held at Bodington Hall, Sept. 22-24, 1980, Leeds, ARC (UK).

HINKLE M.L. and GRAIG T.M., 1980. Monensin as a coccidiostat and its effect on the development of resistance to coccidia by Angora goats, p. 84-88. In: "Sheep and Goat-Wool and Mohair", paper 3712. Texas Agric. Expt. Station, San Angelo (USA).

HORTON G.M.J., BASSENDOWSKI K.A. and KEELER E.H., 1980. Digestion and metabolism in lambs and steers fed monensin with different levels of barley. J. Anim. Sci., 50: 997-1008.

JOUANY J.P. and THIVEND P., 1983. In vitro effects of avoparcin on protein degradability and rumen fermentation. Anim. Feed Sci. and Technol., 15: 215-229.

JOYNER A.E. JR., BROWN L.J., FOGG G.T. and ROSSI R.T., 1979. Effect of monensin on growth, feed efficiency and energy metabolism of lambs. J. Anim. Sci., 48: 1865-1869.

LANDAU S., 1987. Ralgro implant as a growth promoter for Saanen male kids. Ann. Zootech., 36: 342 (Abst.).

MEHREZ A., ABOUL-ELA M.G. and EL-HASHMANY G., 1982. Effect of monensin and trenbolone acetate on the performance of growing kids, p. 571 (Abst). 3rd Intern. Conf. on Goat Production and Disease, Jan. 10-15, 1982. Tucson, Arizona (USA).

NEUENDORF D.A., RUTTER L.H., PETERSON L.A. and RANDER R.O., 1985. Effect of lasalocid on growth and puberall development in Brahman bulls. J. Anim. Sci., 61: 1049-1057.

PATTERSON R.L.S. and SALTER L.J., 1985. Anabolic agents and meat quality: a review. Meat Sci., 14: 191-220.

POOS M.J., HANSON T.L. and KLOPFENSTEIN T.J., 1979. Monensin effect on diet digestibility, ruminal protein bypass and microbial protein synthesis. J. Anim. Sci., 48: 1516-1524.

RANAWEERA K.N.P. and THANGARAJAH P., 1982. Effect of trenbolone acetate on the growth of tropical cattle and goats. Trop. Anim. Health and Prod., 14: 44 (Abst.).

ROCHE J.F. and QUIRKE J.F., 1986. The effect of steroid hormones and xenobiotics on growth of farm animals, p. 39-52. In: BUTTERY P.J., HEYNES N.B. and LINDSAY D.B. (Eds.), Control and Manipulation of Animal Growth. Butterworths, London (UK).

RUSSEL J.B., 1987. A proposed mechanism of monensin action in inhibiting ruminal bacterial growth: effects of ion flux and protonmotive force. J. Anim. Sci., 64: 1519-1525.

SCHELLING G.T., 1984. Monensin mode of action in the rumen. J. Anim. Sci., 58: 1518-1527.

SCHMIDELY P., MORAND-FEHR P., BAS P. and HERVIEU J., 1987. The effect of estradiol-17B combined with trenbolone acetate on growth performance of fattening male kids. Ann. Zootech., 36: 342 (Abst.).

SCHMIDELY P., BAS P. and ROUZEAU A., 1988. Effect of estradiol-17B and trenbolone acetate on nitrogen digestibility and retention in male kids. Seminar of FAO subnetwork on Goat Nutrition and Feeding, Oct. 3-5, 1988, Potenza (ITALY).

SHARPE P.M., HAYNES N.B. and BUTTERY P.J., 1986. Glucocorticoids status and growth, p. 207-222. In: BUTTERY P.J., HAYNES N.B. AND LINDSAY D.B (Eds.), Control and Manipulation of Animal Growth. Butterworths, London (UK).

SILVA M. and BERENGUER F., 1984. Response of growing male kids to repeated zeranol implantation. Anim. Prod., 38: 546 (Abst.).

SNOWDER G.D., SHELTON M., BASSET J.W. and THOMPSON P., 1982. The influence of castration and testosterone or zeranol implants on body weight gain and fiber production of Angora goats,p. 528 (Abst.). 3rd Intern. Conf. on Goat Production and Disease, Jan. 20-21, 1982, Tucson (USA).

THOMPSON W.R. and RILEY J.G., 1980. Protein levels with and without monensin for finishing steers. J. Anim. Sci., 50: 563-571.

THORNTON J.H. and OWENS F.N., 1981. Monensin supplementation and in vitro methane production by steers. J. Anim. Sci., 52: 628-634.

VAN DER WAL P. and BERENDE P.L.M., 1984. Effets des anabolisants sur les animaux de boucherie (Effects of anabolic in beef cattle), p. 77-120. In: Meissonier D. (Ed.), Anabolisants en production animale. Symposium OIE, février 1983, Paris (FRANCE).

VAN NEVEL C.J. and DEMEYER D.I., 1979. Effect of monensin on some rumen fermentation parameters. Ann. Rech. Vet., 10: 338-340.

VAN VUUREN B.G.J. and NEL J.W., 1983. The effect of monensin on the efficency of feed conversion, carcass traits and the occurence of coccidiosis in lambs. S. Afr. J. Anim. Sci., 13: 87-90.

WARDROPE D.D., MACLEED N.S.M. and SLOAN J.R., 1983. Outbreak of monensin poisoning in cattle. Vet. Rec., 45: 560-561.

Part 3

Feeding of adult goats

Chapter 17

INTENSIVE FEEDING OF DAIRY GOATS

M. HADJIPANAYIOTOU and P. MORAND-FEHR

INTRODUCTION

Intensive dairy goat production systems involve animals of high milk potential (Alpine, Anglo-Nubian, Damascus, Saanen, Toggenbourg, etc.), relatively high investment in housing, machinery, veterinary supervision and high inputs on purchased or home-produced feed. Feeding costs comprise the major part (55-75 % on average) of the total production costs (HADJIPANAYIOTOU, 1987a). Research has been undertaken within countries participating in the FAO European sub-network and in other countries (ECONOMIDES, 1986) on the basal features of goat nutrition and the strategies of feeding in various physiological stages.

As feeding is based on the knowledge of animals nutrient requirements (see Chapters 6, 8 and 10) and the nutritive value of feeds (see Chapters 13 and 15) this chapter will deal with the feeding strategy of dairy goats during the cycle of reproduction, the pattern of feed distribution, the response of dairy goats to energy and protein and the effect of diet composition and processing on goat performance.

NUTRITIONAL FACTORS AFFECTING MILK PRODUCTION

Response to energy

The level of energy intake is the major factor affecting milk production. The response to energy deficit being very fast. A decrease in milk yield of dairy goats offered 85 % of their total energy requirements occured within 24 to 48 h (MORAND-FEHR and SAUVANT, 1978). In mid lactation, the correlation between energy intake and milk yield is around 0.83. In early lactation, the effect of energy intake on milk yield is less evident (r = 0.73), and this is associated with the fact that goats like other ruminants rely on body reserves. SKJEVDAL (1981) reported that high yielding goats may produce 0.5 to 1.0 kg of 4 % fat-corrected milk mobilizing body reserves. Similarly, MORAND-FEHR et al. (1987) reported that in goats producing 750 kg milk per year, on average the production of 1.4 kg milk per day is covered by the mobilization of adipose tissues during the first month of lactation and 0.7 kg milk at the beginning of the second month, corresponding to a loss of 1 and 0.5 kg of adipose tissues per week, respectively.

DE SIMIANE (1978) found a negative correlation between milk production and crude fibre content of the forage and a positive correlation between milk yield and the net energy of the forage. Differences in crude fibre content caused differences in feed consumption and milk production by the goats.

Improvement of the organic matter digestibility of the diet fed to goats at the onset of lactation through the inclusion of better quality roughages resulted in higher dry matter and energy intakes and consequently higher milk yields (SAUVANT et al., 1987). In a study by KESSLER (1985) feeding of poor as opposed to good quality hay resulted in reduced feed intake (1.1 vs 1.9 kg DM/day) and milk yield (1.0 vs 1.9 kg/day). High energy (125 % of the NRC, 1981) intake during the last two months of pregnancy and continued in the lactation period had a positive response on litter weight and the milk yield during the first 8 weeks of lactation compared to a 100 % NRC (1981) energy allowance. It must be stressed however, that a response to higher energy intake was obtained in the high yielding, Damascus breed, but not in the Zaraibi breed (GIHAD et al., 1987).

The response of dairy goats to energy supplements is greater in early than in late lactation. Indeed at the end of lactation, it is difficult to increase milk production even if the energy intake increases.

Response to lipids

Dairy goats may have specific lipid requirement because of the low ether extract content of their diets and the inability of the mammary gland to synthesize long chain fatty acids (MORAND-FEHR and SAUVANT, 1987). Supplementary feeding of diets low in ether extract (alfalfa hay, beet pulp, barley grain, soybean meal) with lipids increased milk yield and milk fat content and decreased slightly milk protein content (Table 1). However, in order to maintain cellulose digestion in the rumen at satisfactory levels it is desirable to use saturated and/or protected lipids and the ether extract content should not exceed 5 % of the finished diet. During the declining stage of lactation supplementary feeding of lipids had little effect on milk yield of goats, but increased milk fat content (MORAND-FEHR and SAUVANT, 1987). Further studies by DACCORD (1987) showed that addition of 4 % micronized fat in a concentrate mixture comprising 25 % of the finished diet resulted in a more efficient dietary N-utilization and slightly higher milk yield (No fat = 3.58, fat = 3.73 kg/day) and an increase in the protein and fat content of milk (see Chapter 18).

Table 1

EFFECT OF FAT SUPPLIES ON DIET DIGESTIBILITY, MILK YIELD AND MILK COMPOSITION OF GOATS IN EARLY LACTATION (first 8 weeks)

	Control*	Oil**	Tallow***
Fat content (%) of the finished diet	2.5	4.1	4.7
Dry matter intake (g/day)	2022.0	2381.0	2210.0
Organic matter digestibility (%)	75.3	76.0	-
Milk production (kg/day)	2.9	3.9	4.1
Fat (g/kg milk)	46.0	49.1	48.4
Protein (g/kg milk)	34.4	33.4	33.1

(*) Animals were offered alfalfa hay *ad lib.*, beet pulp silage and concentrates.
(**) 5 % soybean oil added in the concentrate mixture
(***) 5 % beef tallow added in the concentrate mixture.

Response to protein

Provision of protein supplements immediately after parturition has been used to enhance milk yield in dairy cows (FOLDAGER and HUBER, 1979) and ewes (COWAN et al., 1981). Like other ruminants goats respond to increasing dietary nitrogen levels with higher milk yield. The level of dietary protein (100 vs 140 g CP/kg DM) had a positive effect on the post-weaning (56 - 98 days in milk) milk yield of Damascus goats (HADJIPANAYIOTOU, 1987b). In another study with suckling Damascus goats, response to increasing dietary protein was even greater (Table 2). Furthermore, goats suckling twins with high milk production potential responded to a greater extent to high protein intake (high protein, difference between twin suckling and single suckling 0.54 kg/goat/day; low protein, difference 0.27 kg/goat/day). Greater response to increasing dietary protein levels by high compared with low yielding goats has also been reported by MORAND-FEHR and SAUVANT (1980). Finally, in another study with goats at a more advanced stage of lactation (90 - 120 days in milk) increasing dietary nitrogen levels (101, 135, 139, 162 or 183 g CP/kg feed DM) did not affect milk yield of Damascus goats (HADJIPANAYIOTOU, 1987b). In line with the studies of HADJIPANAYIOTOU (1987b) are and those of MORAND-FEHR and SAUVANT(1978) where protein supplementation resulted in only minor responses in milk yield during mid-lactation, but during early lactation, particularly the high yielding goats, were sensitive to the protein content of the diet. Feeding high yielding Norwegian goats (SKJEVDAL, 1981) a concentrate mixture of high protein content (18.5 % digestible crude protein) resulted in 5 to 10 % higher fat corrected milk yield compared to the lower level (12.5 % digestible crude protein). In early lactation, goats like other ruminants are more sensitive to the protein content of the diet because they can rely on body mobilization for energy deficits whereas body protein mobilization is very limited and there is greater need of dietary supply.

Table 2

PERFORMANCE OF SUCKLING DAMASCUS GOATS OFFERED TWO LEVELS
OF DIETARY PROTEIN

Suckling	Treatment				
	High protein*		Low protein*		S D
	Single	Twin	Single	Twin	
No. of goats	11	13	13	11	
Feed intake (kg/day)					
Concentrate		1.932		1.830	-
Barley hay		0.587		0.565	-
Total milk yield (kg/day)	2.39	3.46	2.12	2.92	0.55
Milk suckled (kg/goat/day)	1.36	2.67	1.31	2.31	0.42
Milk composition (g/kg):					
Fat	37.0	39.00	38.0	40.0	9.7
Protein	40.0	40.00	37.0	35.0	3.1
Ash	8.3	8.40	8.5	8.1	0.4
Total solids	125.0	131.00	129.0	128.0	9.5
Liveweight (kg):					
at kidding	55.6	53.40	51.1	55.3	10.4
at weaning (end of trial)	55.7	52.50	51.0	50.4	8.1

* : High protein : 140 g CP/kg DM; low protein : 100 g CP/kg DM.

Feedstuffs

A variety of feedstuffs is used for feeding goats. Forages comprise a major component of the finished diet fed to dairy goats kept in confinement in France, Italy, England and some other countries. A detailed description on the evaluation and utilization of forages by goats is given in Chapter 13.

Straw

Straw as a roughage source makes an important contribution to the nutrition of goats in many countries with limited forage supplies. It is important not only as a sole feed but also as stubble grazing for a considerable part of the year. Apart from its low DM digestibility one of the main factors limiting straw intake is its low nitrogen content.

Supplementation of cereal straws with nitrogen results in a significant improvement in straw intake and digestibility. At the Cyprus Agricultural Reserach Institute where Chios ewes and Damascus goats are kept in mixed flocks, supplementation of straw with 80 g soybean meal and 170 g of barley grain, or 10 g urea and 240 g of barley grain, resulted in higher energy intake than the required energy needs for maintenance (0.412 MJ ME/kgW$^{.73}$), and also in nitrogen equilibrium under stall-fed conditions. Furthermore, in the study of FUJIHARA and TASAKI (1975) addition of urea to poor quality forage fed to goats improved their feeding value and nitrogen retention in goats.

Inclusion of straw (40 % of the finished diet fed to lactating Damascus goats) treated with a 10 % urea solution at the rate of 400 ml/kg resulted in a significant improvement of dry and organic matter and crude protein digestion coefficients in comparison with the untreated straw (HADJIPANAYIOTOU, 1984a).

Cereal grains

There are areas in the world where it is more economical to produce milk on cereal grains rather than on roughages. Studies conducted in Cyprus (ECONOMIDES et al., 1982) showed that sorghum, wheat, corn and barley grain can replace each other in ruminant diets and that consideration should be given to their nutrient content in order to supplement the diet accordingly. Cereal grains comprise a major component of the finished diet in high concentrate feeding systems. Therefore, efficient and economic utilization of cereal grains will maximize economic returns to the farmer.

Processing of cereal grains has been practised with the purpose of ensuring better mixing of the ingredients and better utilization of the diets.However, ORSKOV (1979) reported that any processing of grain given to sheep and goats is likely to be of no value, and suggested that barley should be fed whole. In trials with lactating Chios ewes and Damascus goats (ECONOMIDES et al., 1989) the effect of cereal grain processing (mash (M), the cereal grains were ground and mixed with the other ingredients of the concentrate mixture; pelleted (P), the cereal grains were ground and pelleted, 5 mm cubes, with the other ingredients; whole grains (WG) the whole grains were mixed with pellets made of the other ingredients of the concentrate mixture) on their milk yield and milk composition was studied. Ewes and goats on the WG diets gave similar milk yields with those on the P diets, but their milk was of higher milk fat content. Similarly, animals on the M diet produced milk of lower fat content than those on the WG diet. The higher milk fat content of sheep and goats on the WG diet can be associated with increases in ruminal acetate and decreases in ruminal propionate (ORSKOV et al., 1974). Furthermore, it seems that the structure of the reticulo-omasal orifice in adult goats is similar to sheep allowing regurgitation and rumination of whole grain resulting in an efficient utilization of whole grains.

By product utilization

Under intensive goat feeding systems by-products can be incorporated into concentrate mixtures and/or offered separately. Research work by SAUVANT et al. (1987) has shown that performance of dairy goats fed brewer's grain is in accordance with data reported for sheep and cattle.

Citrus pulp, a very moist (18 % DM) product to store has been ensiled (4 parts citrus pulp and 1 part peanut shells) with peanut shells and fed to dry Damascus goats up to 60 %

of the finished diet (HADJIPANAYIOTOU, 1988a). Sugar beet pulp silage (22 % DM) has been used successfully in diets of lactating goats up to 45-50 % of complete diets (SAUVANT et al., 1988). In addition, dehydrated beet pulp pellets have been used successfully at moderate but not at high levels of inclusion (MIOSSEC and DE SIMIANE, 1981). A too high proportion of beet pulp can increase the risk of digestive disorders (see Chapter 12).

Dried poultry litter has been used successfully in diets of lactating Chios ewes and Damascus goats in Cyprus (HADJIPANAYIOTOU, 1984b). The milk yields were similar among goats or ewes receiving 0 or 15 percent poultry litter. During periods of low productivity (late lactation, dry period, early part of pregnancy) litter can comprise 35 % or more of the finished diet. During pregnancy and the early postpartum period only small quantities of litter (about 5 %) should be fed.

Nitrogen supplements

The quantity and quality of protein reaching the small intestine in early lactation may limit milk yield. Fish meal has been reported to be a protein supplement of high quality and of low degradability in the rumen (ARC, 1984). Two trials, one with 36 twin-suckling Chios ewes and the other with 32 twin-suckling Damascus goats, were conducted to study the effect of protein source (fish meal vs soybean meal) on the pre-weaning milk yield of the dams, and the growth performance of their offspring (HADJIPANAYIOTOU et al., 1988). Ewes on fish meal produced more milk than those on the control diet, whereas milk yield of goats was similar in the two treatments. The difference between species obtained in this study may reflect the higher protein concentration of ewes milk compared with goats milk, the lower level of fish meal inclusion and its higher degradability in the goat trial. This data with goats, however, are in agreement with those of SMALL and GORDON (1985) for dairy cows, where protein source (soybean meal, formaldehyde soybean meal, fish meal or a mixture of fish meal with soybean meal) did not affect milk yield. But other studies with goats (SKJEVDAL, 1981, cited by EKERN, 1982) and with cows (OLDHAM, 1984) have found that fish meal increased milk protein secretion more than soybean meal.

Fish meal addition, on the other hand, gave a significant curvilinear increase in N retention in milk. Optimal level of fish meal inclusion was dependent on fish meal quality, but was in the range 10 to 12 % in the concentrate (LINDBERG and CISZUK, 1987).

In another study (HADJIPANAYIOTOU, unpubl.), 32 Damascus goats were used to test two isonitrogenous concentrate mixtures containing soybean meal (S) or fish meal (F) as protein supplements. The average daily milk production during the preweaning period was similar in both treatments (S = 3.87, F = 3.82 kg). No response to fish meal compared to soybean meal has also been reported in another study with suckling Damascus goats. The average daily milk yield was similar for both treatments (S = 2.3 kg, F = 2.44 kg).

In further studies the effect of feeding formaldehyde-treated soybean meal on the performance of dairy goats and ewes was studied (HADJIPANAYIOTOU, unpubl.). Thirty dairy Damascus goats were used in a change-over design experiment. In the first period, there was difference (P = 0.018) in the rate of milk decline with faster rate on the control (S) than on the treated soybean meal diet. In the second period the decline of milk in the control diet was more steep than that of the treated soybean meal but the differences were not significant (P = 0.067). In another experiment however, feeding formaldehyde-treated soybean meal (PS) had no effect on the preweaning milk yield of goats (Table 3).

Replacement of soybean meal by meat meal of lower degradability did not alter milk yield or milk composition of goats in mid lactation (MORAND-FEHR et al., 1987). In experiments with lactating Swedish Landrace goats (LINDBERG and CISZUK, 1985) it was shown that increasing the proportion of fish meal in the diet reduces feed intake relative to a standard diet, and that differences in nitrogen utilization are related to the degradability of feed nitrogen.

Urea-N replaced soybean meal in diets of dry Damascus goats (HADJIPANAYIOTOU, 1988b). The use of urea resulted in lower N-requirements for maintenance.

Table 3

PRE- AND POST-WEANING PERFORMANCE OF DAMASCUS GOATS
OFFERED UNTREATED (S) OR FORMALDEHYDE TREATED (PS) SOYBEAN MEAL

	Treatment		
	S	P S	S D
No of animals	29	31	-
Pre-weaning period			
Feed intake (kg/day)			
Concentrate	2.79	2.41	-
Lucerne hay	0.58	0.54	-
Barley straw	0.50	0.47	-
Milk yield (kg/day)	3.92	3.93	0.859
Post-weaning period			
Feed intake (kg/day)			
Concentrate	2.23	2.20	-
Lucerne hay	0.35	0.35	-
Barley straw	0.50	0.50	-
Milk yield (kg/day)	2.96	3.16	0.706
Fat corrected (4 %) milk yield (kg/day)	3.27	3.57	0.822

Furthermore, Mudgal (1982) reported that after adaptation to NPN diets the lactating goats can utilize both urea and biuret equally well as plant protein supplements. Further studies by KHAN et al. (1981) revealed no difference in FCM yield between the control (36 % cotton seed cake) and the concentrate containing 1 % urea along with 24 % cotton seed cake. Contrary, the yield of FCM by the experimental goats fed 2 % and 3 % urea was lower from goats fed the control or 1 % urea in the diet.

Pattern of feed distribution

Generally goats are kept in flocks and individual feeding is not common. In a recent work, MORAND-FEHR and DE SIMIANE (1986) presented the results of several French studies comparing various patterns of feed distribution. They concluded that in mid and late lactation, it is possible to offer a fixed concentrate allowance regardless of milk yield of goats without any adverse effect on milk yield. They emphasize however, that fixed levels of concentrate feeding can only be used, when good quality forages are available *ad libitum*, and when there is no big variation in milk yield between animals. On the contrary, fixed concentrate levels are less applicable in the early part of lactation because milk production is higher, there is greater variation in milk yield and feed intake is low relative to requirements. Consequently, it is desirable to have different groups of goats based on their milk yield in the early part of lactation. Finally, the same workers concluded that there is no need to meet the daily nutrient requirements of dairy goats, if animals are fed to requirements on an annual basis. So the distribution of concentrates can be simplified by making use of the ability of the goats to accumulate and mobilize body reserves.

Mixed ration feeding is common in dairy cattle (NOCEK et al., 1986). COPPOCK et al. (1981) outlined several advantages associated with this type of feeding system. Data on goats showed that feeding total mixed ration results in better utilization of the diet than feeding the different components separately (CARASSO et al., 1988).

FEEDING GOATS ON AN ANNUAL BASIS

In intensive dairy goat production systems, feeding is based on the knowledge of the nutrient requirements, the nutritive value of feeds and on diet formulations which meet the requirements of the animals. Feeding is mainly based on purchased or home produced feedstuffs, which are mainly offered indoors. As a rule good quality feedstuffs are offered during periods of high productivity and feedstuffs of lower quality are offered during periods of low productivity. The feedstuffs are selected on the basis of the cost per unit of energy and protein.

Periods of low productivity

During these stages (dry non-pregnant, early pregnancy, low milk production) goats, like other ruminants, are ideally suited to utilize medium to poor quality feedstuffs. The declining stage of lactation is ideally suited to restore energy reserves (Rodstad, 1980, cited by SKJEVDAL, 1981), since the efficiency for body weight gain in a lactating goat is equal to that for lactation (ARMSTRONG and BLAXTER, 1965).

At mating and during pregnancy

In intensive dairy goat production systems the feeding schedule is planned so that goats are continually in good body condition. When goats are in good body condition extra feeding prior to mating is not needed; if however, they are not in good condition, supplementary feeding above maintenance should begin 2-3 weeks prior to mating and continue through the mating season. Because of the minute growth of foetuses that take place in early pregnancy (BLANCHART and SAUVANT, 1974), no additional nutrients are required until 90 days prior to kidding. The energy requirements in the last 6-8 weeks of pregnancy of Damascus goats with twin foetuses is approximately 16 MJ of ME/day (ECONOMIDES and LOUCA, 1981). More liberal prepartum feeding (20.1 MJ ME/day) did not result in higher birth weights of kids or extra milk (ECONOMIDES and LOUCA, 1981). In the same studies, a low level of energy intake (11.3 MJ ME/day) resulted in a high incidence of pregnancy toxaemia (see Chapter 12). Due to the fact that the appetite of the pregnant animal declines, especially in the last 2 weeks of gestation, good quality diets (10.5 to 11.5 MJ of ME and 120 to 140 g CP/kg DM) must be fed during the last part of pregnancy. The level of intake in late pregnancy is declining because rumen volume is reduced due to the growth of foetuses in the abdominal cavity. The strategy of goat feeding at this period must be concentrated on the maintenance of rumen volume through the use of top quality palatable forages. A severe decrease in feed intake in late pregnancy will lead to an intensive body fat mobilization leaving only limited supplies of body reserves during the early post partum period when they are actually needed for insuring a successful onset of lactation. Overfattening should be avoided because in dairy goats the main part of lipid reserves are situated in abdominal tissues resulting in reduced rumen volume leading to lower feed intake and greater risk from pregnancy toxaemia because animals have to rely at a great extent on body fat mobilization. In conclusion, a very good or poor body condition during late pregnancy must be avoided (MORAND-FEHR and SAUVANT, 1987).

During lactation

The most critical period for nutrient supply to the high producing goat is from parturition until peak milk production because maximum dry matter intake occurs after the peak milk yield (MORAND-FEHR, 1981). Feed intake increases by about 30 - 40 % during the first weeks and reaches maximum between 6 and 10 weeks of lactation. Feed intake is closely related to milk yield (HADJIPANAYIOTOU, 1987a). In French studies (MORAND-FEHR, 1981), the maximum intake attained was 180 g/kg $W^{.75}$. In studies conducted at the Cyprus Agricultural Research Institute where a high-concentrate feeding system is applied and the concentrate allowance increases gradually reaching *ad libitum* level 10 to 15 days postpartum the daily DM intake by suckling Damascus goats was 135 g/kg $W^{.75}$. In this study the daily concentrate allowance was increased only when animals were consuming an amount of lucerne and barley hay so that the proportion of roughage was not less than 25 % of the finished diet. MORAND-FEHR and SAUVANT (1980) reported that any plan of goat feeding during early lactation affects milk production during the first weeks of lactation and during mid-lactation. They suggested that long term effects of diets offered to goats should be taken into account in establishing feeding plans during lactation. Substituting grain for roughage will eliminate much of the negative energy balance and subsequently maintain a higher peak milk yield. Considerable quantities of body fat reserves are mobilized to supply energy during the early postpartum period. On the other hand, mobilization of body protein is limited (ARC, 1980). The existence of satisfactory body reserves during the early part of lactation is necessary for insuring a good onset of lactation. Concentrate supply during late gestation (600 vs 150 g/day) and in early lactation (400 vs 200 g/kg milk) had a significant effect on the entire lactation milk yield. Studies in Australia (KONDOS, 1972) and France (SAUVANT and MORAND-FEHR, 1978; cited by MORAND-FEHR and SAUVANT, 1980) showed that the best milk production and most satisfactory economic results in early lactation occured in goats offered higher amount of concentrate in late gestation and early lactation. In early lactation, high levels of concentrates improve milk production to a greater extent than in mid or late lactation. Increasing the energy density of the concentrate mixture in mid lactation resulted in minor changes in milk yield, but in a significant increase of body fat reserves (MORAND-FEHR and SAUVANT, 1980).

In areas where there is lack of good quality roughages is more economical to produce milk from concentrates rather than roughages. It is however, essential to supply animals with a certain roughage allowance to guarantee normal rumen function. During the early part of lactation, concentrate allowance increases gradually reaching control *ad libitum* (the concentrate allowance is increased only when the roughage to concentrate ratio is within the ratio of 3:7) 10 to 15 days postpartum in order to avoid metabolic disorders.

MOWLEM et al. (1985) showed that high rates of concentrate allowance reduce forage intake by goats. Daily intakes of hay offered *ad libitum* were (g/head) 669, 1102 and 1222 for the high (1389 g/head/day), medium (1000 g/head/day) and low (593 g/head/day) concentrate allowance, respectively. The level of forage intake is higher when goats are allowed plenty of time for selection. DE SIMIANE et al. (1983) reported that under temperate conditions to obtain the highest intakes of green fodder the level of residues should be between 15 - 40 % depending on fodder quality and for green corn and sorghum 5 to 20 % residues should be allowed. Forage legumes are more acceptable by goats than other ruminants (MORAND-FEHR and SAUVANT, 1987). Chopping of green fodder decreased selection and consequently green fodder intake by 15 % (HUGUET et al., 1977). Intakes of conserved fodder (hay or silage) are lower than those of green fodder. Under temperate conditions, hay or grass silage offered *ad libitum* can meet the requirements of goats for maintenance and the production of 1 - 2.5 kg milk (GIGER et al., 1986). Green fodder intake can meet the requirements up to 2-3 kg of milk (DE SIMIANE, 1978). This aspect is dealt with in Chapter 13.

In a feeding system largely based on forages, consumption of large quantities of forages during the late pregnancy enhances forage intake during the postpartum period (MORAND-FEHR and SAUVANT, 1978). The same principle might find application in a feeding system largely based on concentrates where enhanced feed intake is sought in the early part of lactation to attain higher peak milk yields and consequently higher total milk yields.

CONCLUSIONS

Feeding costs comprise considerable part of the total production costs in intensive dairy goat production systems in confinement. To improve efficiency of production good quality feedstuffs should be offered during periods of high productivity. By-products can be efficiently used in diets of dairy goats and cereal grains is better to be fed unprocessed. Dietary protein concentration and protein source are factors influencing productivity and should be taken into consideration. Response to dietary energy and protein is more evident in early than in late lactation. Towards the end of lactation it is difficult to increase milk production if there has been a drop in energy intake. This period is ideally suited to restore energy reserves. Improvement in the efficiency of production should be continually sought through the application of new findings.

SUMMARY

A favourable economic climate has resulted in the development of intensive dairy goat production systems. In such systems, animals of high potential are offered a considerable part of their nutrient requirements indoors. In areas where there is shortage of good quality roughages, goats are fed largely on concentrates, whereas in other areas, goats are given large quantities of good quality forages. Generally good quality feedstuffs are offered during periods of high productivity. Rations are formulated using knowledge of the nutrient requirement of goats and the nutritive value of available feeds. The present chapter reviews feeding of intensively fed goats during the different stages of production and how new research findings might improve efficiency of production.

Keywords : Dairy goat, Intensive feeding, Pregnancy, Lactation, Energy, Lipids, Protein, Feedstuffs.

RÉSUMÉ

Un climat économique favorable a conduit au développement des systèmes de production intensifs de lait de chèvre. Dans ces systèmes, les animaux de haut potentiel reçoivent une part importante de leurs besoins alimentaires en chèvrerie. Dans les zones où les fourrages de bonne qualité manquent, les chèvres sont nourries en grande partie par des aliments concentrés. Au contraire, dans d'autres zones, les chèvres reçoivent de grandes quantités de fourrages de bonne qualité. Généralement les fourrages de bonne qualité sont distribués au cours des périodes de haute productivité. Les rations sont formulées en fonction des besoins nutritionnels des chèvres et de la valeur nutritive des aliments disponibles. Ce chapitre passe en revue l'alimentation des chèvres nourries intensivement pendant les différents stades du cycle de production et comment les nouveaux résultats de la recherche peuvent améliorer l'efficacité de la production.

REFERENCES

ARC, 1980. The Nutrient Requirements of Ruminant Livestock. Commonwealth Agricultural Bureaux, Slough (UK).

ARC, 1984. The Nutrient Requirements of Ruminant Livestock. Suppl. No. 1. Commonwealth Agricultural Bureaux, Slough (UK).

ARMSTRONG D.G. and BLAXTER K.L., 1965. Effect of acetic and propionic acids on energy retention and milk secretion in goats, p.59. Proc. 3rd Symp. Energy Metabolism. EAAP Publ. No. 11, London, Acad. Press (UK).

BLANCHART D. and SAUVANT D., 1974. Répercussion de la gestation sur l'évolution de la lactation (Consequences of pregnancy on the onset of lactation), p. 76-87. In: Journée d'études sur l'alimentation de la chèvre laitière, Paris, June 13, 1974, INRA-ITOVIC, Paris (FRANCE).

CARASSO Y., MALTZ E., SILANIKOVE G., SHEFET G., MELTZER A. and BARAK M., 1988. The use of complete diets for Israeli goats. Seminar of FAO subnetwork on Goat Nutrition and Feeding, Oct. 3-5, 1988, Potenza (ITALY).

COPPOCK C.E., BATH D.L. and HARRIS B. Jr., 1981. From Feeding Systems. J. Dairy Sci., 64: 1230-1249.

GOWAN R.T., ROBINSON J.J., McHATTIE I. and PENNIE K., 1981. Effects of protein concentration in the diet on milk yield, change in body composition and the efficiency of utilization of body tissue for milk production in ewes. Anim. Prod. 33: 111-120.

DACCORD R., 1987. Effect of addition of animal or vegetable fat to a hay based diet on digestibility and nitrogen balance in the lactating goat. Ann. Zootech. 36: 329 (Abst.).

DE SIMIANE M., 1978. Utilisation des fourrages par la chèvre et systèmes d'alimentation (Utilization of forages by the goat and systems of nutrition), p. 124. In: MORAND-FEHR P. et al. (Eds.): Données récentes sur l'alimentation de la chèvre, ADEPRINA-INAPG, Paris (FRANCE).

DE SIMIANE M., HUGUET L. and MASSON C., 1983. Comportement alimentaire des chèvres à l'auge et en pâturage, aspects liés au fourrage et à l'animal, conséquences sur les performances zootechniques (Feeding behaviour of goats indoors or on pasture: effect of forages and animals, consequence on performance) p. 71-100. In: 8e Journées de la recherche ovine et caprine, INRA-ITOVIC, Paris (FRANCE).

ECONOMIDES S., 1986. Nutrition and management of sheep and goats, Paper 58, p. 61-73. In: TIMON V.M. and HANRAHAN J.P. (Eds.). Small ruminant production in the developing countries, FAO, Roma (ITALY).

ECONOMIDES S. and LOUCA A., 1981. The effect of the quantity and quality of feed on the performance of pregnant and lactating goats, Vol. 1, p. 319-326. In: MORAND-FEHR P., BOURBOUZE A. and DE SIMIANE M. (Eds.): Nutrition and Systems of Goat Feeding. Symposium International, Tours (FRANCE), May 12-15, 1981, INRA-ITOVIC, Paris (FRANCE).

ECONOMIDES S., PAPACHIRSTOFOROU C. and GEORGHIADES E., 1982. The use of cereal grains in ruminant diets. Miscellanous Reports 2. Agric. Res. Inst. Nicosia (CYPRUS).

ECONOMIDES S., GEORGHIADES E., KOUMAS A. and HADJIPANAYIOTOU M., 1989. The effect of cereal processing on the lactation performance of Chios sheep and Damascus goats and the preweaning performance of their offspring. Anim. Feed Sci. Tech., 26: 93-104.

EKERN A., 1982. Results from feeding trials and practical experience concerning protein feeding of ruminants in Norway, p. 86-102. In: MILLER E.L., PIKE I.H. and VAN ES A.J.H. (Eds.): Protein Contribution of Feedstuffs for Ruminants: Application to Feed Formulation, , Butterworths, London (UK).

FOLDAGER, J. and HUBER J.T., 1979. Influence of protein percent and source on cows in early lactation. J. Dairy Sci. 63: 243-248.

FUJIHARA T. and TASAKI I., 1975. The effect of abomasal infusion of urea on nitrogen retention in goats. J. Agric. Sci., (Camb.) 85: 185-187.

GIGER S., SAUVANT D., HERVIEU J. and DORLEANS M., 1987. Valeur alimentaire du foin de luzerne pour la chèvre (Nutritive value of lucerne hay for goats). Ann. Zootech. 36: 139-152.

GIHAD E.A., EL-GALLAD T.I., ALLAM S.M. and EL-BEDAWY T.M., 1987. Effect of pre and post-partum nutrition on birth weight and early milk yield in goats, Vol. 2, p. 1401-1402 (Abst.). 4th Intern. Conf. on Goats, March 8-13, 1987, Brasilia (BRAZIL).

HADJIPANAYIOTOU M., 1984a. The value of urea-treated straw in diets of lactating goats. Anim. Feed. Sci. Tech. 11: 67-74.

HADJIPANAYIOTOU M., 1984b. The use of poultry litter as ruminant feed in Cyprus. World Anim. Rev., 49: 32-38.

HADJIPANAYIOTOU M., 1987a. Intensive feeding systems for goats in the Near East, Vol. 2, p. 1109-1141. 4th Intern. Conf. on Goats, March 8-13, 1987, Brasilia (BRAZIL)

HADJIPANAYIOTOU M., 1987b. Studies on the response of lactating Damascus goats to dietary protein. J. Anim. Phys. Anim. Nutr., 57: 41-52.

HADJIPANAYIOTOU M., 1988a. The feeding value of peanut hay and silage made from peanut shells and citrus pulp with the addition of urea. Miscellanous reports, 33, Agric. Res. Inst., Nicosia (CYPRUS).

HADJIPANAYIOTOU M., 1988b. Nitrogen requirements of Damascus goats during maintenance. Miscellaneous Reports 32, Agric. Res. Inst. Nicosia (CYPRUS).

HADJIPANAYIOTOU M., GEORGHIADES E. and KOUMAS A., 1988. The effect of protein source on the performance of suckling Chios ewes and Damascus goats. Anim. Prod. 46: 249-255.

HUGUET L., BROQUA B. and DE SIMIANE M., 1977. Factors affecting green forage intake by lactating goat, p. 1549-1552. 13th Intern. Grassland Congr., Section 10, Leipzig (GERMANY).

KESSLER J., 1985. Influence of roughage quality on feed intake milk production and nutrient utilization in the lactating Saanen goat. Ann. Zootech. 34: 482 (Abst.).

KHAN M.A., ALVI T.A., ALI C.S. and GHANDHRY N.A., 1981. Effect of dietary urea on feed intake, digestibility and milk yield in Beetal goats, Vol. 2, p. 615-620. In: MORAND-FEHR P., BOURBOUZE A. and DE SIMIANE M. (Eds.): Nutrition and Systems of Goat Feeding. Symposium International, Tours (FRANCE), May 12-15, 1981, INRA-ITOVIC, Paris (FRANCE).

KONDOS A.C., 1972. Pre- and post-partum nutrition of goat and its effect on milk production, p. 7. Proc. Nat. Goat Breed Conf., Melbourne (AUSTRALIA).

LINDBERG J.E. and CISZUK P., 1985. Nitrogen utilization in lactating dairy goats. Ann. Zootech. 34: 477 (Abst.).

LINDBERG, J.E. and CISZUK P., 1987. Feed intake, digestibility and nitrogen retention in lactating dairy goats fed increments of urea and fish meal. Ann. Zootech. 36: 328-329 (Abst.).

MIOSSEC H. and SIMIANE M. (de), 1981. Utilisation des betteraves fourragères par la chèvre laitière (Use of beet fodders for dairy goats), Vol. 2, p. 643-651. In: MORAND-FEHR P., BOURBOUZE A. and DE SIMIANE M. (Eds.): Nutrition and Systems of goat. Symposium International, Tours (FRANCE), May 12-15, 1981, INRA-ITOVIC, Paris (FRANCE).

MORAND-FEHR P., 1981. Nutrition and feeding of goats: Application to temperate climatic conditions, p. 193. In: GALL C. (Ed.): Goat Production, Academic Press, (UK).

MORAND-FEHR P. and DE SIMIANE M., 1987. Pattern of concentrate distribution in dairy goat , Fourrages, (109), 75-94.

MORAND-FEHR P., HAUZY F. and HERVIEU J., 1987. Use of meat meals for replacing soya oilmeals in dairy goat diets, Vol. 2, p. 1421 (Abst.). 4th Intern. Conf. on Goats, March 8-13, 1987, Brasilia (BRAZIL).

MORAND-FEHR P. and SAUVANT D., 1978. Nutrition and optimum performance of dairy goats. Livest. Prod. Sci., 5: 203-213.

MORAND-FEHR P. and SAUVANT D., 1980. Composition and yield of goat milk as affected by nutritional manipulation. J. Dairy Sci., 63: 1671-1680.

MORAND-FEHR P. and SAUVANT D., 1987. Feeding strategies in goats, Vol. 2, p. 1275-1303. 4th Intern. Conf. on Goats, March 8-13, 1987, Brasilia (BRAZIL).

MORAND-FEHR P., SAUVANT D. and BRUN-BELLUT J., 1987. Recommendations alimentaires pour les caprins (Recommended allowances for goats). Bull. Tech. CRVZ Theix INRA (70) 213-222.

MOWLEM A., OLDHAM J.D. and NASH S., 1985. Effect of concentrate allowance on *ad libitum* hay consumption by lactating British Saanen goats. Ann. Zootech. 34: 474 (Abst.).

MUDGAL V.D., 1982. Protein and non-protein utilization in goats, p. 604. 3rd Intern. Conf. on Goat Production and Disease, Jan. 10-15, 1982, Tucson, Arizona, (USA).

NRC, 1981. Nutrient Requirements of Domestic Animals, No. 15 : Nutrient Requirements of Goats, p. 91. National Academy Press, Washington, DC (USA).

NOCEK J.E., STEELE R.L. and BRAUND D.G., 1986. Performance of dairy cows fed forage and grain separately versus a total mixed ration. J. Dairy Sci. 69: 2140-2147.

OLDHAM J.D., 1984. Amino acid metabolism in ruminants, p. 137-151. Proc. Cornell Conf. for Feed Manufacturers, Cornell (USA).

ORSKOV E.R., 1979. Recent information on processing of grain for ruminants. Livest. Prod. Sci. 6: 335-347.

ORSKOV E.R., FRASER C. and GORDON J.G., 1974. Effect of processing of cereals on rumen fermentation, digestibility, rumination time and firmness of subcutaneous fat in lambs. Brit. J. Nutr. 32: 59-69.

SAUVANT D., GIGER S., HERVIEU J. and DORLEANS M., 1987. Brewer's grain feeding value for dairy goats, Vol. 2, p. 1423. 4th Intern. Conf. on Goats, March 8-13, 1987, Brasilia (BRAZIL).

SAUVANT D., HERVIEU J., GIGER S., TERNOIS F., MANDRAN N. and MORAND-FEHR, 1987. Influence of dietary organic matter digestibility on goat nutrition and production at the onset of lactation. Ann. Zootech. 36: 335 (Abst.).

SAUVANT D., HERVIEU J., KANN G. and DISENHAUS C., 1988. Effects of GRF on dairy goat performances in mid lactation. Seminar of FAO subnetwork on Goat Nutrition and Feeding, Oct. 3-5, 1988, Potenza (ITALY).

SKJEVDAL T., 1981. Effect on goat performances of given quantities of feedstuffs, and their planned distribution during the cycle of reproduction., Vol. 1, p. 300-318. In: MORAND-FEHR P., BOURBOUZE A. and DE SIMIANE M. (Eds.): Nutrition and systems of goat feeding. Symposium International, Tours (FRANCE) May 12-15, 1981, INRA-ITOVIC, Paris (FRANCE).

SMALL J. and GORDON D.J., 1985. The effect of source of supplementary protein on the performance of dairy cows offered grass silage as the basal diet. Anim. Prod. 40: 520 (Abst.).

Chapter 18

INFLUENCE OF FEEDING ON GOAT MILK COMPOSITION AND TECHNOLOGICAL CHARACTERISTICS

P. MORAND-FEHR, P. BAS, G. BLANCHART, R. DACCORD,

S. GIGER-REVERDIN, E.A. GIHAD, M. HADJIPANAYIOTOU, A. MOWLEM,

F. REMEUF and D. SAUVANT

INTRODUCTION

There is an increasing interest in goat milk production, and there are now goat milk development programmes in many countries. In tropical countries it has often proved difficult to meet all domestic milk demand by increasing cow milk production, while the industrialized countries are seeking dietary and gastronomic quality and diversity in their dairy products. Ripened goat cheese still has a good product image as a food for festive occasions, while fresh goat milk is beginning to take a significant share of the market.

Goat milk is very often used for cheesemaking, and while goat farmers generally aim at increase milk output in quantitative terms, they are equally concerned in maintaining or improving its composition. They may try to act on genetic factors by breeding but this will only bring visible results after several years' work. They may also work on dietary factors and flock management methods - factors over which they have at least partial control and which can often have an effect on milk composition within 24 or 48 hours.

This chapter will consider how the feeding of dairy goats can affect milk composition and cheese quality. These investigations are based on the recent (though somewhat limited) data available from studies in Cyprus, Egypt, Switzerland, the United Kingdom and France.

COMPOSITION AND CHARACTERISTICS OF GOAT MILK

There have been articles in several journals on the characteristics of goat milk, and on factors affecting them (PARKASH and JENNESS, 1968; JENNESS, 1981; MORAND-FEHR et al., 1982; JUAREZ and RAMOS, 1986).

In local breeds not bred for milk production, milk composition is often similar to that of milk ewes, having a very high concentration of dry matter (DM) (135 to 175 g/kg), fat (45 to 65 g/kg) and crude protein (CP) (40 to 55 g/kg). Dairy breeds like the Saanen, Alpine and Toggenburg, on the other hand, give milk that is low in DM (110 to 135 g/kg), often owing to low levels of fat (30 to 40 g/kg) and crude protein (27 to 35 g/kg). MORAND-FEHR et al. (1981a, 1982) have shown that milk from dairy breeds of cows and goats is very similar in terms of the composition of its DM, which is far less true of milk from local breeds. In any case, variations in the composition of the dry matter are mainly due to variations in fat and crude protein levels.

Using goat milk, 75 % of all variation in cheese yield is due to variation in CP content and only 25 % to fat content (RICORDEAU and MOCQUOT, 1967). However, fat content determine one important factor for cheese quality : the fat/dry matter ratio. It is therefore especially useful to study these two constituents with goats, which are mainly kept for cheesemaking. It is known that in dairy breeds like the Saanen and Alpine, the low levels of these two milk constituents are a handicap for cheesemaking. Furthermore, the proportions of coagulable proteins in the total crude protein (GRAPPIN et al., 1981) and of αS_1 caseins in the total casein (REMEUF and LENOIR, 1985) are significantly lower than in cow milk. GROSCLAUDE et al. (1987) found a strong correlation between αS_1 casein content and that of rennet-coagulable proteins. REMEUF (1987) has shown that goats possessing variants with high levels of αS_1 casein synthesis produce a better milk for cheesemaking purposes : a higher total casein content, a firmer gel obtained with rennet.

Goat milk from dairy breeds is apparently not very suitable for cheesemaking owing to its low levels of useful constituents, caseins in particular. This is confirmed by the behaviour of such milk in the presence of rennet. The firmness of goat milk coagulum is significantly lower, on average, than that of cow milk; as a result, more useful matter is lost and cheese yield is lower (REMEUF and LENOIR, 1985; REMEUF, 1988). Rennetting-to-clotting time for goat milk is shorter and its rennetting rate is faster. These features are connected with the low true protein content and perhaps also to the size and high ash content of the micella. Moreover, the thermal stability of goat milk is far weaker than that of cow milk, probably because its ionized calcium content is higher. However, all these milk characteristics vary widely from one goat to another and from one flock to another.

The fat in goat milk forms smaller globules than that in cow milk. Although very variable in its fatty acid composition, goat milk fat contains higher proportions of C_6-C_{12} fatty acids than cow milk fat, and lower proportions of C_{16}-C_{18} saturated fatty acids (MORAND-FEHR et al., 1982).

Generally, goat milk has a slightly lower lactose content than cow milk (PARKASH and JENNESS, 1968). Phosphorus and calcium contents are generally very similar in cow and goat milk, but chloride and potassium contents are higher in goat milk (GUEGUEN et al., unpubl.). Goat milk contents only traces of carotenoids if any, which gives it a white appearance. It is also low in vitamins of the B group (B6, B12 and folic acid).

FEEDING AND OTHER FACTORS
AFFECTING GOAT MILK COMPOSITION

It is difficult to isolate the effects of feeding from the other factors affecting milk composition because, as can be seen from Figure 1, these factors are interdependent (MORAND-FEHR et al., 1986b). To take an obvious example, the qualitative and quantitative availability of forage varies according to season and climate. Varying day length varies the time available to grazing goats to satisfy their daily nutritional requirements, and this affects intake levels. Another important factor is the animal's milk potential: the composition of the total intake of a highly productive goat is usually different from that of a low-yielding goat.

In view of this, an effort will be made to describe the effects of dietary factors independently, of other factors. In particular, we shall in all cases distinguish between the direct effects on milk composition and the indirect effects, since dietary factors also affect milk output, and there is generally an inverse correlation between milk output and concentration.

Figure 1

INTERELATIONS BETWEEN FACTORS INFLUENCING MILK COMPOSITION

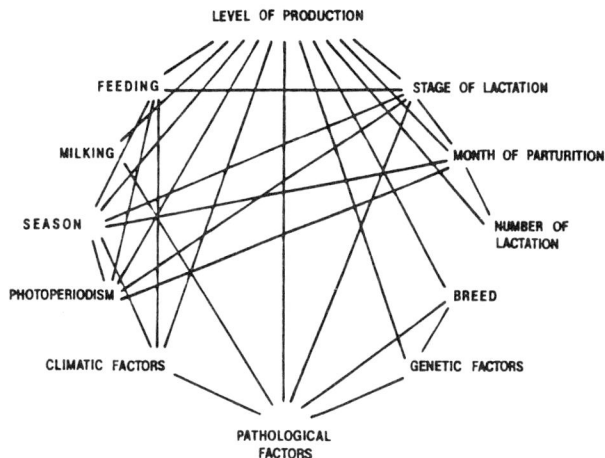

THE INFLUENCE OF NUTRITIONAL AND DIETARY FACTORS

Several workers have attempted to explain the effects of feeding factors on milk composition by linking them to the mechanism by which milk constituents are synthesized (MORAND-FEHR and SAUVANT, 1980; DEVENDRA, 1982; MORAND-FEHR et al., 1981a, 1982, 1986b). This approach does in fact elucidate certain apparently contradictory findings, due to the very variable milk yield potentials of goats and the wide range of rations employed.

Feed intake and energy supply

Fasting rapidly reduces milk output and increases the milk fat and, to a lesser extent, milk CP content. It also raises the proportion of C_{18} fatty acids against short chain saturated fatty acids in the fat (MORAND-FEHR, BAS and SAUVANT, 1988, unpubl.; MASSART-LEEN and PEETERS, 1985; Table 1). While the impact on the CP content seems to be essentially due to the drop in the quantity of milk secreted, fat content and the composition of the fatty acids are directly affected. Fasting causes a massive mobilization of lipids stored in the adipose tissue, mainly C_{18} fatty acids. The inflow of these C_{18} fatty acids into the secretary tissue of the mammary gland tends to significantly reduce the *de novo* synthesis of shorter chain fatty acids in the udder (MORAND-FEHR et al., 1981b, 1982; CHILLIARD, 1985).

In mid-lactation, an improvement in energy supply in the ration due to increased distribution of concentrates has the opposite effect; fat content declines (0.15 to 0.35 percentage points) but the proportion of C_4-C_{16} saturated fatty acids, but casein and protein contents (0.1 to 0.15 percentage points) rise (FEHR and DELAGE, 1973; FEHR and LE JAOUEN, 1977; MORAND-FEHR and SAUVANT, 1980). However, GIHAD et al. (1987) showed that in goats with low milk yield potential, when feed supply was raised from equilibrium to 125 % of energy requirement, milk production was not improved and the milk fat content did not fall, while SNF and particularly CP rose significantly (Table 2). This effect on CP content is a direct effect, since additional energy intake increases the degradable energy available for synthesizing microbial proteins in the rumen. Although additional energy supply often raises CP content, restricting energy supply does not always lower milk CP contents. BLANCHART et al. (1986, unpubl.) showed that a 20 % reduction in organic matter supply did not reduce casein or total CP contents even though milk production fell by 15 %.

Table 1

EFFECT OF FASTING ON THE COMPOSITION OF GOAT MILK
A. Main components (1) : From MORAND-FEHR and HERVIEU (1969, unpubl.)

	Before fasting	After 24 hours fasting	Level of significance
Milk production (kg/day)	2.1	1.7	+
Fat content (%)	3.1	3.6	+++
Protein content (%)	2.7	2.9	++
Total dry matter content (%)	11.1	11.7	++

B. Composition of fatty acids in milk fat (2) : From MASSART-LEEN and PEETERS (1985)

	Before fasting	After 48 hours fasting	Level of significance
Fatty acids (%) :			
$C_{10:0}$	10.3	3.7	+++
$C_{12:0}$	5.6	1.7	+++
$C_{14:0}$	11.9	3.4	+++
$C_{16:0}$	33.7	23.1	++
$C_{18:0}$	5.9	10.5	+++
$C_{18:1}$ + isomers	21.9	37.3	+++

(1) experiment on 8 one year old Alpine goats in 4th month of lactation
(2) experiment on 5 lactating, non pregnant goats of a fawn-colored breed
+++ : $P \le 0.001$;   ++ : $P \le 0.01$;   + : $P \le 0.05$

Table 2

EFFECT OF ENERGY INTAKE AND ROUGHAGE LEVEL ON MILK COMPOSITION OF ZARAIBI DOES DURING THE FIRST 32 WEEKS OF LACTATION
(From GIHAD et al., 1987)

	Level of NRC (1981) energy requirements (%)				Level of significance		
	100		125		(A)	(B)	(C)
Level of roughage (%)	20	40	20	40			
Milk production (kg/d)	0.69	0.69	0.72	0.72	NS	NS	NS
Fat content (%)	4.39	4.04	4.37	4.11	NS	++	NS
SNF content (%)	8.18	8.11	8.55	8.50	+	NS	NS

The average milk production of goats was 0.66 kg/day
NS: Non significant; + : $P \le 0.05$; ++ : $P \le 0.01$.
(A): Level of energy; (B): Level of roughage; (C): Interaction of these two factors.

However, it does seem that the lower the energy balance (energy supply minus energy requirements), the more visible the above-mentioned effects due to augmenting energy supply. This has been demonstrated in early lactation, with the greater energy deficit and the resulting mobilization of lipid reserves, the higher milk fat content and proportion of C_{18} fatty acids, and the lower proportion of short chain fatty acids (SAUVANT, 1982; SAUVANT et al., 1987a).

Amount and type of dietary carbohydrates

Cell carbohydrates such as starch are mainly provided by cereal grains or other concentrates and cell wall carbohydrates by forage feeds. Researchers have therefore sought to identify what proportions of the two types of feed, or of the two types of carbohydrates, will give optimum output of goat milk while keeping lipid and CP levels high.

From earlier experiments (MORAND-FEHR et al., 1981a; MOWLEM et al., 1985), it would seem that for a given energy supply, a lower forage / concentrate ratio tends to reduce fat content and increase the proportion of C_{18} fatty acids in milk, while CP content is not greatly affected (Table 3). This has since been confirmed by GIHAD et al. (1987) with Zaraibi goats of low milk yield potential. When the proportion of forage was reduced from 40 to 20 % of the ration during the first 32 weeks of lactation, milk fat content fell significantly (0.26 to 0.35 percentage points). Similarly, when forage / concentrate ratio dropped below 20/80, the milk fat content fell rapidly by 0.4 to 1.5 percentage points, while CP content changed little (CALDERON et al., 1984). These results are at least partly due to reduced availability of acetic acid (the main precursor for the fatty acids synthesized in the udder), which is produced more readily in the rumen when cell wall carbohydrates are fed. However, supplying bicarbonate in the ration can partly or entirely restore milk fat levels and maintain a satisfactory level of acetic acid in the rumen fluid (HADJIPANAYIOTOU, 1982; Table 4). In fact the finding of CALDERON et al. (1984) and HADJIPANAYIOTOU (1982), and also certain unpublished observations, seem to suggest that the goat is less sensitive than the cow to a fibre-deficient diet and that its milk fat content drops less.

Table 3

EFFECT OF THE RATIO ROUGHAGE/CONCENTRATE OF DIETS ON MILK PERFORMANCES
OF LACTAGING GOATS
(From MOWLEM et al., 1985)

Level of intake (g DM/ day) :			
Concentrate	593	1000	1389
Hay	1222	1102	669
Milk production (kg/ day)	2.69	2.94	3.21
Fat content (%)	2.68	2.63	2.59

Recent experiments (SAUVANT et al., 1986) show that if the forage/concentrate ratio stays above 20/80, the goats' nutritional energy status is a more essential factor than the relative proportions of concentrate to forage. The ruminant energy balance, however, is far more strongly influenced by the digestibility of the organic matter in the ration (DMO). In early lactation, increasing DMO from 71 to 77 % by switching from a lucerne hay based diet to one based on beet pulp, led to an increase in goat milk production and secretion of $C_{4:0}$ to $C_{16:0}$ fatty acids and to a decline in milk fat content and secretion of C_{18} fatty acids (SAUVANT et al., 1986). The amount of total fat secreted tends to fall once DMO levels reach about 75-76 % of the ration, which corresponds to an energy concentration of 0.89

UFL (Milk Feed Unit)/kg of DM for a ration containing 2-3 % fat. On the other hand, with a ration based on maize silage or lucerne hay, and a total DMO of around 80 %, the type of concentrate has no effect on the fat or CP content of the milk. The fatty acid composition is only slightly altered (GIGER et al., 1986; Table 5). Similarly, in early lactation the effect of a concentrate rich in cell wall carbohydrates or starch seems relatively minor compared to the effect of the energy balance.

Table 4

EFFECT OF SODIUM BICARBONATE SUPPLIES TO DIETS CONTAINING 15 OR 30 % OF ROUGHAGE ON RUMEN CHARACTERISTICS AND ON MILK PRODUCTION AND COMPOSITION OF GOATS

(From HADJIPANAYIOTOU, 1982)

	Without NaHCO$_3$		With NaHCO$_3$		Level of significance	
					Hay %	NaHCO$_3$
Hay percentage in the diet	15 %	30 %	15 %	30 %		
Dry matter intake (kg/ day):						
Concentrate	1.44	1.26	1.57	1.26	+	NS
Barley hay	0.22	0.45	0.24	0.46	++	NS
VFA concentration in rumen liquor (mm/l)	139$_1$	57$_2$	123$_1$	59$_2$		
Molar proportion of :						
acetate	63.5$_{1,2}$[b]	71.3$_2$[a]	67.2$_{1,2}$[b]	70.3$_{1,2}$[ab]		
propionate	21.7$_1$[a]	14.4$_2$[b]	17.9$_{1,2}$[ab]	15.5$_2$[b]		
Rumen pH	6.1[b]	6.9[a]	6.2[b]	7.1[a]		
Milk yield (kg/day)	2.22	2.20	2.17	2.07	NS	NS
Fat content (%)	4.02	4.22	5.00	4.77	NS	++
Protein content (%)	4.20	3.93	3.97	4.13	NS	NS

NS: non significant; + : P < 0.05; ++ : P < 0.01.
Means on the same line with unlike superscripts differ (P ≤ 0.05) and with unlike subscripts differ (P ≤ 0.01).

GARCIDUENAS (1978), comparing rations based on green forage, hay and maize silage, observed no significant effect on milk composition due to the nature of the feed, but found that energy intake levels had a slight effect. Furthermore, cheese yield was unaffected by the type of forage in the ration. These results were explained by the fact that there was little difference in the fat and coagulable protein contents of the milk from goats fed on the different types of forage.

Milk composition is slightly affected by an increase in concentrate supply, at least where small quantities are offered (0.2 to 0.5 kg) and provided the concentrate comprises 30 % or less of total DM intake. This was observed by GARMO (1986) on mountain pastures and confirmed by some observations made in France.

Table 5

INFLUENCE OF TYPE OF FORAGE AND CONCENTRATE
ON GOAT MILK AND FAT COMPOSITION
(From GIGER et al., 1986)

	Lucerne hay			Lucerne hay + maize silage			Maize silage		
Number of observations*	54			48			22		
	(A)	(B)	(C)	(A)	(B)	(C)	(A)	(B)	(C)
Fat production (g/day)	1.26	NS	+	13.56	NS	++	- 8.63	NS	NS
Fat content (%)	0.79	NS	+	2.56	NS	++	- 1.53	NS	NS
Protein content (%)	0.96	NS	NS	0.83	NS	++	1.15	NS	NS
Fatty acid (% in total fat):									
$C_{4:0}$ - $C_{14:0}$	- 1.63	NS	++	0.92	NS	++	- 1.26	NS	NS
$C_{16:0}$	3.48	++	++	0.90	NS	++	0.15	NS	NS
$C_{18:0}$ + $C_{18:1}$	- 2.52	NS	++	- 1.56	NS	++	1.93	NS	+

Level of significance : NS : non significance; + = < 0.05; ++ = P < 0.05.
* : one observation = average value for one goat which ate R and P concentrates in Latin square experiments.
(A): R-P; R: concentrate rich in digestible fiber; P: concentrate poor in digestible fiber; R-P: difference.
(B): Level of significance; (C): Covariate : the covariate is the measured energy balance expressed as g. of DMO.

So, it seems that unless there is a very marked imbalance in the starch/cell wall carbohydrate ratio, the nature of the carbohydrate in the ration has only an indirect effect on goat milk composition, through its impact on energy intake. Nevertheless, ECONOMIDES et al. (1989) observed that milk fat content is higher in goats fed a whole-grain diet than mash or pelleted grain diets.

On the other hand, a forage/concentrate ratio of less than 20/80 upsets rumen fermentations; less acetic acid is produced, and this has a clear, direct effect, lowering the milk fat content and the amount of fatty acids synthesized in the udder. However, these effects tend to be less marked than in the cow.

Feed additives

It has been shown quite clearly that monensin sodium added to goat ration favors the production of propionic acid in the rumen at the expense of acetic acid, reducing milk fat content by 0.15 to 0.45 percentage points and increasing CP levels (BROWN and HOGUE, 1985).

Sodium bicarbonate, on the other hand, which tends to raise the proportion of acetic acid against propionic acid in the rumen, lowers the CP content of the milk and significantly raises the fat content, while milk output is not significantly altered (HADJIPANAYIOTOU, 1988).

Lipid feed supplements

As with cows, addition of lipids to dairy goat rations has given variable results. Fats added to a forage-plus-concentrate ration resulted in some cases in an increased lipid secretion, associated with an increase in milk output,milk fat levels, or both. Tables 6 and 7 give two examples of the latter instance in early lactation. There are a number of possible feedstuffs for fat supply: lupin (MASSON, 1981), brewers' grains (SAUVANT et al., 1987b), soya beans (SAUVANT et al., 1983; MORAND-FEHR et al., 1984a; DACCORD, 1987). Animal fats or vegetable oil can also be fed (DACCORD, 1985, 1987; MORAND-FEHR and HAUZY, 1986; MORAND-FEHR et al., 1981b, 1986a, 1987a).

The effectiveness of added lipids in altering milk composition depends on many factors, especially the lipid content of the rest of the ration. Obviously, a high level of added lipids (over 6 % of the DM), or a lower level (4-5 % of the DM) when the ration is already rich in ether extract, may have harmful effects on milk output and even on the milk fat content. High levels of added lipids can upset rumen fermentation, reducing production of acetic acid in the rumen. Reduced synthesis in the udder of saturated acids with 4 to 16 carbon atoms cannot be entirely compensated by feeding long chain fatty acids as a supplement in the ration.

On the other hand, where the diet is low on lipids, the adipose tissue has proved to be incapable of synthesizing enough C_{18} fatty acids to cover udder requirements. Consequently, the milk fat content can be reduced by 0.3 to 0.5 percentage points, and the amount of fat secreted by 8 % to 20 %. Moreover, the proportion of C_{18} acids falls by 20 to 50 % (MORAND-FEHR and SAUVANT, 1980). This situation has recently been observed in high yielding dairy goats on rations containing less than 2 - 2.5 % of ether extract in the DM (MORAND-FEHR et al., 1982); this would apply to diets of hay, grass silage, tubers and roots such as beet, and concentrates like barley and solvent-extracted oilcakes (Tables 6, 7). Milk fat of goats fed rations low in lipids is low in C_{18} fatty acids and rich in C_4 - C_6 acids.

Table 6

EFFECT OF PROTECTED OR UNPROTECTED FAT SUPPLIES ON THE LEVEL OF FEED INTAKE, MILK PRODUCTION, COMPOSITION AND TECHNOLOGICAL CHARACTERISTICS OF GOATS IN EARLY LACTATION*
(From MORAND-FEHR et al., 1987a))

Diets **	G 0	G 5	D 5	D 10
Level of dry matter intake (g/day)	2254	2157	2461	2204
Energy balance (Kcal NE/day)	- 659[a]	- 466[ab]	- 167[b]	- 302[ab]
Corrected milk production*** (kg/day)	3.63[a]	3.69[a]	4.02[b]	4.01[b]
Fat content (%)	3.95[a]	4.43[b]	4.46[b]	4.35[b]
Corrected fat production*** (g/day)	148[a]	153[a]	177[b]	180[ab]
True protein content (%)	2.47	2.80	2.75	2.68
Casein content (%)	2.00	2.14	2.15	2.16
Maximal gel firmness	43	54	49	54
Time for gel hardening	2.6	3.3	4.0	3.0
Time for clotting (min)	27	24	23	24

* : early lactation : first 8 weeks of lactation; ** : diets composed of luzerne hay, beet pulp and concentrates. From 2nd week of lactation : G0 : concentrate without fat; G5 : concentrate + 5 % tallow; D5 : concentrate + 5 % dairy fat prills; D10 : concentrate + 10 % dairy fat prills. *** : the milk and fat production were adjusted using the value of 1st week of lactation as covariate. For the other parameters, the covariates were not significant. Values not followed by same letter differ significantly (P ≤ 0.05).

As with cows, goats on a starch-rich diet respond less well to a fat supplement than goats on a diet rich in cell wall carbohydrates (MORAND-FEHR and HERVIEU, unpubl.). This is probably because starch and added lipids together have an effect on rumen fermentation.

The impact on milk secretion also depends on the type of fat fed and how it is processed. Animal fats and solid oils like palm oil have proved more satisfactory than unsaturated vegetable oils (MORAND-FEHR et al., 1984b; MORAND-FEHR and HAUZY, 1986; DACCORD, 1987). Fats protected in crystalized or saponified form (MORAND-FEHR et al., 1984b, 1987a; DACCORD, 1987; DE MARIA GHIONNA et al., 1987) are less likely to upset rumen fermentation processes, provided they are properly protected (MORAND-FEHR et al., 1981b); as a result, they are in some cases more effective in raising milk production and milk fat content. However, unprotected fats and even unsaturated fats can give comparable results when incorporated in fairly small proportions in the ration.

Table 7

EFFECT OF ADDITION OF ANIMAL OR VEGETABLE FAT TO HAY BASED DIET
ON THE PRODUCTION AND COMPOSITION OF GOAT MILK
(FromDACCORD, 1987)

	Standard	Level + fodder beet + 4 % micronized animal fat	+ 20 % soybeans
Level of intake (kg DM/day)	2.15	2.06	2.03
Digestibility of fibre (%)	62.0	59.1	60.7
Fat (3 %) corrected milk yield (kg/ day)	3.58	3.73	3.15
Fat content (%)	2.83	3.20	3.28
Protein content (%)	2.67[a]	2.81[b]	2.86[b]

Values not followed by same letter differ significantly (P ≤ 0.05).

The fatty acid composition of the milk fat is affected by that of the lipid supplement in the feed; the longer the carbon chains of the fatty acids fed, the more this applies.

A fat supplement in the ration sometimes tends to reduce CP content in the milk, by an average of 0.1 points (MORAND-FEHR et al., 1984b). This phenomenon could be connected with insulin secretion and alterations of protein synthesis by the rumen microflora. But the more saturated or protected the dietary fat, and the smaller the amounts fed, the less marked is the effect (see Table 6). Generally, this effect can be explained by an effect of dilution. Moreover in some cases, CP content can be significantly improved (DACCORD, 1987; Table 7).

Unlike such other energy sources such as cell and cell wall carbohydrates, dietary lipids apparently have a direct impact on the fat and CP contents of goat milk. They may have positive or negative effects depending on how they are used.

In view of the tendency for dietary fat supplements to lower milk CP content, it would be interesting to know what effect they would have on the technological characteristics of the milk. Unfortunately we have the results of only one experiment on this question (Table 6); caution is therefore called for in the interpretation of these findings. In this experiment, the time for clotting was shorter, the gel firmer and rennetting rate faster with goats which had been fed a lipid supplement. These results should be considered in the light of the higher casein content of milk from these goats. However, milk casein content is probably not always improved, as it was in this experiment, by feeding a fat supplement with the ration.

Intake and source of crude protein

There have been few studies analysing the effects of the amount of CP in the ration on goat milk composition . HADJIPANAYIOTOU (1987) varied the total CP content of the ration between 8 and 16 %, and observed that an increase in the dietary CP content can improve milk production, but that milk composition is unaltered. In particular, milk CP content does not increase following addition of CP to the feed; this is in accordance with observations on cows (REMOND, 1985) and sheep (ROBINSON et al., 1974).

Studies on the effects of the type of CP fed on the composition of goat milk have given contradictory results. With Alpine goats in mid-lactation, soya meal was replaced by meat meal (a protein source that is richer in rumen undegradable protein) in a ration that remained isoenergetic and isonitrogenous in terms of protein digestible in the intestine (PDI), and just met energy and protein requirements. The change did not significantly alter milk protein production or true protein content in milk (MORAND-FEHR et al., 1987b; Table 8). By contrast, partial replacement of soya meal by fish meal improved the CP content of the milk of Damascus goats by 0.3 percentage units, without altering milk output (HADJIPANAYIOTOU et al., 1987; Table 9). Replacing untreated soya and rapeseed oil cakes with formaldehyde-treated cakes and feeding additional urea (BLANCHART et al., 1986, unpubl.), or replacing soya meal with protein-rich seeds (MASSON, 1981), while the rations remained isoenergetic and isonitrogenous, did not lead to any significant change in the CP content of the milk or the composition of the proteins.

Table 8

EFFECT OF DIETARY NITROGEN SOURCE ON THE PRODUCTION AND COMPOSITION OF GOAT MILK
(From MORAND-FEHR et al., 1987b)

	Soya bean meal	Meat meal A	Meat meal B	Level of significance
Level of intake :				
Net energy (UFL/ day)	1.83	1.84	1.83	NS
Protein (g PDI/ day)	171	174	171	NS
Milk production (kg/ day)	2.83	2.87	2.82	NS
Corrected milk production (3,5 % fat milk kg/ day)	2.63	2.64	2.61	NS
Fat content (%)	3.03	3.00	3.03	NS
True protein content (%)	2.78	2.73	2.77	NS

Twenty four goats in latin square design experiment from 4th to 6th month of lactation. Isoenergetic and isonitrogenous diets composed of lucerne hay, beet pulps and concentrates. Meat meal A and B contained 59 and 67 % true protein in dry matter, respectively.
UFL : milk feed unit; PDI : protein digestible in intestine; NS : non significant.

In most cases then, with isoenergetic and isonitrogenous diets, the CP and casein contents of goat milk are not very sensitive to a dietary CP supplement, or to changes in the type of protein source in the diet. It is not surprising, therefore, that the technological characteristics should show little change when meat meal is fed, although the rennetting rate is significantly faster with diets containing meat meal.

218

On the other hand, non-protein nitrogen (NPN) levels in milk, urea (MUC) especially, largely depend on the diet composition and on the nitrogen digestion and metabolization in the rumen (see Chapter 8). A high MUC may mean too much degradable nitrogen, a lack of available energy for the microorganisms in the rumen, or both. The MUC may also be high if there is much excess crude protein in the ration, even if supplies of degradable and undegradable nitrogen are well balanced. In fact, because of the low values involved (0.4 g/ kg), variations in MUC have only a limited influence on milk CP content.

Table 9

EFFECT OF THE DIETARY NITROGEN SOURCE ON THE MILK PRODUCTION
AND COMPOSITION OF DAMASCUS GOATS
(From HADJIPANAYIOTOU et al., 1987)

Nitrogen source	Soybean meal	Fish meal
Milk production (kg/day)	3.87	3.82
Fat percent (%)	3.20	3.60
Protein percent (%)	3.80	4.10

Diet : 2.18 kg 20 % CP concentrate/day + 0.31 kg lucerne hay /day + 0.61 kg barley hay /day.

Inspite of the different dietary CP concentrations and protein sources used in the experiments referred to, the milk fat content was altered in only three cases: fish meal raised it by 0.4 percentage units (HADJIPANAYIOTOU et al., 1987), field beans lowered it by 0.45 and lupins increased it by 0.12 (MASSON, 1981), although in the latter case the effect was probably due to the high fat content of the lupin.

In conclusion, it seems that dietary CP source and concentration have only a limited effect on milk composition.

PRACTICAL CONSEQUENCES OF THE EFFECTS OF DIET ON MILK COMPOSITION

The higher the milk output of the goat, the greater the impact of diet on milk composition. With dairy breeds,high levels of energy supply are essential to obtain satisfactory contents of CP and coagulable protein in the milk. Depending on the level of milk production, the minimum DOM threshold that must be met, lies between 70 and 75 %.

It would perhaps also be helpful to avoid unbalanced diets rich in degradable protein and poor in energy available to the rumen flora. Diets of this kind induce high MUC, which reduce the quality of the curd as has been shown with cow milk. Although this effect has not been demonstrated in goats, it is advisable to avoid such diets, especially in summer, when cheesemaking mishaps are frequent.

Although the milk fat content has less impact on cheese yield than the protein content, it is advisable not to produce low fat milk since this affects the proportion of fat to DM and the flavour and creaminess of the cheese. Care should therefore be taken not to provide a diet too low in fibre or fat, as this has an adverse effect on the secretion of C_4 - C_{16} fatty acids due to a lack of the necessary precursor, acetic acid, and also on the production of C_{18} fatty acids which cannot be synthesized in the udder.

To conclude, the diet must be well balanced,with neither too much starch (which often in practice means too little fibre), nor too little starch (which may lead to an energy deficit), and neither too rich nor too poor in lipids (optimum 4 - 5 % of total DM) and, lastly, with a good balance between sources of rumen degradable and undegradable nitrogen.

CONCLUSIONS

As diet has a limited effect on the contents of the different milk nitrogen fractions apart from urea, it is better to make use of breeding techniques as well to improve the protein content of goat milk. On the other hand, diet can play a significant role in determining the quantity and composition of secreted fats. Priority should therefore be given to diet when seeking to raise a low milk fat content or modify unsatisfactory fatty acid composition. The quantity and type of energy supplies are the dietary factors that have most impact on milk composition.

In order to produce milk with optimum cheesemaking potential, it is best to combine a high level of output with high concentrations of fats and, above all, of coagulable protein. Goats must consume large amounts of net energy, as this will enhance both milk production and CP content. High energy intake can be assured by maintaining a high digestibility of organic matter in the ration. Further, for a satisfactory milk fat content, adequate amounts of dietary lipids and cell wall carbohydrates must be fed.

Goat milk composition will be satisfactory as long as care is taken to maintain a good, well-balanced diet and, more particularly to ensure supplies of starch, fibre and lipids that are neither too high nor too low.

SUMMARY

An improvement of the energy density of the ration due to concentrate supply reduces the fat content and increases protein content of goat milk but the effect of concentrate supply has little effect on milk composition when it comprises one third or less of total dry matter intake.

The type of carbohydrates in the ration seems to have only an indirect effect on the composition of the goat milk, through its impact on energy intake. Generally fats added to rations result in an increased lipid secretion. Generally, the negative effect on protein content can be explained by an effect of dilution. Frequently protein supplies and the type of protein sources have a very limited effect on milk composition.

To produce goat milk with high fat and protein contents, the diet must be well balanced, in termes of starch, lipids and rumen degradable and undegradable protein.

Keywords : Dairy goat, Milk composition, Technological characteristics, Energy supply, Dietary carbohydrates, Feed additives, Fats, Protein sources.

RÉSUMÉ

Une amélioration de la densité énergétique du régime par un apport de concentrés réduit la teneur en lipides du lait et augmente la teneur en protéines. L'effet de l'apport de concentrés a peu d'effets sur la composition du lait quand il représente un tiers ou moins de la matière sèche totale ingérée.

La nature des glucides alimentaires semble avoir seulement un effet indirect sur la composition du lait de chèvre par l'intermédiaire de son impact sur l'ingestion d'énergie. Généralement les matières grasses ajoutées aux régimes provoquent une augmentation de la sécrétion de lipides. L'effet négatif sur la teneur en protéines peut généralement être expliqué par un effet de dilution. Fréquemment les apports protéiques et le type de sources protéiques ont un effet très limité sur la composition du lait.

Pour produire un lait à haute teneur en lipides et en protéines, le régime doit être bien équilibré sans trop ni pas assez d'amidon ni trop riche ni trop pauvre en lipides et enfin, avec un bon équilibre entre les sources d'azotes fermentescibles et non fermentescibles dans le rumen.

REFERENCES

BROWN D.L. and HOGUE D.E., 1985. Effects of feeding monensin sodium to lactating goats: milk composition and ruminal volatile fatty acids. J. Dairy Sci., 68: 1141-1147.

CALDERON I., DE PEETERS E.J., SMITH N.E. and FRANKE A.A., 1984. Composition of goat milk: changes within milking and effects of high concentrate diet. J. Dairy Sci., 67: 1905-1911.

CHILLIARD Y., 1985. Métabolisme du tissu adipeux, lipogenèse mammaire et activités lipoprotéine-lipasiques chez la chèvre au cours du cycle gestation-lactation (Adipose tissue metabolism, mammary lipogenesis and lipoprotein lipase activities in goats during gestation and lactation). Thèse de doctorat d'État, Université Paris 6 (FRANCE).

DACCORD R., 1985. Effect of type of concentrate on digestibility and nitrogen utilization of a forage based diet in lactating goats. Ann. Zootech., 34: 480 (Abst.).

DACCORD R., 1987. Effect of addition of animal or vegetable fat to a hay based diet on digestibility and nitrogen balance in the lactating goat. Ann. Zootech., 36: 329 (Abst.).

DE MARIA GHIONNA G., BARTOCCI S., TERZANO G.M. and GORGHESE A., 1987. Acidi grassi satificati con calcio nella dieta de capre in lattazione. Effetto sulta produzione, sul contenuto di grasso e di proteine del latte cio nella dieta de capre in lattazione. Ann. Sperim. Zootech., 20: 231-242.

DEVENDRA C., 1982. Goat. Dietary factors affecting milk secretion and composition. Intern. Goat Sheep Res., 2: 61-76.

ECONOMIDES S., GEORGHIADES E., KOUMAS A. and HADJIPANAYIOTOU M., 1989. The effect of cereal processing on the lactation performance of Chios sheep and Damascus goats and the pre-weaning growth of their offspring. Anim. Feed Sci. Technol., 26: 93-104.

FEHR P.M. and DELAGE J., 1973. Effet du niveau énergétique de la ration sur l'utilisation par la glande mammaire de chèvre de l'acétate comme précurseur des acides gras du lait (Effect of energy level of diets on the utilization of acetate as precursor of milk fatty acids in goat mammary gland). C.R. Acad. Sc. Paris, 276, Série D: 3449-3452.

FEHR P.M. and LE JAOUEN J.C., 1977. Influence de divers facteurs alimentaires sur la composition du lait et les caractéristiques du fromage de chèvre (Effects of various dietary factors on milk composition and on goat cheese characteristics). Rev. Lait. Fr., 338: 39-55.

GARCIDUENAS A., 1978. Étude de l'incidence de certains types de rations alimentaires sur la composition du lait de chèvre et les rendements en fromages (Effect of various kinds of diets on goat milk composition and cheese yield). Document ITOVIC, Paris (FRANCE).

GARMO T.H., 1986. Dairy goat grazing on mountain pastures. 1 - Effect of supplementary feeding. Meldinger fra Norges Landbrukhogskole, 85 (26), 19 p.

GIGER S., SAUVANT D. and HERVIEU J., 1986. Influence of the kind of component feed on goat milk production and composition. Ann. Zootech., 36: 334-335.

GIHAD E.A., EL-GALLAD T.T., ALLAM S.M. and EL BEDAWY T.M., 1987. Performance of Zaraibi and Damascus lactating does fed high and low energy rations. Ann. Zootech., 36: 336-337.

GRAPIN R., JEUNET R., PILLET R. and TOQUIN A., 1981. Étude des laits de chèvre. I - Teneur du lait de chèvre en matière grasse, matière azotée et fractions azotées (Study on goat milks. I - Fat and protein contents and protein composition of goat milk). Le Lait, 61: 117-133.

GROSCLAUDE F., MAME M.F., GRIGNON G., DI STASIO L. and JEUNET R., 1987. A Mendelian polymorphism underlying quantitative variations of goats αS_1 casein. Génét., Sél., Evol., 19: 399-412.

HADJIPANAYIOTOU M., 1982. Effect of sodium bicarbonate and of roughage on milk yield and milk composition of goats and on rumen fermentation of sheep. J. Dairy Sci., 65: 59-64.

HADJIPANAYIOTOU M., 1987. Studies on the response of lactating Damascus goats to dietary protein. J. Anim. Phys. Anim. Nutr., 57: 41-52

HADJIPANAYIOTOU M., 1988. Effect of sodium bicarbonate on milk yield and milk composition of goats and on rumen fermentation of kids. Small Rum. Res., 1: 37-47.

HADJIPANAYIOTOU M., KOUMAS A. and GEORGHIADES E., 1987. Effect of protein source on performance in lactation of Chios ewes and Damascus goats, and degradability of protected and unprotected soybean meal in the rumen of goats. Ann. Zootech., 36: 326-327.

JENNESS R., 1980. Composition and characteristics of goat milk: Review 1968-1979. J. Dairy Sci., 63: 1605-1630.

JUAREZ M. and RAMOS M., 1986. Physico-chemical characteristics of goat's milk as distinct from those of cow's milk. In: Production and utilization of ewe's and goat's milk. Bull. Intern. Dairy Fed., (202), p. 55-67.

MASSART-LEEN A.M. and FEETERS G., 1985. Changes in the fatty acid composition of goat milk fat after 48 hours fast. Reprod., Nutr., Dévelop., 25: 873-881.

MASSON C., 1981. Utilisation des graines protéagineuses dans l'alimentation de la chèvre en début de lactation (Utilization of proteagenous grains in the diets of goats in early lactation). Ann. Zootech., 30: 435-442.

MORAND-FEHR P. and SAUVANT D., 1980. Composition and yield of goat milk as affected by nutritional manipulation. J. Dairy Sci., 63: 1671-1680.

MORAND-FEHR P., LE JAOUEN J.C., CHILLIARD Y. and SAUVANT D., 1981a. Les constituants du lait de chèvre, synthèse et facteurs de variation (Components of goat milk, synthesis and factors of variation), p. 234-270. 6e Journées de la Recherche ovine et caprine, INRA-ITOVIC, Paris (FRANCE).

MORAND-FEHR P., BAS P. and SAUVANT D., 1981b. Influence des lipides protégés sur les performances et le métabolisme des chèvres en début de lactation (Effects of protected lipids on the performance and metabolism of goats in early lactation). Vol. 2, p. 594-611. In: MORAND-FEHR P., DE SIMIANE M. and BOURBOUZE A. (Eds.): Nutrition and systems of goat feeding. Symposium International, Tours (FRANCE),May 12-15, 1981, INRA-ITOVIC, Paris (FRANCE).

MORAND-FEHR P., CHILLIARD Y. and SAUVANT D., 1982. Goat milk and its components: secretary mecanisms and influence of nutritional factors, p. 113-121. 3rd Intern. Conf. on goat production and disease, Jan. 10-15, 1982, Tucson, Arizona (USA).

MORAND-FEHR P., SAUVANT D. and BAS P., 1984a. Effect of soybeans and soya oil on milk performance in dairy goats. 35th EAAP Annual Meeting, Aug. 6-9, 1984. The Hague (THE NETHERLANDS).

MORAND-FEHR P., SAUVANT D. and BAS P., 1984b. Utilisation des matières grasses chez les ruminants. Expériences sur chèvres laitières (Utilization of fats in ruminants: experiments on dairy goats), p. D1-D21. In: CAAA: Peut-on et comment utiliser les matières grasses dans les rations des vaches laitières, Nov. 8, 1984, ADEPRINA, Paris (FRANCE).

MORAND-FEHR P. and HAUZY F., 1986. Graisses chez les ruminants (Fats in ruminants). Revue de l'alimentation animale, (339), p. 30-34.

MORAND-FEHR P., BAS P., SAUVANT D., HERVIEU J. and CHILLIARD Y., 1986a. Influence de la nature de l'aliment concentré sur le métabolisme des chèvres en fin de gestation et en début de lactation (Effect of the kind of concentrate feeds on goats metabolism during late pregnancy and early lactation). Reprod., Nutr., Dévelop., 26: 349-350.

MORAND-FEHR P., LANCHART G., LE MENS P., REMEUF F., SAUVANT D., LENOIR J., LAMBERET G. and LE JAOUEN J.C., 1986b. Données récentes sur la composition du lait de chèvre (Recent data on goat milk composition), p. 253-298. 11e Journées de la Recherche ovine et caprine, INRA-ITOVIC, Paris (FRANCE).

MORAND-FEHR P., BAS P., SAUVANT D., TERNOIS F. and HERVIEU J., 1987a. Effect of dairy fat prills on milk performance of goats in early lactation, p. 1420 (Abst.). 4th Intern. Conf. on Goats. March 8-13, 1987, Brasilia (BRAZIL).

MORAND-FEHR P., HAUZY F. and HERVIEU J., 1987b. Use of meat meals for remplacing soya oilmeals in dairy goat diets, Vol. 2, p. 1421 (Abst.). 4th Intern. Conf. on Goats, March 8-13, 1987, Brasilia (BRAZIL).

MOWLEM A., OLDHAM J.D., NASH S., 1985. Effect of concentrate allowance on *ad libitum* hay consumption by lactating British Saanen goats. Ann. Zootech., 34: 474 (Abst.).

PARKASH S. and JENNESS R., 1968. The composition and characterization of goat's milk, a review. Dairy Sci. Abst., 30: 67-87.

REMEUF F., 1988. Contribution à l'étude des aptitudes fromagères du lait de chèvre (Contribution to studying the capacity of goat milk for cheese making). Thèse de Docteur Ingénieur, INA Paris-Grignon (FRANCE).

REMEUF F., LENOIR J., 1985. Caractéristiques physico-chimiques de laits de chèvre et leur aptitude à la coagulation par la présure (Physio-chemical characteristics of goat milk and its capacity for clotting by rennet). Rev. Lait. Fr., (446), 32-40.

REMOND B., 1985. Influence de l'alimentation sur la composition du lait de vache. 2 - Taux protéique, facteurs généraux (Effect of feeding on cow milk composition. 2 - Protein percentage, main factors). Bull. Tech. CRZV, Theix, INRA (62) 53-67.

RICORDEAU G. and MOCQUOT G., 1967. Influence des variations saisonnières de la composition du lait de chèvre sur le rendement en fromage. Conséquences pratiques pour la sélection (Effect of seasonnal variations of goat milk composition on cheese yield. Practical consequences for selection). Ann. Zootech., 16: 165-181.

ROBINSON J.J., FRASER C., GILL J.C. and MAC HATTIE J., 1974. The effect of dietary crude protein concentration and time of weaning on milk production and body-weight change in the ewes. Anim. Prod. 19: 331-339.

SAUVANT D., 1982. Alimentation énergétique des caprins (Energy feeding for goats). Les dossiers de l'élevage, 5: (1) 31-48.

SAUVANT D., MORAND-FEHR P., BAS P., 1983. L'intérêt des lipides dans les aliments concentrés (Interest of fat supplies in concentrate feeds), p. K1-17. In: CAAA: Quels aliments concentrés pour les hautes productrices de lait. ADEPRINA, Paris (FRANCE).

SAUVANT D., HERVIEU J., GIGER S., TERNOIS F., MANDRAN N. and MORAND-FEHR P., 1986. Influence of dietary organic matter digestibility on goat nutrition and production at the onset of lactation. Ann. Zootech., 36: 335-336.

SAUVANT D., GIGER S., HERVIEU J. and DORLEANS M., 1987a. Brewer grain feeding value for dairy goats, Vol. 2, p. 1423. 4th Intern. Conf. on Goats, March 8-13, 1987, Brasilia (BRAZIL).

SAUVANT D., HERVIEU J., TERNOIS F., MANDRAN N., BAS P. and MORAND-FEHR P., 1987b. Secretion of short chain fatty acids at the onset of lactation, Vol. 2, p. 1472. 4th Intern. Conf. on Goats, March 8-13, 1987, Brasilia (BRAZIL).

Chapter 19

GOAT BREEDING AND FEEDING SYSTEMS
IN MEDITERRANEAN SYLVO-PASTORAL AREAS

H. NARJISSE, M. NAPOLEONE, B. HUBERT and P. M. SANTUCCI

INTRODUCTION

Goat husbandry is still an important socio-economic activity in most marginal zones of the Mediterranean region. This activity is based on agro-sylvo-pastoral resources produced in areas with substantial constraints. In these areas the predominance of mountains and the variability and harshness of the climate, in addition to access difficulties and lack of infrastructure, often impart to goat husbandry an extensive character and limit the possibilities of its intensification.

MEDITERRANEAN SYLVO-PASTORAL ZONES

We will characterize these zones using information taken from available literature on the northern rim southeastern France) and southern rim (Moroccan High Atlas) of the Mediterranean Sea.

The Natural Milieu

The Mediterranean region is characterized by the dominance of hills and mountains. Altitudes are relatively low in southeastern France, rarely exceeding 1600 m, while they vary in the Moroccan High Atlas from 1500 m on valley floors to over 4000 m at the peaks. In terms of vegetation, this mountainous countryside presents marked contrasts arising from the differentiation of the land into strata defined along an altitude gradient, and delimiting a succession of ecozones and therefore of different plant communities. These communities ranging from low shrubs to high elevation grassland, offer a seasonal distribution of feed availabilities. They also enable the producers to design a feeding calendar in which each of these communities occupies a sequence and whose implementation requires some herd movement.

The Mediterranean climate is characterized by rainfall variability and a very pronounced opposition of wet and dry seasons. The combined effects of low temperatures in high altitudes and of insufficient water availability allow a brief growing period of 3 to 5 months. Rain falls generally from October to May. In the Var department of France, precipitation exceeds 700 mm/year, while it reaches only 500 mm/year in the High Atlas. Part of this precipitation falls in the form of snow, contributing to the isolation of these zones, especially on the southern rim of the Mediterranean sea, and setting the timing of summer grazing in the mountains.

Agricultural activities

The general characteristics of the region we have outlined determine human activities in these areas. North of the Mediterranean Sea, we have observed since the late 19th century a continuing decline of traditional mountainous systems resulting from their inability to insure the economic and ecological conditions of their renewal (BAZIN, 1985). This has resulted in an alarming frequency of fires and a thinning of the social fabric. Presently, active programs are undertaken to ensure the revival of pastoral activities, and prevention of forest fires (HUBERT et al., 1985).

South of the Mediterranean sea, especially in the Moroccan High Atlas, traditional mountainous agricultural systems are well preserved despite a significant rural exodus compensated by high demographic growth. In these systems, subsistence agricultural operations exist in an extremely hostile environment. Income generated from these operations is rarely adequate to meet the basic needs of the family. These are then secured by revenues earned within the region or elsewhere. This results in an overutilization of the sylvo-pastoral resources due to excessive use of the forest (encroachment of cropping, illegal wood cutting, branch lopping) in order to satisfy domestic needs as well as those of the herd. In this context, the major concern of the government agencies in these regions is to protect these resources threatened by desertification.

Plant Production

The steep topography of marginal zones of the Mediterranean region limits cropping, often concentrated on valley floors. The valleys of the High Atlas are thus carefully arranged in extremely small irrigated terraces. BOURBOUZE (1986) reported that the mean size of an agricultural operation does not exceed 0.50 hectare. Faced with this miniaturization, farmers compensate by a remarkable intensification reflected by the quasi-general practice of two crops per year on the same plot and by obtaining very high yields. The most common speculations are cereals (barley and corn), which cover more that 80 % of the cultivated land, followed by vegetable gardening. Irrigated prairies (Agdal) are located on land difficult to terrace on the edges of rivers.Forage crops (clover and alfalfa) are widely used, primarily to satisfy the needs of cattle.

By contrast, in the North of the Mediterranean Sea goats are produced primarily for the market, and are considered the most profitable way of using dense woodland. In these regions, intensive crops (orchards, gardening, cereals) are concentrated in the plains, except on rough surfaces covered by scrub growth and chaparral.

Animal production

Animal production in hilly Mediterranean regions is largely dominated by herds of small ruminants, often managed as mixed herds except in specialized operations. The excessively rough topography and presence of shrub tree cover reduces the benefit of cattle raising.

In the South of the Mediterranean sea, goat production is directed primarily toward meat, except for a few places where milk tradition exists. In this case, herd size ranges from a few to several hundred goats. By contrast, we observe in the north of the Mediterranean sea movement toward specialization favoring the production of goat milk for farm-made cheese. Along these lines there is a progressive introduction of highly productive breeds such as the Alpine chamoisé, and a better control of feeding practices (GAUTIER, 1983).

One of the essential characteristics of extensive goat husbandry in the Mediterranean region is its significant use of sylvo-pastoral resources. Thus, there are systems in which rangeland alone provides more than 90 % of the feed resources. These systems are widespread, especially on the south rim of the Mediterranean sea (BOURBOUZE, 1982; NARJISSE et al., 1983). In the north, however, this practice is progressively being replaced by a combination of pastoral resources, mown and/or grazed grasslands, and concentrates, most often purchased.

This production practice necessitates the use of animals adapted to seasonal changes in feeding systems and variable availability of feed resources. This results in the predominance of hardy, native breeds of goats, especially in southern Morocco and Corsica, where they account for 100 (NARJISSE et al., 1983) and 95 % (PROST et al., 1985) of the herds, respectively.

GOAT BREEDING STRATEGY IN THE MEDITERRANEAN ZONE

The limitations of sylvo-pastoral zones discussed above require special strategic choices on goat farmers concerning both herd conception and husbandry practices.

Conception of the herds

Type of herd

The producer can organize his herd via the choice of one or several species, the choice of one or several breeds with different potentials and the choice of one or several products within the same herd. These different strategies result from local cultural traditions, technical and financial capabilities of the farmer and the economic environment.

When the farmer has access to forage resources, the operation is often oriented toward milk production. In this case, the herd is composed of animals most capable of generating efficient milk production from this resource. In Morocco, this is achieved by high milk-producing cattle, while in France intensive milk production can be performed in units including both goats and cattle.

Inversely, when the only resource is rangeland with drought tolerant vegetation, the herd is generally oriented toward meat production through:

-   sheep and goat herds composed of native breeds in transhumant systems located in the foothills of Morocco or the Crau in Provence (France);

-   herds composed only of local goat breeds in regions dominated by woody vegetation (Moroccan mountains, Mediterranean back-country in Provence).

Between these two extreme situations (forage calendar based on the use of grasslands, versus native drought-resistant vegetation), there are intermediate situations in which the farmer seeks production output regularily by combining milk and meat production and forming mixed herds including groups of animals having different capacities.

A portion of the herd is composed of dairy animals whose role is to insure animal production in the spring. After raising their kids for 1 or 2 months, these animals are milked (1 or 2 times daily) during their lactation period lasting about 8 months.

A second sub-herd is composed of animals oriented toward the production of 5 to 10 month old kids. This is usually seen as a supplement rather than as a production goal, and it may supply income when females dry up. Although the output of this sub-herd is low (one kid per animal), its presence in the herd is justified by the little extra work it requires, and the higher return from time spent taking care of the herd.

A third sub-herd present in all operations plays a fundamental role. It is composed of animals selected for both conformation and production level. These animals must have sufficient flexibility to be milked or suckled. These goats are generally milked after 3 to 4 months of suckling and therefore maintain the lactation output of the dairy herd when production begins to decline. This group plays the role of a reserve herd. This mode of organization, based on the combination of breeds having different production potential (Table 1) and on the changes in numbers of goats milked during the season, enables the farmer to maintain herd production at a stable level in spite of inevitable changes in individual production.

Table 1

GOATS BREEDS AND PERFORMANCE LEVELS
IN THE MEDITERRANEAN REGIONS OF MOROCCO AND FRANCE

Site / Production system	Breeding stock	Production				
		Milk			Meat	
		Estimated milk production kidding-drying	Milking duration	Duration of suckling before milking	Slaughter age	Slaughter weight
MOROCCO						
High Atlas Extensive rangeland	Local population	40 kg		4 months*	3-24 months	7-25 kg
Rif	Cross (Murcia/ Local)	170 kg	6 months			
FRANCE (Provence)						
Rangeland + supplem.	Alpine	350-600 kg	7-9 months	1-2 months		
Rangeland + supplem.	Rove	200-300 kg	5-7 months	3-5 months		
Rangeland	Rove				4-12 months	8-25 kg
Rangeland + supplem.	Cross: Alpine x Rove	300-500 kg	5-9 months	1-5 months	1-5months	8-25kg

(*) These herds were not milked.

Herd size

From the highly productive dairy animal to the native breed producing kids, there are large differences in individual productivity. By adjusting the size of the herd, the farmer regulates these differences in order to obtain the desired herd production level.

Thus, dairy herds on the plains of Provence are usually of small size (about 50 heads). Large herds of more than 100 heads, are often made of hardy animals that rely primarily on rangelands with low productivity.

Although the mean herd size targeted is a function of family objectives and animals' potential, this size may undergo considerable variations from one year to another depending on feed availability and the immediate financial needs of the family. In Morocco, for example, a period of drought may lead to a high mortality (10 to 50 % mortality in young animals) and the sale of a large proportion of the herd (100 goats may be sold from a herd of 150). These substantial variations in herd size necessarily affect the strategy of herd renewal.

Conception of husbandry practice

Within the context defined by operation objectives, the farmer seeks a sufficiently flexible husbandry mode through which he can modulate herd requirements and forage supply.

BOURBOUZE (1982) and VALLERAND and SANTUCCI (1989) stressed the importance of this management flexibility for the herd survival during periods of shortages and its tolerance of alternations of feed surplus and restriction. In the latter case, requirements are adjusted to available resources by:

- mobilization of body reserves (GIBON, 1981);

- regulation of production periods (circulating females in Corsica which "miss" a reproductive cycle (SANTUCCI and MAESTRINI, 1985);

- regulation of production levels depending on forage supply (maximum grass feeding throughout lactation, and compensatory growth of young).

By using these adjustment capabilities of the animal, the farmer can use the production cycle to prolong or reduce critical sequences in which requirements are high.

For meat production, in Morocco as well as in Provence, the farmer will rely more on slaughter age (3 to 12 months) than on average daily gain to obtain a marketable weight. The feeding plan of the kids is therefore dictated by the slaughter age chosen. Kids of a hardy breed such as Rove reach practically the same weight at the age of 7 months regardless of their husbandry system. In this case, the most likely variable to be affected by feeding is growth rate. For example, the farmer can carefully select the feeding regime to delay the slaughter age through a succession of start-up and growth decline phases. This method may be advantageous to the farmers who desire to sell the kids at 3 months of age, and to milk the mother subsequently. On the other hand, if the target slaughter age is 7 months, extensive mother-kid management is then necessary.

In milk production the farmer must predict the response of the animal in order to define husbandry practices:

- in the group of specialized dairy animals, the farmer defines the feed supplementation by combining dry forage and concentrate to allow the animal to produce as regularly as possible for more than 7 months;

- in the group of hardy mixed breeds, the farmer increases the supplementation level in order to limit the importance of mobilization at certain periods of the year. In addition, depending on feed availability, the animal may dry up after 3 to 7 months of milking.

In an attempt to influence the performance of certain animals in the herd, the farmer may change his husbandry practices in the course of the season. Sometimes, the result is, however different from that expected: the dairy response of the hardy Rove breed in Provence to a momentary change in husbandry is low (NAPOLEONE and HUBERT, 1986). Selected breeds, on the other hand, react to changes with an intensity and in a direction often difficult to control. Supplementation may therefore be considered as a tool to increase production, insure coverage of all requirements, or to maintain production level. Depending on the physiological stage of the animal, the impact of this change on feeding systems varies: it is low after 4 to 5 months of lactation, but may be high at the onset. This change in production at the start-up phase may adversely affect the persistance and duration of lactation.

FEEDING SYSTEMS

Major systems observed

As we have seen, goats do not occupy the same place in the different rural societies dealt with here. South of the Mediterranean sea, they are often managed as mixed herds with sheep, directed toward family meat consumption or restricted marketing. Some females may nevertheless be kept for milk production for the household, but in Morocco most milk is obtained from cattle. North of the Mediterranean sea, particularly in France, goats are mostly oriented toward the production of milk to be processed into cheese, industrially or on the farm.

Feeding systems can thus be relatively different depending on the production goals of the farmers, their know-how, technical capacity and the physical environment and resources available to feed their animals, as well as the socio-cultural endowment of the region.

In Morocco, goats must find at least 90 % of their feed resources on rangelands: 3 to 10 % of these resources are provided by the agricultural operation itself (flattening barley in February, grazing on stubble May - June). In winter, when the climatic conditions are harsh, some farmers feed hay to their sheltered animals along with freshly cut oak or juniper foliage harvested from the forest (BOURBOUZE, 1982; NARJISSE et al., 1983), or even with barley if foliage gathering is impossible. In the French Mediterranean zone, there is a gradient from feedlots to operations distributing no more than 50 MFU (milk feed unit corresponds to the equivalent of the net energy content of 1 kg of barley for lactation) per goat and per year. Different types of goat husbandry practices can thus be distinguished as a function of production goals (delivery of milk, or cheese-making on the farm), the tools used (combination between individual performance and herd size), the breeds raised (native vs. improved), targeted feeding costs, etc. Some of these types identified in recent studies (GAUTIER, 1983; LAGACHERIE, 1988; MADANI, 1988) are presented in Table 2.

In most of these operations, supplementation is regularly provided to grazing goats, sometimes even beyond drying, in order to maintain milk production at a given level. This supplementation consists of concentrates (cereals, commercial feed, etc.) or variable quality hay. The effects of these different supplements on the increased value of feed ingested on rangeland has been discussed in Chapter 13. A basic feed resource such as *Quercus ilex* leaves can have different feeding values depending on the type of supplement used.

Organization of these feeding systems in space and time

The farmer uses natural vegetation by maximizing the diversity of potentially grazable resources during different seasons of the year, i.e. by adjusting the cycle of the herd's nutritional requirements (related to reproduction and production management, if such a cycle can indeed be identified at the level of the herd) to phenological stages of the vegetation which depend on climatic events.

Thus, in Morocco most mountain herds move among winter ranges consisting of evergreen oak located at 1200-1800 m from November to April, and summer pastures located above 2000 m and grazed from May to October. In this transhumance process, the herder may move from one fixed point to another ("azib") or migrate with his tent (BOURBOUZE, 1982; NARJISSE et al., 1983). The feeding regime of the animals includes foliage of ash trees or *Quercus ilex* (evergreen oak) in spring, grasses in summer, and juniper and acorns (up to 70 % of the diet) in the fall. In Corsica, it is considered that 70 % of the herds move to the mountains in summer (1500-2000 m) where milking continues until the end of July, then to low hills from October to April for farmers quartered at 800 m (VAN DEN DRIESCH, 1987; SANTUCCI and VALLERAND, 1988).

Table 2

GOATS BREEDS , FEEDING SYSTEM AND PERFORMANCE LEVELS IN SOUTHERN FRANCE

Feeding system	Herd size (head)	Genetic type	Average milk product. (kg/goat)	Supplementation MFU (1) Hay	Cereals	Summer supply of rangelands (%)	Authors
Mostly grazing	70	local	350	80	40	80	GAUTIER (1983)
Grazing + suppl.	30	Alpine	600	300	200	40	
Feed lots	60	Alpine	850	500	300	0	
Occasional grazing	30	Alpine	750	700		3-10	MADANI (1983)
Summer grazing	40-70	Alpine	650	500		15-20	
Grazing + suppl.		Alpine + loc.	450-800	350		35-50	
Pastoral	40	local + Alp.	200-500	250		40-70	
Extensive small herds	<50	local	milking + suckling	30	0	>90	LAGACHERIE (1988)
Extensive large herds	100-200	local	milking + suckling	50	0	>90	
Large grazing herds supplem.	100	local	kids	90	170	35-75	
Exclusively dairy	55	loc. + alp.	only milking	70	30	70	

(1) Milk Feed Unit

Herds located at lower altitudes do not move as much (NARJISSE et al., 1983; MOREAU, 1988), but use different grazing areas around a relatively fixed site (evergreen oak woodland, low shrubland, stubble). Most herds in the Mediterranean region of France fall into this category.

Thus, different ranges and even well identified pastures are used in different seasons, according to a rotation pattern, with frequency ranging from several weeks to several months. Gradients of altitude, exposure, soil types, even the history of plant formations thus contribute to provide diversified resources at different periods of the year (Figure 1, MADANI, 1988). In this case, the animals feed on *Quercus ilex* in winter;grasses in spring, *Quercus pubescens, Coronilla* and *Arbutus* in summer. During the fall season, the main components of their diet consist of grass, shrubs and acorns. Many herds also combine grazing on cultivated forages (alfalfa, sainfoin) with the use of rangelands, either at turning out or more regularly throughout the year, particularly in the summer. These movements or rotations allow the development of vegetation between two grazing periods separated by several weeks to one year. This organization can be facilitated by fencing, or even as in the case in Morocco, by social restrictions as the prohibition of high altitude ranges from April to July before their opening to grazing (MENDES and NARJISSE, 1987).

In Morocco and Corsica, where these collective uses of rangelands are common, their regulations are largely codified by social agreements which allocate grazing land or periods of grazing to different farmers. Sometimes, the allocation process specifies the animal species that are allowed to graze.

Managing the feeding systems

The fine organization of grazing throughout the year as well as the supplementation level take into account the farmer's goal, his resources and space availabilities, and his desire to direct his animals toward a feeding behavior appropriate to different situations (choice, grazing standing, etc.). We may thus distinguish practices related to land use and others related to herd management, in order to adjust feeding practices during the breeding season. These include supplementation, or as an alternative to unavailable sylvo-pastoral resources, synthesis and mobilization of body reserves, compensatory growth of young, etc.

Figure 1

UTILIZATION AND CONTRIBUTION OF GRAZED FORAGE RESOURCES

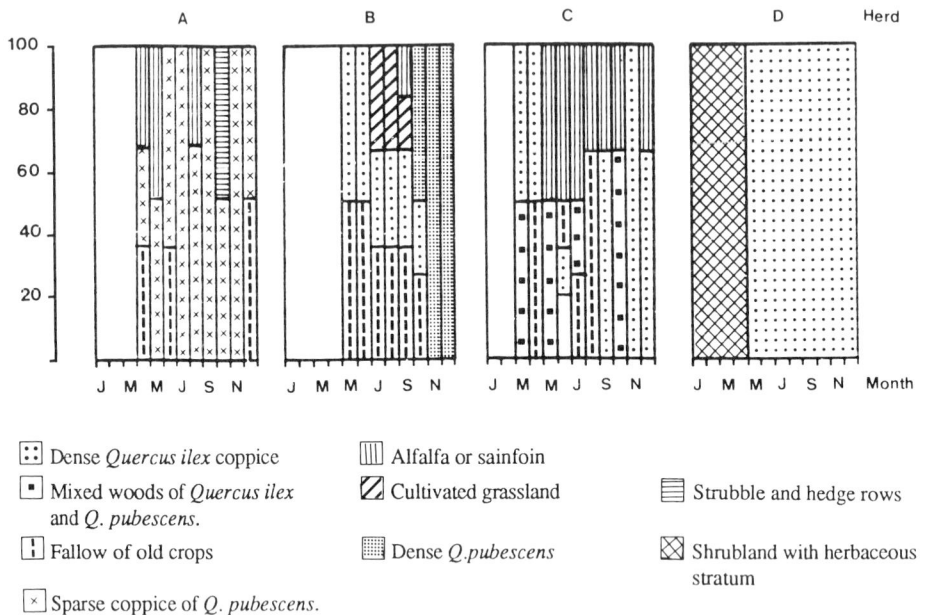

Legend:
- :: Dense *Quercus ilex* coppice
- ▪ Mixed woods of *Quercus ilex* and *Q. pubescens.*
- ⊡ Fallow of old crops
- ⊠ Sparse coppice of *Q. pubescens.*
- ⦀ Alfalfa or sainfoin
- ⊘ Cultivated grassland
- ▤ Dense *Q.pubescens*
- ▤ Strubble and hedge rows
- ⊠ Shrubland with herbaceous stratum

The overall dimension of this feeding system should not be understood as a mere succession of independent technical sequences. Instead, it is more relevant to consider the organization and the management of this grazing practice, in order to identify the practical problems and attempt to resolve them. A sectorial approach to such an integrated system might indeed lead to confusions.

An illustration of this system suggested by MENIER and GUERIN (1987) and GUERIN and BELLON (1989) involves identifying "functions" within a "pasture grazing sequence" (Figure 2) showing land use as a function of time throughout the growing season. These functions are characterized by the season of use of each forage resource: they integrate seasonal variations of the vegetation as well as those related to changes in animal status and requirement. They also point out the type of management which is carried out (high intensity versus low stocking rate allowing the grazing animals to select their diet among available forage resources).

Figure 2

PASTURE GRAZING SEQUENCE IN A GOAT FARM IN SOUTH EASTERN FRANCE
(GUERIN and BELLON, 1989)

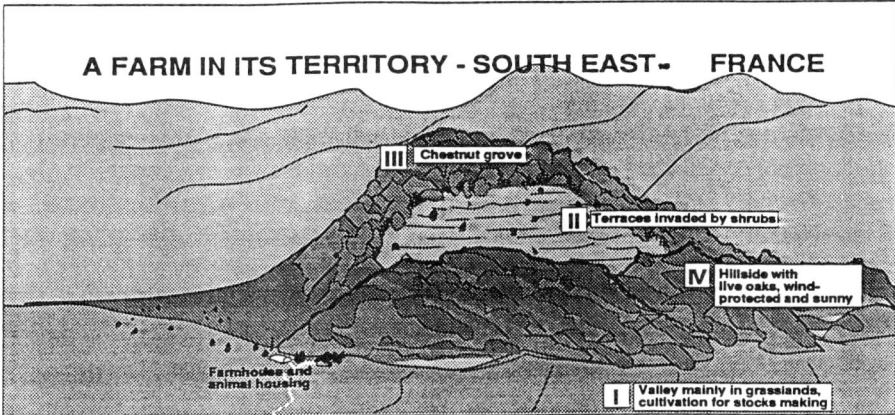

A FARM IN ITS TERRITORY - SOUTH EAST - FRANCE

- III Chestnut grove
- II Terraces invaded by shrubs
- IV Hillside with live oaks, wind-protected and sunny
- Farmhouses and animal housing
- I Valley mainly in grasslands, cultivation for stocks making

REPRESENTATION OF THE PASTURE GRAZING SEQUENCE (PLANNING FOR 1989 -1990)

LEGEND I.P. : Intake pattern

M ✗ Mowing periods

S ✗ Sowing periods (for sorghum as an intercrop)

■ 3 Number of grazing days per area

✶ Daily combination of several grazed areas ("soupade")

MAIN FUNCTIONS OF THE VARIOUS FARM AREAS

A Goats turning out to grass

B Spring grazing regulation

C Goats lactating during spring (I.P. I in table 1)

D Late spring goats lactation maintenance

E Spring - summer transition

F Summer lactation

G Autumnal goats'lactation

H Goats' " recovery" and flock maintenance during late autumn (I.P. II in table 1)

I Winter pregnancy and kidding (I.P. III)

This representation is a better way of understanding how a pastoral area is structured by farmers through grazing. It points out the contribution of the different portions of this space to the entire feeding system. This method may be adapted to herded grazing animals, for which the circuits are probably less clearcut, where the combination of successive grazing sequences plays a role, and where the distances covered depend on day length, temperature, etc. Many authors have investigated such circuits and reported their highly structured nature (BOURBOUZE, 1982; NARJISSE et al., 1983; MADANI, 1988; BOURBOUZE and DONADIEU, 1987; VERTES, 1983).

FEEDING BALANCES

Methodological considerations

The analysis of the feeding balance has a twofold objective. It enables the comparison of animal requirements and feed supplies during different periods of the year. It also leads to an estimation of secondary production of rangeland. Forage consumption is, in fact deduced by difference of the calculated requirements and known feed supplies. The establishment of the feeding balance of the herd is based on the idea of a zootechnical unit, composed of one reproducing female and her litter. It is derived from assumptions involving the reproductive efficiency of the herd, the annual distribution of births, as well as the variation in herd size resulting from mortality, sale and possible consumption.

In compliance with these hypotheses and taking into account the performance level monitored regularly on a sample of herds, the energy requirements of the herd are estimated on a monthly basis. In parallel, the periods of use of feed resources purchased and/or provided by the farm are noted monthly. Similarly, the energy value of these controlled supplies is quantified and compared to that of the calculated requirements leading to the deduction of the feed supply attributed to the sylvo-pastoral land.

This analysis, however, does not explain effects related to climatic changes. In addition, there are several sources of error attached to it and so it has only an indicative value. Although some estimations arise from measurements, others result from declarations or from more or less fragile hypotheses, especially those related to available standards of energy requirements of hardy breeds and the feeding value of sylvo-pastoral resources. This determination can therefore be useful only as a way of comparing supply and demand cycles in a given site and under particular conditions. It can also be considered as a base for determining the relative variations between seasons, years, and operations of the animal requirements and the grazed resources contribution.

Contribution of different forage resources

These balances allow to identify adjustment practices in terms of supplementation or herd management. Thus, concentrate is often used to bolster milk production. This is the case of relatively extensive operations, such as the "large supplemented pastoral goat herds" described by LAGACHERIE (1988), and illustrated in Figure 3: concentrate covers almost one-third of the requirements during the lactation period. Another highly demonstrative situation is that of dairy goat herds in the region of Chaouen (Morocco). Figure 4 shows the contribution of the different feed sources in three categories of herds: III - herds using improved goats and close to the city; II - herds composed of native breeds, and located in the mountains; and I - herds using mixed breeds and located in foothills and hilly zones. Nevertheless, in this example, as shown in Figure 5, concentrate is strictly distributed during lactation periods as a supplement to evergreen oak branches distributed in winter in the trough. It is interrupted in the spring when vegetation begins to grow. This figure also shows that herd organization as illustrated by its structure is not particularly linked to the seasonal cycle of resource availability. Instead, it seems more related to the production goals throughout the year.

When we attempt to further define the respective roles of concentrate and pastoral resources in the entire season of milk production, different strategies of supplementation can be seen. Thus, some farmers rely on pastoral resources to cover maintenance requirements. Some achieve this objective by regularly adjusting quantities of concentrate distributed according to changes in animal requirements and forage availability in rangelands. Others provide a fixed quantity of concentrate (a sort of "forage insurance"). Concentrates can thus be used as a fine tool for piloting milk production by controlling supply/demand adjustments (NAPOLEONE and HUBERT, 1987; MADANI, 1988).

Figure 3

INDIVIDUAL DAILY REQUIREMENTS AND FEED INTAKE THROUGHOUT THE YEAR IN EXTENSIVE PASTORAL FARMING SYSTEM

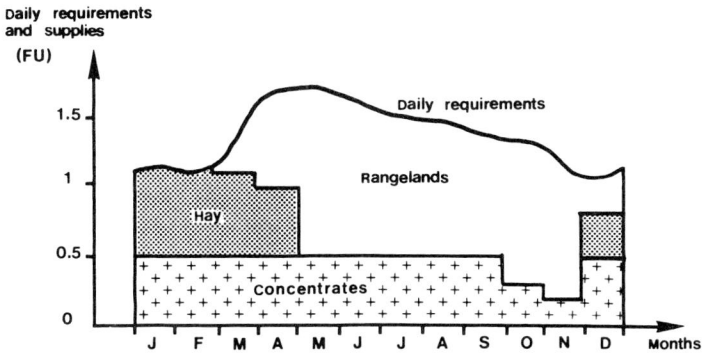

Figure 4

CONTRIBUTION OF DIFFERENT RESSOURCES TO SATISFY ENERGY REQUIREMENTS IN ZONES I, II and III

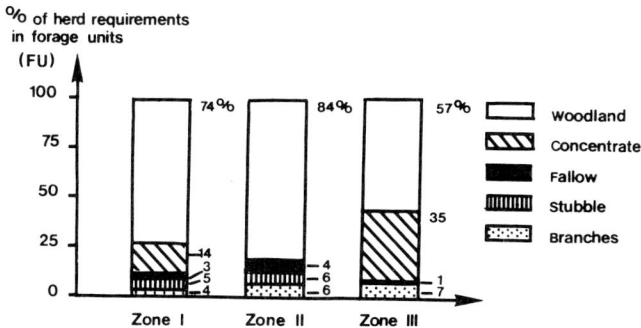

The practice of supplementation is far from being the general case. Thus, in the Moroccan High Atlas, where goat raising is directed toward meat production, feeding is largely dominated by sylvo-pastoral resources with a significant contribution of summer grassland for herds located at high altitudes. For those in the middle and low valleys, the difficulty of access to high altitude ranges is compensated by continuous use of forest resources and grazing on stubble in the summer (Figure 6).

Figure 5

FORAGE BALANCE OF ZONE III CLOSE TO THE CITY OF CHAOUEN

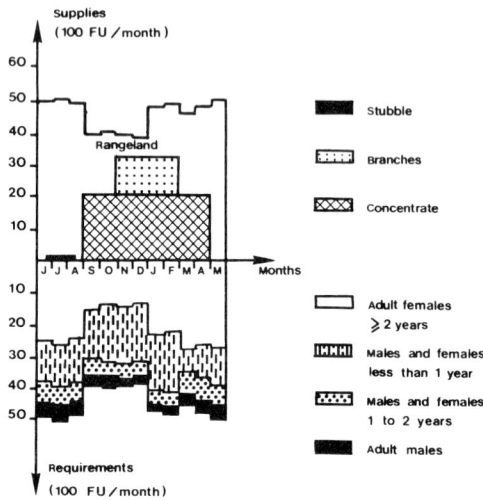

CONCLUSION

Faced with constraints related to the supply/demand cycle and its adjustment and taking into account the cultural and economic environment of the farm, the farmer reserves a number of degrees of freedom by organizing the herd and its management in a specific way.

This diversity of herds, management and production systems, far from being experienced as a handicap, makes the farmer capable of adaptation to a variable environment. Thus, the objective is not to seek a homogenization of management practices in order to promote an intensification plan, but to understand the relationships between the components of herd management (organization, feeding, animal reserves, production, etc.) in order to maximize benefit from effects induced in short and medium term (response of individuals and the herd, repercussion on feed reserves, etc.) in a given economic environment. The latter very often determines technical choices through opportunities offered by the market.

This understanding should lead to the implementation of new production plans (even though substantially inspired by traditional techniques) based on productivity concepts different from those classically used in northern Europe. These goat farming systems must function and produce in spite of (or taking into account) the harshness of climatic, social and economic conditions.

Figure 6

ANNUAL DISTRIBUTION OF FEED RESOURCES ACCORDING TO ALTITUDE
(BOURBOUZE, 1982)

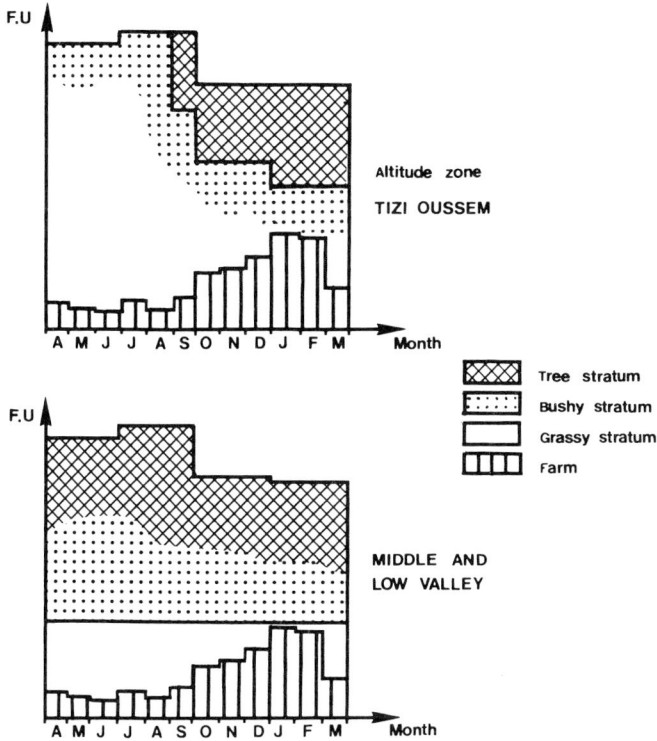

SUMMARY

Goat raising in Mediterranean sylvo-pastoral areas operates in a specific physical, social and economic environment. These specificities and the resulting constraints lead to a herd organization involving the selection of appropriate breeding stock, type of product, and production system. Husbandry practices are dictated by these choices which take into account the food resource level and the market requirements. This enables the farmer to run his operation by adjusting the size of the herd, the reproduction calendar and supplementation scheme. Goat feeding systems differ on the north and south rims of the Mediterranean Sea, as a result of differences in farmers' objectives and the physical and economic environment in which they operate. In general, these systems rely on a diversified range of feed resources, including the sylvo-pastoral resources. The relative contribution and utilization of these resources are determined by farmers' goals and the factors related to land use and herd management.

Key words : Goat, Feeding systems, Feeding balance, Mediterranean area, Forage resource, Sylvo-pastoral systems.

RÉSUMÉ

L'élevage caprin dans les zones sylvo-pastorales méditerranéennes tire son originalité des spécificités de l'environnement physique, social et économique dans lequel il opère. Face à ces spécificités et aux contraintes qu'elles engendrent, l'éleveur s'organise en jouant sur le choix du matériel génétique et du type de production. De surcroît viennent se superposer sur ce choix des conceptions de conduite qui, tenant compte du niveau des ressources alimentaires de l'exploitation et des exigences du marché vont permettre à l'éleveur de piloter son élevage en intervenant sur la gestion des effectifs, le calendrier de reproduction et la pratique de la complémentation. Les systèmes d'alimentation des caprins dans la zone méditerranéenne sont souvent contrastés au Nord et au Sud de la Méditerranée en raison des différences dans les objectifs visés par les éleveurs et de leur environnement physique et économique. Dans tous les cas, ces systèmes s'appuient sur une gamme diversifiée de ressources alimentaires, et notamment les ressources sylvo-pastorales dont l'importance relative et le mode d'utilisation sont dictés par les objectifs de l'éleveur et des contraintes liées à l'utilisation de l'espace et à la conduite du troupeau.

REFERENCES

BAZIN M., 1985. Quelles perspectives pour les agriculteurs montagnards ? (Future projects for agriculture in mountainous areas ?), Thèse de docteur-ingénieur, INA-PG, Paris (FRANCE).

BOURBOUZE A., 1982. L'élevage dans la montagne marocaine. Organisation de l'espace et utilisation des parcours par les éleveurs du Haut Atlas (Animal husbandry in mountainous areas of Marocco. Organization of space and using of rangelands by High Atlas breeders). Thèse de docteur-ingénieur INA-PG, Paris (FRANCE).

BOURBOUZE A., 1986. Adaptation à différents milieux des systèmes de production des paysans du Haut-Atlas (Adaptation of production systems of High Atlas peasants to various environments). Techniques et Culture, 7: 59-94.

BOURBOUZE A. and DONADIEU P., 1987. L'élevage sur parcours en régions méditerranéennes (Animal raising on rangelands in Mediterranean areas). Options méditerranéennes. Série Étude, Nov. 1987. IAM de Montpellier (FRANCE).

GAUTIER L.A., 1983. L'élevage caprin dans le Sud Varois. Analyse de systèmes de production (Goat farming in the South of Var country in France). Analysis of production systems). Mémoire de fin d'étude ENSAM, Montpellier (FRANCE).

GIBON A., 1981. Pratiques d'éleveurs et résultats d'élevage dans les Pyrénées Centrales (Sheep and cattle farming in the Pyrenees mountains : farmers practices and animal production results). Thèse de docteur-ingénieur, INA Paris-Grignon (FRANCE).

GUERIN B. and BELLON S., 1989. Approche et fonctionnement des surfaces fourragères et pastorales de l'exploitation agricole en zone méditerranéenne française, p. 147-156. In: CAPILLON A.: Grassland systems approachs ; some French research proposals. Études et recherches sur les systèmes agraires et le développement, n° 16, Oct. 1989, INRA, Paris (FRANCE).

HUBERT B., GUERIN G., BOURBOUZE A. and PREVOST F., 1985. Problèmes posés par l'utilisation des ressources sylvopastorales par les ovins et les caprins (Problems set by using of sylvo-pastoral resources in sheep and goat farming), p. 131. 9e Journées de la recherche ovine et caprine. INRA-ITOVIC, Paris (FRANCE).

LAGACHERIE M., 1988. Eleveurs caprins utilisateurs de la race Rove (Goat breeders using Rove breed). INRA Ecodéveloppement - ITOVIC, Paris (FRANCE).

MADANI T., 1988. Élevages caprins laitiers du Sud-Est. Analyse de l'organisation des calendriers fourragers en fonction des objectifs de production et des pratiques pastorales (Analysis of the organization of pasture grazing sequences according to production aims and grazing practices). Mémoire DEA, ENSAM, Montpellier (FRANCE).

MENDES L. and NARJISSE H., 1987. Range-Animal Ecology and Agropastoralists in Morocco's Western High Atlas. In: Mac CORKLE C. (Ed.): Plants, Animals and People: Crop / Liverstock systems research in the SR-CRSP Rural Sociology Program, University of Missouri, Columbia (USA).

MENIER D. and GUERIN G., 1987. Les fonctions des surfaces dans l'exploitation agricole en région méditerranéenne (Fonctions of pastoral areas in Mediterranean farms). Document ITOVIC, Paris (FRANCE).

MOREAU F., 1988. Identification d'un projet d'amélioration de l'élevage caprin dans le Haouz de Marrakech (Identification of a improving project of goat farming in Marrakech Haouz, Morocco). Mémoire de fin d'étude: IAV Hassan II, Rabat (MAROC), ENSSAA, Dijon, and IAM, Montpellier (FRANCE).

NAPOLEONE M. and HUBERT B., 1987. Caractérisation et évaluation de systèmes de production caprine utilisateurs de parcours dans le Sud-Est de la France (Characteristics and assessment of goat production systems using rangelands in Southeastern France), p. 72-84. In: FLAMANT J.C. and MORAND-FEHR P. (Eds.): L'évaluation des ovins et des caprins méditerranéens. CEE Rapport EUR 11893 (LUXEMBOURG).

NARJISSE H., BERKAT O., MERZOUGUI L. and JEBARI M., 1983. Analyse du système des productions animales dans la vallée de l'Ourein du Haut-Atlas (Analysis of animal production systems in Ourein Valley in High Atlas). Document ronéotypé. Direction du Développement rural, IAV Hassan II, Rabat (MAROC).

PROST J.A. and VALLERAND F., 1985. Développer l'élevage ovin et caprin dans les zones marginalisées. Quelques enseignements de la situation corse (Development of sheep and goat farming in marginal areas. Some information on Corsican situation), p. 513-538. 9e Journées de la recherche ovine et caprine, INRA-ITOVIC, Paris (FRANCE).

SANTUCCI P. and MAESTRINI O., 1985. Body conditions of dairy goats in extensive systems of production : method of estimation. Ann. Zootech. 34: 473-474 (Abst.).

SANTUCCI P.M. and VALLERAND F., 1988. Le relazioni gregge - territorio nell'allevamento caprine. Agricultura Ricerca, 91: 87-94.

VALLERAND F. and SANTUCCI P.M., 1989. Conduite des animaux et équilibration des systèmes fourragers très saisonnés (Animal management and balances between very seasonnal forage production). In: XANDE A. and ALEXANDRE G. (Eds.): Symp. sur l'alimentation des ruminants en zone tropicale humide. Guadeloupe (French West Indies), INRA (FRANCE).

VAN DEN DRIERSCH T., 1987. Enquête sur l'élevage caprin en Corse (Animal farming survey in Corsica). Mémoire de BTS. INRA, Corte (FRANCE).

VERTES C., 1983. L'utilisation du territoire pastoral par les chèvres corses (Utilization of pastoral areas by Corsican goats). Mémoires de DAA. INRA, Corte (FRANCE).

Chapter 20

BODY CONDITION SCORING OF GOATS
IN EXTENSIVE CONDITIONS

P.M. SANTUCCI, A. BRANCA, M. NAPOLEONE, R. BOUCHE,
G. AUMONT, F. POISOT and G. ALEXANDRE

INTRODUCTION

Animal body condition considered as an indicator of fat reserves (RUSSEL et al., 1969) has mainly been studied in cows and sheep. To our knowledge research in this field on goats, has only been developed in recent years. Only one paper (MORAND-FEHR et al., 1987) provides an analysis of the various methods of assessing body condition score or fat content in goats, whatever the production system used. However, the physiological mechanisms regulating the storage and mobilization of body reserves have mainly been studied in intensively fed dairy goats (SAUVANT et al., 1979; CHILLIARD, 1985...).

In extensive rearing systems, goats which are privileged users of forage areas give rise to specific problems mainly linked to flock management, ecological equilibrium and economic viability of production systems. Moreover, the rearing conditions often complicate the quantification of conventional production parameters (live weight, milk yield, level of feed intake...) and any experimental protocol generally represents a major constraint to the farmer. Therefore, using simple measures such as body condition scoring by palpation, seems to be a means better adapted to these rearing conditions; the practical advantages of this method were studied in several farming situations and in particular in meat producing sheep and dairy cattle. Knowledge of the body condition facilitates the establishment of feeding programmes during the production cycle, contributes to improving the reproductive performance and leads to a better definition of the optimum stage for sale of meat animals.

At the present time, body condition scoring of goats is regularly included in the research programmes of several countries (SPAIN, FRANCE, ITALY, PORTUGAL). Therefore, it seems interesting in this chapter to summarize work carried out in Mediterranean areas, on extensive production systems (CORSICA, SARDINIA, PROVENCE) as well as in GUADELOUPE (WEST INDIES) which are regions with relatively high farming activity.

Two questions concerning body condition of goats in extensive production systems will be treated in this chapter.

The role of body reserves in animals subjected to strong constraints is first indicated in order to show the relationship between forage availability, body condition, and production performance of the flock; then, the work on the definition of a scoring scale is presented and an analysis of the relationship between body condition scoring involving palpation and the volume of sub-cutaneous and internal fat reserves is given.

The second part deals with factors of variation in body condition, as well as their relationship to lipomobilization indicators and the reproductive performances of the animals.

The results of two series of experiments carried out either on a research station (SARDINIA, GUADELOUPE), or on private farms (CORSICA, PROVENCE, SARDINIA) are given in this paper.

ROLE OF BODY RESERVES

In countries with rangeland farming, and hence a seasonality of forage resources, the animal is subjected to a series of alternating phases of feed scarcity or abundance, the duration and intensity of which determine the performance of the animals and the organisation of the production systems (VALLERAND and SANTUCCI, 1987).

In addition to this seasonality of feed resources, there are other cyclic effects due to mesological (for instance excessive water deficiency) or anthropic factors (fire...), which modulate the forage potential and make the forage availability kinetics even more irregular.

These different feeding phases affect the nutritional status of the animals and modify their nutritional balances; thus, the energy balance is subject to the same alternation (positive/negative) of feed supplies although it does not directly depend on these fluctuations since the kinetics of the curve of nutritional requirements are specific for each animal. Thus, according to the date of kidding the goat may for a given period of time show a negative energy balance during the period of feed abundancy and a positive balance during a period of feed restriction.

In extensive production systems the milk yield or the daily mean gain of the goats do not seem to be reliable instantaneous indicators of a moderate feed restriction; variations in the daily milk yield of the whole flock or of each goat are often found, especially during bad or hot weather; these variations are never very large (generally 15 % in Corsican goat flocks). The animal maintains its milk production mainly by using energy derived from fat reserves. This physiological phenomenon, usually called the "buffer effect", which has been well studied in dairy animals especially in early lactation, is one of the production characteristics of the animal and reveals its ability to cope with all kinds of hazards (e.g. environment, health status...).

At a given moment of the production cycle there certainly are limits to this buffer effect of the body reserves, beyond which other physiological mechanisms are disturbed, i.e. immune defense or ovarian activity as shown in high yielding dairy cattle (HANSEN et al., 1983) and in meat producing sheep (GUNN and DONEY, 1975; PARAMIO and FOLCH, 1985; GIBON et al., 1985). Moreover, in extensive goat production the composition of the flock at a given moment (categories of females with different kidding intervals) and a highly variable inter-year fertility (from 75 to 90 %) are indicators of the role played by the relationships between the body reserve status and the reproductive performance (SANTUCCI, 1985).

A METHOD FOR BODY CONDITION SCORING OF GOATS

On the basis of Scottish work (RUSSEL et al., 1969), a scale for assessing the body condition was developed for Corsican goats (SANTUCCI and MAESTRINI, 1985, see appendix).

Principle of the method

Using a scale from 0 to 5, the score given to an animal is obtained by palpation of 2 anatomical regions:

- the sternum;
- the lumbar vertebrae.

The volume of sternal fat (thickness, width, length) and the thickness of surrounding tissue layers covering the chondrosternal joints are assessed by touch.

Sternal fat is the only sub-cutaneous adipose tissue which is generally well distinguished in a form of a "fat pad" palpable over some ten centimeters. The sternal score on a scale of 0 to 5 can rapidly be given because the "fat bread" can be easily detached and examined.

The covering of the lumbar vertebrae is then assessed as in sheep by using several reference marks, e.g. transverse, spinous and articular processes, enabling the establishment of the lumbar score.

In both cases, fingers are used to:

- act like a grip (sternal fat, transverse and articular processes);

- exert a fixed pressure (chondrosternal, transverse and articular processes);

- exert a movement with the tip of the forefinger on the line formed by the spinous processes.

The total score takes into account the palpation of these two sites.

Application and limits

In adult animals, the assessor does not meet any major difficulty in applying this method. During training he/she should practise on groups of animals with large score differences (a minimum of 2 points) in order to determine the corresponding individual body conditions and define the intermediate conditions. Well trained assessors are able to distinguish between different conditions with an accuracy of 1/4 of a point. Each assessor must naturally define his/her own reference marks during palpation depending on the genotype considered, especially for 1/2 point adjustments.

Because of the anatomical particularities of young females (e.g. difficulty in defining sternal fat) and old animals (e.g. sternal callosity) it is more difficult to allocate a body score to animals of eight years and more and to animals under 18 months, according to the previously described principle. A scoring method specific to young growing animals was thus defined for Creole goats (POISOT, 1988).

A monthly control seems sufficient to supply information on variations in animal body condition. However, during key periods like mating or onset of lactation, the control frequency can be set at 21 days.

The reproductibility of the scoring method was evaluated in Creole goats (POISOT, 1988). A heavy experimental device was adopted to test the variability within and between assessors, as well as the influence of certain factors like the effect of the last animal scored. Mean deviations within and between assessors were 0.4 and 0.6, respectively. The repeatability was 0.88. The reproductibility was also high, i.e. 0.80. On the other hand, it was shown that the body condition of the last goat scored did not influence the score given to the following animal.

PHYSIOLOGICAL SIGNIFICANCE OF BODY CONDITION SCORING

Relationship between body condition score and adipose tissue weight

Slaughtering of goats of different genotypes as well as carcass cutting have led to establishment of a certain number of relationships between body condition scores and main adipose tissue weights. Results shown in Table 1 are rather disparate probably because of tissue sampling methods. The low correlations obtained in Corsican goats were certainly due to the heterogeneous groups of animals coming from flocks of different geographical origins.

The weight of sternal, subscapular and lumbar fat tissues was generally quite well correlated with the body condition score, the highest correlations being obtained with sternal fat, which confirms the advantage of using this tissue for determining animal body condition. Correlation coefficients were on average higher with visceral adipose tissues in lactating Sardinian goats (r = 0.91; P < 0.01).

Table 1

CORRELATION COEFFICIENTS BETWEEN VALUES OF BODY CONDITION
AND ADIPOSE TISSUES WEIGHT OF GOATS

Authors	Genotype (number of goats)	Subcutaneous adipose tissues			Internal adipose tissues
		Sternum	Lumbar	Sub-scapular	
POISOT (1988)	Creole (n=18)	0.84***	-	-	0.76***
BRANCA (1987)	Sardinian (n=15)	0.76**	0.68**	0.72**	0.91**
SANTUCCI (1984)	Corsican (n=41)	0.62**	0.59*	-	-

* : P ≤ 0.05; ** : P ≤ 0.01; *** : P ≤ 0.001.

Biochemical indicators of variations in adipose reserves

Other methods and in particular biochemical indicators in the blood and milk were used in flocks on rangeland in order to give supplementary information on the body condition of goats.

Blood parameters

Several authors using animals placed in metabolism crates have shown that blood concentrations of non esterified fatty acids (NEFA) are closely linked to the energy balance (MORAND-FEHR et al., 1977; GIGER and SAUVANT, 1982; CHILLIARD, 1985). This method was applied in Provence on two rangeland farms and on one intensive indoor farming system. The blood levels of NEFA were low even at kidding and the variation range was small in the case of feed allowance adjusted to requirements (intensive indoor farming). In contrast, in rangeland farming the ranges of variation in NEFA and levels of mobilization may in some periods be very large; animals receiving less additional feed when grazing rangeland exhibited much higher NEFA levels at the moment of kidding (Figure 1). Moreover these animals which received an energy supplement one month before kidding, in 1985 (contrary to 1984), exhibited lower levels of lipomobilization. In these flocks, which were great users of rangeland and for which it was difficult to control feeding conditions, NEFA kinetics can be integrated into a series of indicators (e.g. body condition, milk yield) the analysis of which enables the farmers to adjust the flock management systems.

Milk fat

The milk fat content and proportion of fatty acids with 18 carbon atoms are also indicators of lipomobilization (MORAND-FEHR et al., 1977). No significant difference was observed in groups of goats exhibiting large differences in the body condition score (from the 3rd to the 7th month of lactation) (Table 2). This may be explained by the advanced stage of lactation, during which lipomobilization was low; at that stage the influence of feeding factors was probably dominant.

Figure 1

VARIATION OF BLOOD PLASMA NEFA (mEq/l)
CONCENTRATION IN TWO GOAT HERDS OF SOUTH EAST OF FRANCE
(NAPOLEONE, unpubl.)

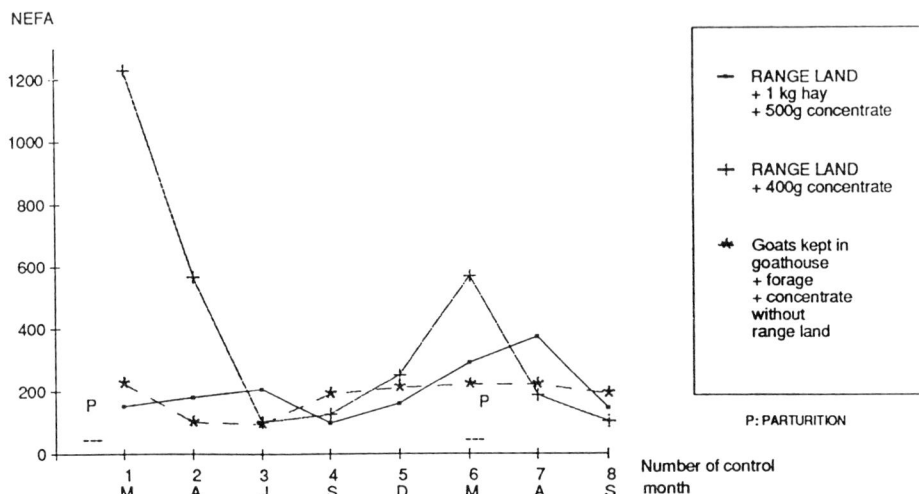

The body condition score at a moment, clearly reflects the reserves, but may well have no direct relationship to the indicators of lipomobilization, which is a dynamic notion. Besides, the relationship between energy balance and body condition is not as close as might be assumed; sometimes the correlation between these two parameters is not significant (MORAND-FEHR, personal communication).

However, the consideration of the body condition dynamics of goats during a relatively short period (40 days) yielded interesting results concerning the indicators of the level of lipomobilization . During the phase of reserve accumulation, the NEFA values, fat content and quantity of fatty acids with 18 carbon atoms in milk were different from those of goats with a reduced body condition (Table 3). This seemed to indicate that variation in the animal body condition certainly was a more interesting parameter to analyse than the condition score itself.

Table 2

CONTENT OF C18 FATTY ACIDS IN MILK FAT IN TWO GROUPS OF CORSICAN GOATS
WITH DIFFERENT LEVELS OF BODY CONDITION
(SANTUCCI, unpubl.)

Month of lactation	Milk yield (kg/day)	FP** (%)	$C_{18:0}$ (%)	$C_{18:0}$ (g)	$C_{18:1}$ (%)	$C_{18:1}$ (g)	TOTAL C_{18} (%)	TOTAL C_{18} (g)
7th Month								
(n=16) BC* = 2.3	1.0	3.9	10.3	4.1	17.4	6.8	32.8	12.9
(n=9) BC = 3.1	1.2	4.0	10.4	4.9	17.5	8.1	33.2	15.5
3rd Month								
(n=12) BC = 1.6	1.4	3.9	10.8	5.7	16.9	8.8	33.3	17.4
(n=18) BC = 2.5	1.2	3.8	11.9	5.14	17.0	7.3	34.4	14.7

(*) BC : Average value of body condition; (**) FP : Fat percentage.

Table 3

VARIATION OF THE VALUE BODY CONDITION AND PARAMETERS OF LIPOMOBILIZATION
(SANTUCCI, unpubl.)

Group (1)	Variation BC (2)	Aver. BC (3)	Milk (kg)	F.P. (%) (4)	$C_{18:0}$ (%) (5)	$C_{18:0}$ (g) (5)	$T_{C:18}$ (%) (6)	$T_{C:18}$ (g) (6)	Plasma (mEq/l) (7)
I	- 0.6	2.3	1.2	4.1	11.6	5.8	34.3	16.9	737
(n=24)		0.6 (+)	0.3	0.6	2.6	1.9	3.2	4.4	197
II	0	2.3	1.2	3.7	10.4	4.4	32.5	13.8	438
(n=21)		0.5 (+)	0.3	0.5	1.8	1.2	4.2	3.9	64
III	+ 0.5	2.5	1.2	3.7	10.7	4.8	33.0	14.6	314
(n=9)		0.6 (+)	0.5	0.7	1.7	2.4	4.1	5.9	65
Level of significance									
I/II	NS	NS	NS	**	*	****	*	**	****
II/III	**	NS	NS	NS	NS	NS	NS	NS	****
I/III	*	NS	NS	*	NS	NS	NS	NS	****

(+) *Standard deviation*
(1) Group of goats at the 7th month of lactation
(2) Variation of body condition (BC) values between Day 0 and Day 40
(3) Average value of body condition (BC) between Day 0 and Day 40.
(4) Fat percentage (FP) of goat milk
(5) Stearic acid: % percent in total fatty acids of goat milk : (g) : quantity secreted per day
(6) Total C18 fatty acids : % percent in total fatty acids of goat milk; (g) : quantity secreted per day
(7) Non esterified fatty acids (NEFA) concentration in blood plasma.
(*) : NS : non significant; P ≤ 0.1; (**) : P ≤ 0.05; (***) : P ≤ 0.01; (****) : P ≤ 0.001.

FACTORS INFLUENCING GOAT BODY CONDITION DURING THE YEAR

Influence of plant cycle

The body condition of goats on rangeland is very sensitive to the availability of forage resources. In periods of underfeeding (january-february) the body condition of animals decreases. The phase of recovery is very significant in spring and at the beginning of summer, a period which corresponds to the highest forage availability during the year (Figure 2).

Figure 2

EVOLUTION OF BODY CONDITION OF GOATS (o) DURING THE CYCLE OF NATIVE FORAGES
(from BRANCA and SANTUCCI, unpubl.)

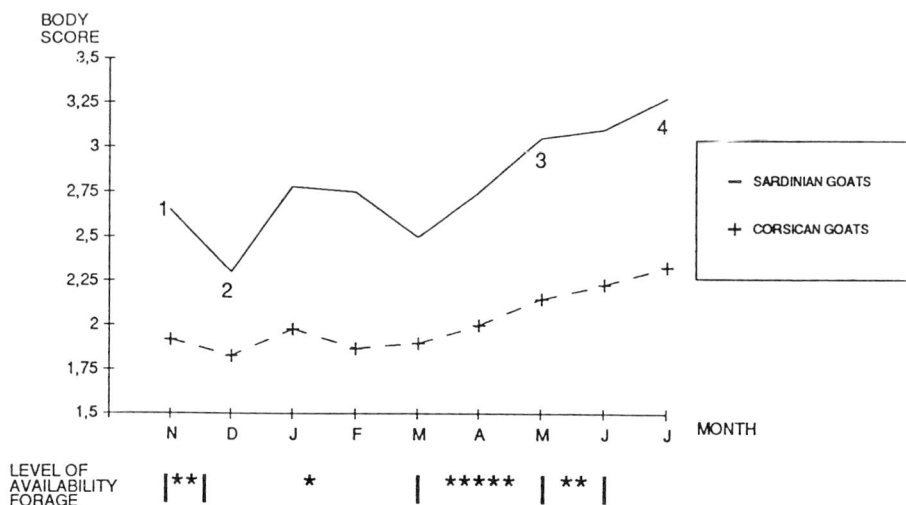

(o) Goats of two herds in Corsica and Sardinia using only maquis.
(1) Parturition; (2) Start of milking; (3) Mating period; (4) Drying off period
* Low availability forage; ** Moderate availability forage; ***** high availability forage.

Influence of the feed supply

The body condition of goats is considerably improved by the supply of concentrates. In farms where goats are given concentrates the flock body condition is more regular and exhibits less variation compared to that of flocks fed only on rangeland (Figure 3a, 3b). Moreover, goats fed with concentrates always exhibit a higher body condition during the production cycle (Figure 3a). This was confirmed by a feeding test conducted at a station with homogeneous groups of Creole goats (Figure 3c).

Figure 3

EFFECT OF CONCENTRATION SUPPLIES ON GOAT BODY CONDITION

3a. Sardinian goats (herd using rangeland)
(BRANCA and CASU, 1987)

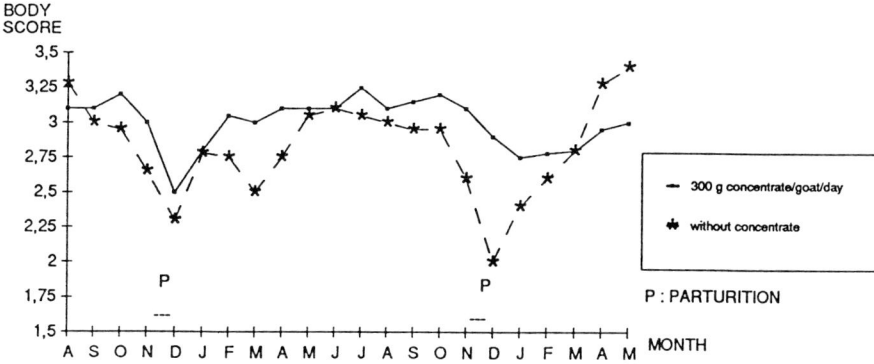

3b. Corsican goats (herd using rangeland)
(SANTUCCI, unpubl.)

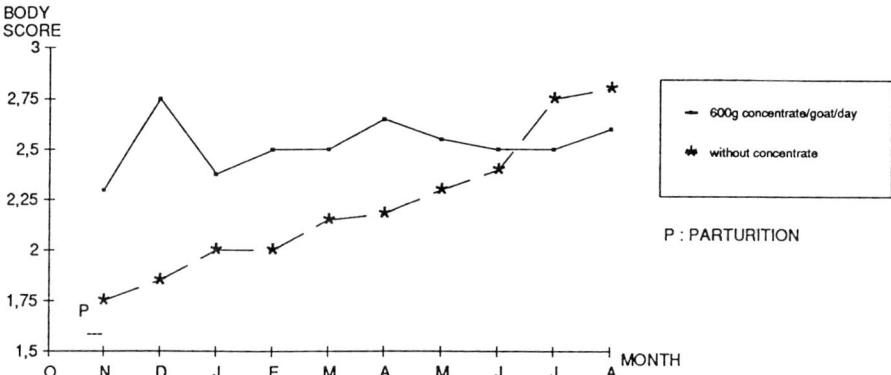

3c. Creole goats (experiment in research station)
(POISOT, 1988)

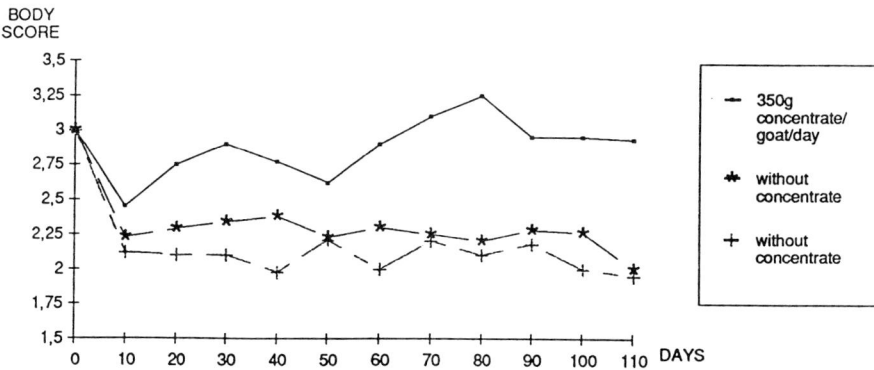

Influence of the physiological stage

During the last phase of gestation goats mobilize intensely their body reserves (Figure 3a); however, during that period forage availability on rangeland is rather favourable (october-november). This mobilization continues during early lactation (1 month) corresponding to the phase of suckling. From the very beginning of milking, the animals recover and their body condition returns to the level reached during the kidding period.

In a Corsican goat flock where kiddings took place two months before the optimum forage availability, significant differences were observed between the body condition of milking and suckling goats (Figure 4). The body condition of suckling goats was always lower than that of milking goats probably because of a larger milk excretion. In the same herd, unproductive goats (females remaining non-pregnant during the year) showed a higher body condition score than that of goats after kidding.

Figure 4

EFFECT OF SUCKLING DURATION ON BODY CONDITION OF CORSICAN GOATS
(SANTUCCI, 1987)

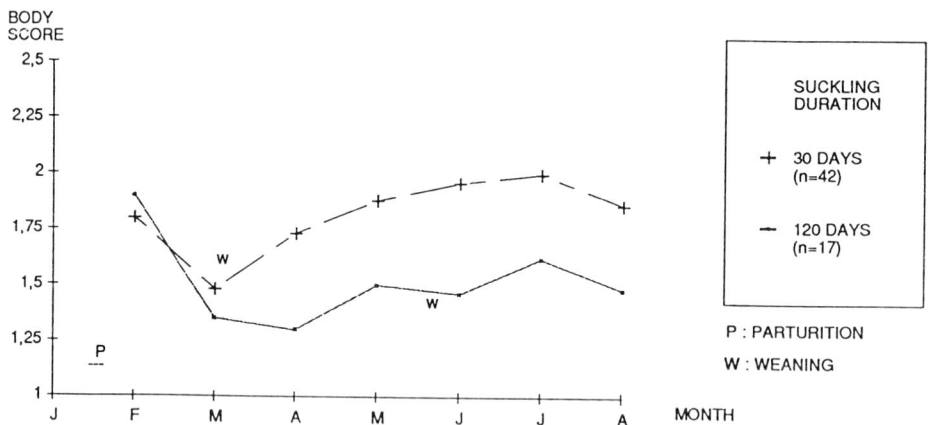

Influence of kidding interval

Two gestations in less than one year considerably influences the body condition and milk yield of goats (Figure 5). Thus, in a flock at the beginning of a new production cycle, goats with a short kidding interval (9 months) exhibited a lower body condition score at kidding than that of goats with a longer kidding interval.

248

Figure 5

BODY CONDITION AND MILK PRODUCTION OF TWO GOAT GROUPS
FROM THE SAME HERD WITH DIFFERENT KIDDING INTERVALS (KI)

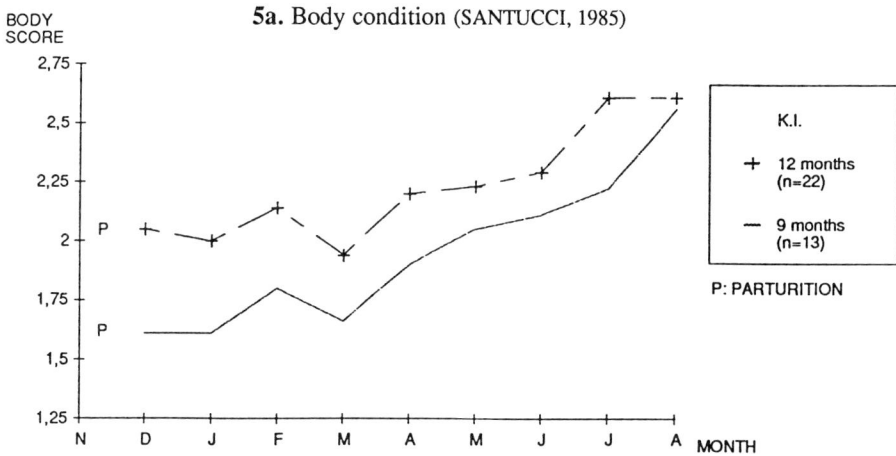

5a. Body condition (SANTUCCI, 1985)

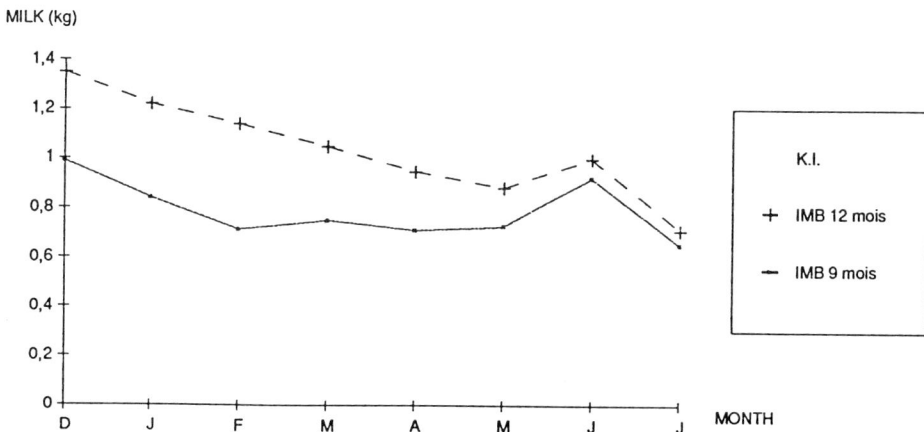

5b. Milk production (SANTUCCI, 1985)

BODY CONDITION SCORE AND PERFORMANCE IN GOATS

Body condition score and milk yield

Considerable differences in body condition have been observed in the same flock between females exhibiting different milk yields. During the whole year, the body condition of high milk-yielding goats was always lower than that of goats with a lower milk yield. Such results have been obtained in Sardinia (BRANCA and CASU, 1987) as well as in Corsica (SANTUCCI, unpubl.); this may lead one to assume that fat reserves of the high yielding goats are restored with more difficulty since the proportion of nutrients transformed into milk is larger than in low-yielding goats.

Condition score and fertility

Observations in Corsican and Sardinian goats showed that the fertility of females with poor body condition at the moment of mating was low (Tables 4 and 5). Table 5 indicates that the percentage of goats with reproductive disorders was twice as high when the body condition at mating was low. Results of this kind are in keeping with those obtained in sheep (GUNN et al., 1972; FOLCH et al., 1983; PARAMIO and FOLCH, 1985).

Table 4

EFFECT OF BODY CONDITION ON FERTILITY, PROLIFICACY AND FECUNDITY
OF SARDINIAN GOATS (in %)
(BRANCA and CASU, 1987)

Values of body condition		Number of goats	Fertility	Prolifilacy	Fecundity
Low	(2.0)	37	83.8	1.64	1.38
Moderate	(3.0)	34	88.2	1.73	1.53
High	(4.3)	16	87.5	1.64	1.44

Table 5

RELATION BETWEEN BODY CONDITION AND REPRODUCTIVE PERFORMANCES OF
CORSICAN GOATS
(SANTUCCI, unpubl.)

Groups	Average BC	F 21 (%)	F 42 (%)	Success at kidding (%)		
				Non pregnant goats	Goats with delay	Normal goats
	(1)	(2)	(3)	(4)	(5)	(6)
I: (n=30) BC < 2.2	1.75 0.26*	86.7	96.7	10	10	80
II: (n=32) BC > 2.2	2.61 0.33*	90.6	100	3	6	91

(*) Standard deviation.
Group I : Goats with a value of body condition lower than the average value of the herd.
Group II : Goats with a value of body condition higher than the average value of the herd.
(1) Average BC : Average of two values of body condition around mating
(2) F 21 : Fertility at 21 days estimated by pregnancy diagnostic
(3) F 42 : Fertility at 42 days estimated by pregnancy diagnostic
(4) Percentage of non pregnant goats
(5) Percentage of goats with a kidding interval higher than 13 months
(6) Percentage of goats with a kidding interval of 12 months.

However, it is difficult to interpret the data in Tables 4 and 5 because all the physiological factors involved in reproduction are not controlled and the results are not repeatable from one year to another; it is often noted in herds that some goats exhibiting a very poor score at mating (approximately 1.5) do not have fertility problems. This may probably be explained by the role played by internal adipose reserves at that moment.

Thus, based on the two condition scores before and after mating (Day 0 and Day 40, length of two sexual cycles) three body condition profiles were defined corresponding to a decrease (profile I), an increase (III) or a stability (II) of the condition score. In Corsican goats without any abortive diseases and whose ovarian activity was controlled by a plasma progesterone assay, a relationship was found between the body condition profile, fertility and length of kidding interval. Fertility at 21 days was higher in animals in the phase of body reserve accumulation; it seemed that females with a rising body condition profile have shorter kidding intervals whereas the intervals of the other females (falling and stable profiles) exceeded 13 months and thus exhibited out-of-season kiddings (Table 6).

Table 6

RELATION BETWEEN EVOLUTION OF BODY CONDITION
AT MATING BLOOD PLASMA NEFA CONTENTS AND PERFORMANCES OF CORSICAN GOATS
(SANTUCCI, unpubl.)

Number of goats (n=62)		EVOLUTION OF BODY CONDITION (1)		
		I (n=14)	II(n=25)	III(n=23)
NEFA (2) (mEq/l)		204	182	163
		145 (+)	99	116
Average BC (3)		2.23	2.24	2.10
		0.58	0.52	0.49
F 21 (4) (%)		78.6	96.0	86.9
F 42 (5) (%)		92.9	100.0	100.0
Success at mating (%)	0 (6)	7	12	0
	1 (6)	7	4	13
Mating (%)	2 (6)	86	84	87
Kidding		13.0	13.4	12.0
Interval (month)		3.70	4.25	2.22

(+) *Standard deviation*
(1) Variation of body condition score during mating period (40 days). I: Decreasing value (- 0.6 point); II: Constant value; III: Increasing value (0.5 point).
(2) Non esterified fatty acids.
(3) Average of the two values of body condition before and after mating.
(4) Fertility at 21 days after mating.
(5) Fertility at 42 days after mating.
(6) 0: non pregnant goats; 1: kidding interval higher than 13 months; 2: kidding interval of 12 months.

CONCLUSION

The body condition of goats in extensive production systems is a measurable parameter. The present method seems satisfactory because of its physiological significance and its reproductibility.

Scoring the amount of subcutaneous adipose tissue (e.g. thickness, covering) is possible on a scale involving 1/4 of a point decisions; the training required for use of method is easy and several works (EVANS, 1978; POISOT, 1988) show the repeatable character of the technique applied.

Each body condition score has a precise physiological meaning linked to the quantity of subcutaneous or visceral adipose tissue. The body condition score is closely correlated with carcass fat contents ($r = 0.91$) in lactating Sardinian goats (BRANCA and CASU, 1987). Relationships between the score and lipomobilization indicators are less clear.

The scoring method seems to be a precious tool for monitoring animal performance in the flock; it is noted that the body conditions of goats follow a law which is generally repeated at each production cycle, i.e. the animals lose 0.5 to 0.75 point after kidding, gain 1 to 1.5 points during spring, lose another 0.25 point at the end of the cycle and then show a loss of 0.5 to 1 point at the end of gestation. The monitoring of body condition shows between - and within - year fluctuations of environmental forage resources and may be used as a reference mark for defining feeding conditions of animals using rangeland. Moreover, this allows a partial explanation of the animals' performance. For this reason, the body condition profile seems to be an important indicator of the dynamics of external body reserves and seems to explain the relationships between the state of the reserves and better fertility, than the condition score itself.

At the present stage of research, it is difficult to make recommendations for critical body condition thresholds. However, it seems important to consider the kinetics of goat body condition during reproduction because a decrease in the condition score may delay ovulation and subsequent conception.

SUMMARY

A method of body condition scoring of goats in extensive production systems was proposed in this paper. The purpose was to develop a diagnostic tool, easy to use and reliable enough to explain the production parameters of the animals.

Slaughtering of goats of different genotypes (Corsican, Sardinian, Creole) showed that sternal fat was correlated with the body condition score and the total quantity of visceral adipose tissues.

During the production cycle, a clear negative relationship was found between body condition and milk yield. Using the body condition profile established around the period of mating it was possible to make a partial interpretation of the causes of subfertility observed in extensive rearing systems.

Keywords : Goat, Body condition, Scoring method, Extensive production systems, Body reserves, Reproductive performance.

RÉSUMÉ

Une méthode pour l'évaluation de l'état corporel des chèvres en élevage extensif est proposée. L'objectif recherché est la mise au point d'un outil de diagnostic, simple d'emploi et suffisamment fiable, pour expliquer les performances zootechniques des animaux.

L'abattage de chèvres de génotypes différents (Corse, Sarde, Créole) fait apparaître que le gras sternal est corrélé à la note d'état corporel et à la quantité totale de tissus adipeux viscéraux.

On constate, au cours du cycle de production, une liaison négative très nette entre l'état corporel et la quantité de lait produite. Le profil d'état corporel autour dela période de lutte permet d'avancer une interprétation partielle des origines de la subfertilité observée dans les troupeaux extensifs.

Acknowledgments

The authors wish to acknowledge F. VALLERAND, P. MORAND-FEHR and Elisabeth BERNARD for their co-operation during the preparation of this article.

Appendix

Score 0 - Animal in skeleton conditions on the point of death: no subcutaneous tissue can be seen.

Score 1

Aspect of the animal : Emaciated animal: the backbone is highly visible and forms a continuous ridge, the flank is hollow. The ribs can be seen and the croup stands out.

Sternum region : The sternal fat can easily be seized with the fingers; it is very flat and slightly hard. It moves with the hand's movement from the right to the left (a sternal fat weight of approximately 20 g at slaughter). The chondrosternal joints and the beginning of the ribs can be felt by a slight finger touch.

Lumbar region : The lumbar vertebrae are seized with the whole hand; it is a rough and prominent mass. No muscle or fat thickness is noted between the skin and the bones. The fingers can easily reach and seize the transverse processes which are clearly outlined. The prominent articular processes can be felt perfectly well with the fingertips.

Score 2

Aspect of the animal : Slightly raw-boned: the backbone is still visible with a continuous ridge. Protuberant croup.

Sternum region : The sternal fat can still be seized with the fingers but it is thicker (1 to 2 cm). It can easily be detached throughout its whole length (average weight at slaughter: 50 g) with the fingertips, a small tissue layer can be seen between the skin and the chondrosternal joints.

Lumbar region : The vertebrae can still be seized with the whole hand, but a tissue mass appears under and over the transverse processes; the outlines of the transverse processes are difficult to perceive with the fingertips. Spinous processes are less prominent and the articular processes can be felt by slight pressures.

Score 3

Aspect of the animal : The backbone is not prominent; the croup is well covered.

Sternum region : The sternal fat can be well distinguished; it is thick and little mobile. It is difficult to seize it because the surrounding mass of tissue (fat and muscles) is not very large (weight of sternal fat at slaughter: 80 g). A thorough palpation is needed to perceive the chondrosternal joints.

Lumbar region : The tissue layer covering the lumbar vertebrae is very thick but it can be seized with 3 fingers. When running a finger over the spinous processes a slight hollow is felt. The articular processes and the outlines of the transverse processes are no longer perceived.

Score 4

Aspect of the animal : No particular signs.

Sternum region : It is difficult to seize the sternal fat because of its thickness. It is almost intermingled with the mass of fat and muscle covering the chondrosternal joints and the ribs.

Lumbar region : It is difficult to put the fingers under the transverse processes, which can no longer be seized: they are wrapped in a thick layer of tissue. The spinous processes can no longer be felt with the fingers. They form a continuous line.

Score 5

Aspect of the animal : No particular signs.

Sternum region : The sternal fat cannot be identified. It cannot be seized. A thick mass of tissue covering the ribs and sternum evenly is felt between the fingers.

Lumbar region : The thickness of the tissue mass is so large that reference marks on the transverse and spinous processes are lost. Moreover, it is impossible to put the fingers under the transverse processes.

REFERENCES

BRANCA A. and CASU S., 1987. Variation of body condition score during a year and its relationship with body reserves in Sarda goats, p. 221-236. In: FLAMANT J.C. and MORAND-FEHR P. (Eds.): L'évaluation des ovins et des caprins méditerranéens. Symposium Philœkios, Fonte Boa (PORTUGAL), sept. 23-25, 1987. Rapport EUR 11893, OPOCE (LUXEMBOURG).

CHILLIARD Y., 1985. Métabolisme du tissu adipeux lipogenèse mammaire et activités lipoprotéine-lipasique chez la chèvre au cours du cycle gestion-lactation (Adipose tissue metabolism, mammary lipogenesis and lipoprotein lipase activities in goats during gestation and lactation). Thèse de doctorat d'État, Université Paris 6 (FRANCE).

EVANS D.G., 1978. The interpretation and analysis of subjective body condition scores. Anim. Prod., 26: 119-125.

GIBON A., DEDIEU B. and THERIEZ M., 1985. Les réserves corporelles des brebis. Stockage, mobilisation et rôle dans les élevages en milieu difficile (Storage, mobilisation and part of ewes body reserves in difficult environment), p. 178-212. 10e journée de la Recherche ovine et caprine, INRA-ITOVIC, Paris (FRANCE).

GUNN R.G., DONEY J.M. and RUSSEL A.J.F., 1972. Embryo mortality in Scottish Blackface ewes as influenced by body condition at mating and by post-mating nutrition. J. Agri. Sci. (Camb.), 79: 19-25.

GUNN R.G. and DONEY J.M., 1975. The interaction of nutrition and body condition at mating on ovulation rate and early embryo mortality in scottish Blackface ewes. J. Agri. Sci. (Camb.), 85: 464-470.

HANSEN L.B., FREEMAN A.E. and BERGER P.J., 1983. Association of heifer fertility with cow fertility and yield in cattle. J. Agri. Sci. (Camb.), 66: 306-314.

MORAND-FEHR P., SAUVANT D., BAS P. and ROUZEAU A., 1977. Paramètres caractérisant l'état nutritionnel de la chèvre (Parameters of nutritional status in goats), p. 195-203. Symposium on Goat Breeding in Mediterranean Countries, Oct. 3-7, 1977, Malaga, Grenada, Murcia (SPAIN).

MORAND-FEHR P., BRANCA A., SANTUCCI P.M. and NAPOLEONE M., 1987. Méthodes d'estimation de l'état corporel des chèvres reproductrices (Methods for estimating goat body conditions), p. 202-220. In: FLAMANT J.C. and MORAND-FEHR P. (Eds.): L'évaluation des ovins et des caprins méditerranéens. Symposium Philœkios, Sept. 23-25, 1987, Fonte Boa (PORTUGAL). Rapport EUR 11893, OPOCE (LUXEMBOURG).

NAPOLEONE M., 1987. Les acides gras non estérifiés: Un "outil" de prédiction du niveau alimentaire des caprins (Non esterified fatty acids: a tool for predicting the level of goat intake). Document interne INRA, Avignon (FRANCE).

PARAMIO M.T. and FOLCH J., 1985. Puntuacion de la condicion corporal en la oveja rasa Aragonesa y su relacion con las reservas energeticas y los parametos reproductivos. Informacion Tecnica Economica Agraria, 58: 29-44.

POISOT F., 1988. Méthodes d'appréciation des réserves corporelles des caprins créoles de Guadeloupe (Methods for estimating body reserves of Creole goats in French West Indies). Mémoire de fin d'étude ENITA, Clermont-Ferrand (FRANCE).

RUSSEL A.F.J., DONEY J.M. and GUNN R.G., 1969. Subjective assessment of body fat in live sheep. J. Agri. Sci., (Camb.), 72: 451-454.

SANTUCCI P. and MAESTRINI O., 1985. Body conditions of dairy goats in extensive systems of production: method of estimation. Ann. Zootech., 34: 473-474 (Abst.).

SANTUCCI P., 1985. L'élevage caprin extensif: indicateurs de la conduite du troupeau (Goat farming in extensive conditions: indicators for head management). 36th EAAP Annual Meeting, Sept. 30 - Oct. 3, 1985, Kallitea (GREECE).

SAUVANT D., CHILLIARD Y. and MORAND-FEHR P., 1979. Goat adipose tissue mobilisation and milk production level. Ann. Rech. Vet., 10: 404-407.

VALLERAND F. and SANTUCCI P.M., 1987. Conduite des animaux et équilibration des systèmes fourragers très saisonnés (Animal management and gestion of forage production during a year), p. 259-296. Proc. Symp. on the Feeding of the Ruminants in tropical zones. Pointe-à-Pitre, 1987. Série : Les Colloques de l'INRA, Paris (FRANCE).

Part 4

Feeding of young goats

Chapter   21

MILK FEEDING SYSTEMS OF YOUNG GOATS

Ø. HAVREVOLL, M. HADJIPANAYIOTOU, M.R. SANZ SAMPELAYO, Z. NITSAN and P. SCHMIDELY

INTRODUCTION

Milk is an essential feed for the newborn kid. Kids can be reared naturally or artificially. The milk period may last only 3-4 weeks or up to 5-6 months (MORAND-FEHR, 1981; MORAND-FEHR et al., 1982). Natural rearing is most common in meat producing herds (GALL, 1981a, 1981c). Early weaning, partly suckling for some hours each day and artificial rearing increase the saleable milk in dairy herds (MORAND-FEHR, 1981, 1987b; ECONOMIDES, 1982; HADJIPANAYIOTOU, 1984, 1986).

The success in artificial rearing depends on many factors such as breed of animal, quality and access to feeds and pastures, environment and last but not least the person who takes care of the animals. The goal in any rearing system is to ensure satisfactory health and performance of the animal at a minimum cost.

This chapter deals with liquid feeding and feeding systems for young goat kids. The following experiences are mainly based on research in connection to FAO sub-network on Goat Nutrition conducted during the period 1981-1988 in Western Europe and in Mediterranean countries (MORAND-FEHR, 1985, 1987a). Experiences from rearing of calves or lambs are also useful in planning of artificial rearing of kids (OWEN, 1974; BURGKART and BAUER, 1974; PENNING, 1979; DAENICKE, 1980; ROY, 1980; STOBO, 1981; GRØNDALEN et al., 1984).

FEEDING AND PHYSIOLOGICAL ASPECTS OF YOUNG RUMINANTS

In any rearing system it is necessary to take into consideration the physiology of the young animal, particularly the development and the capacity of the digestive tract (RADOSTITS and BELL, 1970; CHURCH, 1976; GIHAD and MORAD, 1977; ROY, 1980; THIVEND et al., 1980). The milk feed consumed by the kid will normally pass through into œsophageal groove directly into the abomasum. The œsophageal groove reflex is stimulated by suckling milk and even by "feed signals" from the stockman. This physiological phenomenon is essential for digestion of milk feed. If milk passes into the rumen of newborn animal, digestive disorders may occur. The œsophageal groove reflex continues to function for several months if milk feeding continues (CHURCH, 1976; THIVEND et al., 1980; ØRSKOV, 1987).

The digestive tract of newborn mammals is well adapted to digest milk. Chymosin, pepsin and HCl-acid in the abomasum are responsible for milk clotting which separate milk into whey and clots of casein, calcium and fat. Thus, noncasein protein in the liquid diet for young kids will not clot in the abomasum. The levels of chymosin and lactase decrease with increasing age in calves and lambs, but may maintain for a longer period if a high level of milk is fed in the ration (RADOSTITS and BELL, 1970; CHURCH, 1976; ROY, 1980; THIVEND, 1980; GUILLOTEAU et al., 1983).

Little information is available about enzymes in the digestive tract and the intensity of lipase, glycolytic, amylolytic and proteolytic activities as influenced by age and diet composition in goats (MORAND-FEHR et al., 1982). The dry matter digestibility usually increases during the first month of age and mainly with milk replacers. This increase with age is greater with highly saturated long-chain fatty acids and poorly digestible proteins in calves (THIVEND et al., 1980).

In Spain, SANZ SAMPELAYO et al. (1987) carried out four trials with goats of Granadia breed to examine the effects of milk type (milk and milk replacer), intake level (1.86 and 2.48 times maintenance) and slaughter age (15, 30, 45 and 60 days) on the digestive tract. The kids were weaned at 31 and 45 days of age. No differences in growth and feed utilization were observed between kids on different milk type. Effects of intake level and age are shown in Table 1.

Table 1

EFFECT OF INTAKE LEVEL AND KID AGE ON WEIGHT OF TOTAL STOMACH COMPARTMENT AND PROPORTIONAL WEIGHTS OF RETICULORUMEN, ABOMASUM AND OMASUM

(SANZ SAMPELAYO et al., 1987)

	Intake level		Age (days)			
	1.86 M*	2.48 M*	15	30	45	60
Weight of stomach compartments, g	125	133	60^d	70^d	103^d	284^c
Proportional weights of :						
- Reticulorumen, (g/kg)**	583^a	547^b	$384^{d,b}$	$474^{d,a}$	636^c	760^e
- Abomasum, (g/kg)**	372^b	404^a	$564^{d,b}$	$470^{d,a}$	317^c	199^e
- Omasum, (g/kg)**	46	49	47^b	$56^{c,a}$	47^b	41^d

(*) M = Metabolisable energy for maintenance.

(**) g/kg total stomach compartment. Means with one common superscript are not statistically significant. a, b: $P \leq 0.05$; c, d, e: $P \leq 0.01$.

Postnatal development of the ruminants' stomach is related to live weight, age and diet. Reticulorumen increases with age, but a high proportion of liquid feed delays the growth in terms of tissue thickness, mobility of the rumen wall and papillary development (SKJEVDAL, 1974; CHURCH, 1976; GIHAD and MORAD, 1977; STOBO, 1981; SANZ SAMPELAYO et al., 1987).

COLOSTRUM FEEDING AND IMMUNITY

Colostrum feeding is a main factor in limiting kid losses (MORAND-FEHR, 1987b). Colostrum is rich in fat which is a good energy source and enables the kids to improve their thermoregulation and to adapt themselves easily to environmental conditions. Colostrum is also rich in minerals and vitamins essential for nutrition of the young animal (ROY, 1980; GALL, 1981b; RENNER, 1983).

Colostrum has a high content of immunoglobulins or antibodies which play a major role in protecting newborn animals against bacterial and viral agents present in its new environment (SHELDRAKE and HUSBAND, 1985). The kids have no circulating antibodies at birth. Consequently the immunity is to be derived passively from absorption of immunoglobulins in the colostrum and later by an active production of antibodies against pathogens to which the animal has been exposed (HAVREVOLL et al., 1988; Figure 1).

Figure 1

SERUM LEVEL OF IG G IN LAMBS FED COLOSTRUM FROM DIFFERENT SOURCES
DURING THE FIRST DAY POSTPARTUM
(HAVREVOLL et al., 1988)

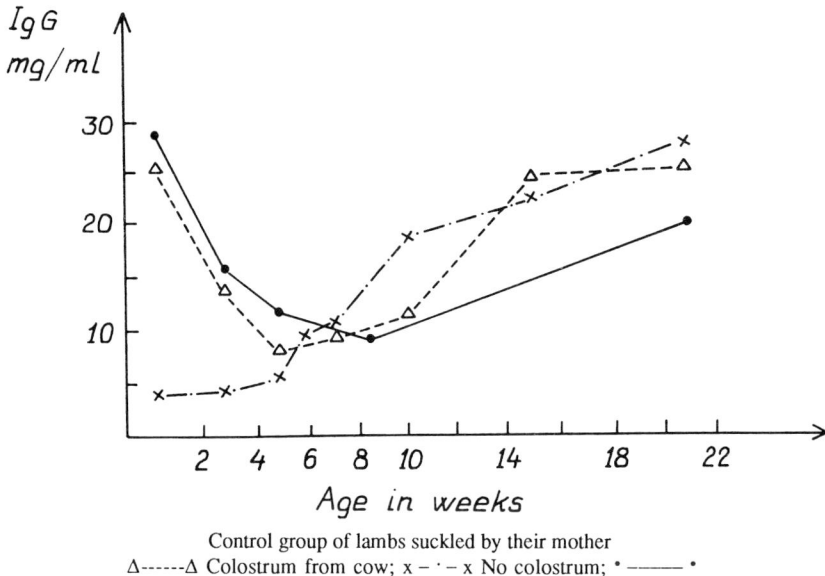

Control group of lambs suckled by their mother
Δ------Δ Colostrum from cow; x – · – x No colostrum; •——— •

The effectiveness of the transfer of immunoglobulins from colostrum to kids plasma, is a function of antibody concentration in the milk, level of colostral intake by the kid and time of consumption in relation to birth (ROY, 1980; NORRMAN, 1987). Kids may loose the ability to absorb immunoglobulins from colostrum 20-28 hours after birth, but there is an evidence that the ability persists longer in starved kids. The concentration of immunoglobulins in colostrum depends on breed, season, dam age, litter size and number of sucklings (HALLIDAY, 1968; CABELLO and LEVIEUX, 1981; NORRMAN, 1987). When the dam dies at parturition, has mastitis or no colostrum it is possible to give its kids colostrum from other goats. Colostrum can be stored in a deep freezer for up to two years and remain immunologically effective (MORAND-FEHR, 1987b). Cow colostrum is also efficient for lambs and kids (LEVIEUX, 1984, cited by MORAND-FEHR, 1987b; EALES et al., 1982; AL-JAWAD and LEES, 1985; HAVREVOLL et al., 1988). However, severe anemia in 2-3 week old lambs fed cow's colostrum has occasionally been observed (ØVERÅS et al., 1988). For kids not receiving colostrum from their dam it is advisable to give 100 ml colostrum per kg live weight in 2 or 3 feedings, fresh or stored in a freezer (MORAND-FEHR, 1987b). Immunoglobulins are thermosensitive. Therefore, during thawing, the temperature of colostrum should not be raised above 50°C.

MILK FEEDS FOR KIDS

Different types of milk in kid diet

It is essential to feed newborn kids colostrum and milk. The older the kids, the greater the possibilities to choose feeds according to price or availability. The content of nutrients in different feeds are shown in Table 2.

Table 2

THE AVERAGE CONTENT OF NUTRIENTS IN DIFFERENT MILK FEEDS

(HAVREVOLL et al., 1987)

	Goat milk	Cow milk	Skim milk	Whey	Permeat
Dry matter (DM), %	11.8	12.7	8.9	5.0	4.6
Protein (% DM)	26.8	26.6	36.2	13.2	4.5
Fat (% DM)	29.5	30.0	0.6	0.7	0.0
Lactose (% DM)	36.7	37.8	54.2	77.5	86.5
Minerals (% DM)	7.0	5.6	9.0	8.6	9.0
Feed Unit per kg feed	0.20	0.23	0.11	0.06	0.04

Colostrum should be offered to the youngest kids either fresh, frozen or acidified (HAVREVOLL et al., 1987; MORAND FEHR, 1987b). Young kids should get whole milk from goat or cow or milk replacer until at least 4-5 weeks of age. Skim milk in the rations for 6-10 week old kids may be favourable for their growth. However, to avoid digestive disorders, an restricted amount of bacteriologically acidified skim milk should be used (SKJEVDAL, 1974; HAVREVOLL et al., 1985). Whey can be fed to kids from two months of age with 1-2 litres per day and per animal (HAVREVOLL et al., 1987). Permeat is made by ultra filtration of whey or skim milk and contains mainly lactose and minerals (Table 2) (SCHINGOETHE, 1976; WAYNE MODLER, 1986). No information is available about permeat in rations for young goats. However, it is advisable to offer less permeat than whey due to a higher content of lactose.

Milk replacer

A milk replacer is normally used in order to save goat milk for sale and to reduce milk feeding cost. The performance of kids fed milk substitute may be similar to kids fed milk. The result depends on the composition and quality of the milk replacer, level of intake, feeding system and management (RADOSTITS and BELL, 1970; SKJEVDAL, 1974; MOWLEM, 1979, 1981, 1982; OWEN and DE PAIVA, 1980, 1982; MORAND-FEHR et al., 1982; NITSAN et al., 1985, 1987; SANZ SAMPELAYO et al., 1987, 1988b).

The fat content of the milk replacer may vary from 15 to 25 % and protein from 20 to 25 % of dry matter (MORAND-FEHR, 1981; MORAND-FEHR et al., 1982). High energy content in the milk replacer requires high quality fat substitutes. We also have to consider the relationship between energy and protein content in the milk feeds. TEH and ESCOBAR (1987) concluded that a milk replacer based on skim milk with no less than

24 % crude protein and 20 % crude fat was adequate and suitable for optimum performance of goat kids. Veal-type kids require more energy in the milk diet compared to goat kids reared for herd replacement (SANZ SAMPELAYO et al., 1988a).

The apparent digestibility of milk substitutes increases with age of the animal particularly during the first month of life (RADOSTITS and BELL, 1970; ROY, 1980; THIVEND et al., 1980). Protein from whey, soybean or fish meal results in lower weight gain of preruminants compared to skim milk based milk replacers (ROY, 1980; MOWLEM, 1979, 1981; SANZ SAMPELAYO et al., 1987). A high level of lactose in milk replacers may cause fermentative diarrhoea (RADOSTITS and BELL, 1970; ROY, 1980; MORAND-FEHR et al., 1982; HAVREVOLL et al., 1985). Heat-treated starch can also be used in milk replacers. However, raw corn starch will lower live weight gain and feed utilization in kids compared to lactose-based milk replacers (NITSAN et al., 1985).

Dry matter content in liquid milk replacers may vary between 12 and 24 % (FEHR and SAUVANT, 1974; SKJEVDAL, 1974; ROBSTAD and SKRØVSETH, 1979; MOWLEM, 1981, 1982). For young kids the recommended level of dry matter is 14-18 %. Below this concentration the young kids can hardly compensate by consuming more liquid (MORAND-FEHR, 1982; HAVREVOLL et al., 1987).

Several experiments have been carried out in order to compare goat milk, cow milk and milk replacer for growing kids. OPSTVEDT (1968) obtained similar weight gains in kids of Norwegian Dairy Goat with goat milk, cow milk and milk replacer. However, due to a lower fat content of the milk replacer and a slightly lower digestibility, the amount of milk replacer offered had to exceed 50 % of that of the goat milk. Experiments conducted by FEHR (1971) showed no significant difference in growth rates between kids fed goat milk or milk replacer. In the case of early weaning, however, goat milk reduced the post-weaning growth check (SKJEVDAL, 1974). MOWLEM (1979, 1981) obtained better growth rates in a group of kids receiving milk replacer compared to kids fed goat milk. The high concentration of the milk replacer used (18 ans 20 % DM) can easily explain this result. SANZ SAMPELAYO et al. (1988b) showed that the intake of goat milk or milk replacer was the most influential factor on the energy retained in the body as protein or fat. The daily ME requirements for maintenance and growth of kids on goat milk or milk replacer were 444 and 427 kJ/kg $W^{.75}$, respectively and 16.12 and 17.91 kJ/g empty body weight gain (see Chapter 7).

Acidified milk feeds

Milk can be acidified bacteriologically by adding 1-2 % starter culture or 0.2-0.5 % organic acid (FOLEY and OTTERBY, 1978; DAENICKE, 1980; HAVREVOLL, 1984). For young animals fermented milk is favourable in order to reduce digestive disorders. During fermentation the increasing number of bacteria will partly split the nutritive components in the milk and produce lactic acid. The pH will decrease from 6.4 to below 5.0. The milk may coagulate (at pH 4.6) with small-size curd formation which is favourable to digestion. Milk replacers based on whey or whey protein or other casein substitute proteins will not coagulate by fermentation (RENNER, 1983; HAVREVOLL, 1984).

Chemically acidified milk replacers are now commercially available in many countries. This type of milk replacer is recommended in automatic feeding systems based on simple teat-feeders. Acidified milk can be stored for a maximum of three weeks at a temperature of 10-12°C or below before feeding to young kids. The palatability of warm, fresh milk is higher compared to acidified and cold milk (HAVREVOLL et al., 1985).

In Norway, good results have been obtained by souring of the milk replacer after mixing with water. Fermented milk replacers may decrease the frequency of abomasal bloat and other gastrointestinal disorders (HAVREVOLL et al., 1985, 1987). For practical purposes, fermentation of milk replacers can be obtained by mixing 200 g powder per litre of water (40-50°C) and by adding about 2 % cultured milk as a starter culture. The milk should be allowed to ferment for 1-2 days at 20°C before feeding.

NATURAL SUCKLING VERSUS ARTIFICIAL REARING

The young kid is entirely dependent on milk for its early growth when the rumen has not been fully developed. In meat producing animals weaning may be delayed until the age of six months, whereas in dairy or dual purpose breeds weaning takes place earlier so that commercial milk yield is maximized (MORAND-FEHR et al., 1982; ECONOMIDES, 1982; HADJIPANAYIOTOU, 1984, 1986; (see Chapter 24)).

The increase in consumption of commercial milk is achieved by decreasing the amount of milk suckled by the kids through reduction of the weaning age and/or the suckling time. However, such an increase in the commercial milk yield should not adversely affect the post weaning growth of kids. LOUCA et al. (1975) obtained no significant differences between kids reared naturally by their dams or artificially to 70 days.

Experiments in Cyprus showed that the total milk yield of Damascus goats was not affected by the length of suckling period, but the commercial milk yield increased with either early weaning or partial suckling of kids. However, the growth rate of kids, either weaned at 35 days of age or partially suckled from 20 to 70 days of age, was inferior compared to kids suckled *ad libitum* until the age of 70 days (ECONOMIDES, 1982). Probably these results from Damascus goats cannot be applied to all breeds. HADJIPANAYIOTOU (1986) showed that single suckling kids gained more weight until weaning than twin suckling kids or kids reared artificially.

Natural rearing is common in extensive production systems particularly when the goats are mainly kept for meat production. However, natural rearing with early weaning may also be an advisable method to raise young animals in highly intensive dairy herds. The quota-system (low price for milk delivered to the dairy over a fixed quota) will probably increase the interest for this feeding system. Natural rearing will also save labour compared to artificial rearing. Farmers, however, should examine the udder of their dairy goats and remove any extra milk once a day in the suckling period.

MILK FEEDING : BUCKET VERSUS TEAT

When kids are separated from their dams, the milk feed may be offered by using various feeding devices: group feeding in a groove, individual feeding in small pails, feeding by teat-bottle feeder or by an automatic feeder. All these devices give satisfactory results. However, according to a French observation, swelling of the abomasum is less frequent when the number or meals is high (teat- or self-feeder) compared to one or two meals per day (bucket feeding) (MORAND-FEHR et al., 1982). When using an automatic feeder, milk may be available at any time. Under such conditions, milk intake is higher compared to bucket feeding with 2 or 3 meals a day offered *ad libitum*. The growth rate is generally higher, but the feed efficiency is reduced by 8-12 % (MORAND-FEHR et al., 1982). These data are similar to those from Norwegian (HAVREVOLL et al., 1985) (Figure 2) and other French experiments (SCHMIDELY, unpubl.). *Ad libitum* teat feeding of sour milk at ambient temperature will increase intake and performance of kids compared with bucket feeding. This method will also save labour. However, individual control of feed intake and health is more difficult with *ad libitum* teat feeding of kids in a pen compared to individual bucket feeding. Hungry, young kids (2-3 days) adapt themselves easier to a self service teat feeding system than older kids which have suckled their dams for several days (HAVREVOLL et al., 1985). It is important to ensure that all kids are adapted to teat-feeding during the 2-3 days of training before they are placed in pens with older kids.

Figure 2

COMPARISON BETWEEN RESTRICTED BUCKET FEEDING OF A WARM MILK REPLACER
IN TWO MEALS A DAY ⬜ AND *AD LIBITUM* TEAT FEEDING AT A SELF SERVICE
SYSTEM▨ RELATIVE VALUE, BUCKET FEEDING = 100
(HAVREVOLL et al., 1985).

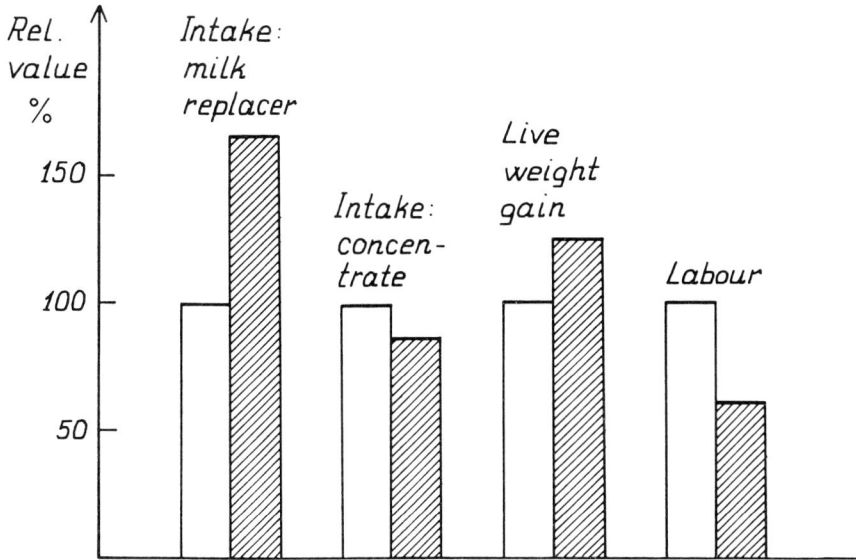

For bucket-feeding, warm milk (38°C) offered in 2 or 3 meals a day is recommended, particularly for kids housed in non isolated stables in cold climate (SKJEVDAL, 1974; ROBSTAD and SKRØVSETH, 1979; EIK, 1988). A high concentration of milk replacer, 17.5 % dry matter may reduce frequency of bloat among bucket-fed kids. Risks for overfilling the abomasum will be less compared to a lower concentration and a higher amount of liquid feed per meal.

Feeding one meal instead of two reduces labour significantly. Kids seem to become rapidly accustomed to one meal per day, when the milk is highly concentrated and offered at body temperature (FEHR and SAUVANT, 1974; MORAND-FEHR et al., 1982). For kids younger than three weeks of age, at least two meals a day should be recommended in order to obtain satisfactory health and growth performance (FEHR and SAUVANT, 1974; ROY, 1980; MORAND-FEHR et al., 1982).

When feeding milk from a reservoir, the milk should be acidified before feeding by fermentation or by adding organic acid. Sour cold milk offered *ad libitum* decreases intake when fed *ad libitum* compared to non acidified warm milk (HAVREVOLL et al., 1985). Individual daily inspection of kids in a pen is essential in order to control feed intake and health of the animals.

RESTRICTED VERSUS *AD LIBITUM* MILK FEEDING

The pre-weaning weight gain in goat kids is always highly correlated ($r = 0.75$) with milk intake (MORAND-FEHR et al., 1982). Figure 2 shows growth performance of kids receiving a milk replacer *ad libitum* or restricted (1 kg/day) (HAVREVOLL et al., 1985). SKJEVDAL (1974) and OWEN and DE PAIVA (1980, 1982) observed similar results on the basis of comparisons between diets offered *ad libitum* and more or less restricted. However, SKJEVDAL (1974) observed intestinal inflamations in the group of kids subjected to a high feeding level and duodenal ulcers in the cases of mortality. The level of milk dry matter intake is the main factor influencing the growth rate whatever the mode of distribution. Kids will consume more solid feed when milk is offered in restricted quantities. ECONOMIDES (1982) examined the effects of age at weaning (31 to 90 days), system of feeding (constant quantity of milk throughout versus restriction of milk intake two weeks prior to weaning) and quantity of milk suckled (50, 70 or 90 kg/kid) on the performance of kids reared artificially. The kids offered milk *ad libitum* grew faster until weaning and slower from weaning to 140 days of age compared to kids on the other two feeding systems. However, the daily gain of kids on the three feeding systems was similar from birth to 140 days of age.

However, because of the high cost price of feed during the milk feeding period, the minimum amount of milk or milk powder necessary for satisfactory growth of young goats should be determined. In France, at least 7 kg of milk powder per female Alpine kid is recommended (FEHR and HERVIEU, 1975). ECONOMIDES (1982) suggested weaning of kids at 52-56 days of age with a restriction of milk intake two weeks prior to weaning.

CONCLUSION

Colostrum and milk are essential as feeds for neonatal kids. The kid capacity to get adapted to different feeds depends on age and physiological development of the digestive tract. All changing of diets in young animals should be done gradually. The choice of milk feeding methods depends on economical conditions and labour availability. The preweaning feeding system should promote high postweaning feed intake and growth (see Chapter 23). It is absolutely necessary, however, that the diet, the feeding system and the environment meet the requirements for nutrients and behaviour of the kids in any rearing system.

SUMMARY

This chapter deals with liquid feeding of goat kids. The milk feeding period may last only 4-5 weeks or up to several months. In a newborn kid the digestive tract with its enzymes is particulary adapted to digest and absorb milk nutrients. During the preruminant period the young animal gradually becomes able to digest solid feed.

Colostrum should be fed *ad libitum* to all kids as soon as possible after birth. Colostrum is essential to obtain immunity and positive energy balance in the body. Liquid feed for kids could be whole milk from goat or cow, milk replacer, skim milk or whey. Skim milk as the only milk feed can be used from 4-5 weeks of age and whey from two months of age (1-2 l/day). The milk replacer should contain 15-25 % good-quality fat and 20-25 % protein on a dry basis. Replacement of milk proteins by soyabean and fish protein in the milk replacer may reduce the kid performance, particularly before 3-4 weeks of age. Milk replacers for calves or lambs can also be used in kids. When a milk replacer is fed, the concentration can vary from 14 to 18 % dry matter. Bacteriologically acidified milk may be favourable to young kids compared to non acidified milk.

Natural rearing is recommended in meat producing herds. Also in dairy herds natural rearing with early weaning or partly suckling is used if the price of goat milk substitute is high or if the milk replacer is not available and if the management is poor. Individual milk feeding from a small pail in 2 or 3 meals a day with warm milk is useful for individual controls. *Ad libitum* teat feeding, a method more adapted to natural rearing, may save labour but will increase consumption of milk compared to bucket feeding. Restricted milk feeding will enhance solid feed intake and reduce postweaning growth check.

Keywords: Young goat, Development of digestive tract, Liquid feeding, Milk feeding systems, Colostrum, Immunity, Milk replacer, Acidified milk feeds.

RÉSUMÉ

Ce chapitre traite de l'alimentation liquide des chevreaux. La période d'alimentation lactée peut durer de 4 à 5 semaines seulement jusqu'à plusieurs mois. Le chevreau nouveau-né est particulièrement adapté à digérer et à absorber les nutriments du lait. Au cours de la période préruminant, le jeune animal devient progressivement capable de digérer des aliments solides.

Le colostrum doit être distribué à volonté à tous les chevreaux aussi rapidement que possible après la naissance. L'ingestion de colostrum est essentielle pour obtenir une bonne immunité et une balance énergétique positive. L'aliment liquide pour chevreaux peut être du lait entier de chèvre ou de vache, du lait de remplacement, du lait écrémé ou de petit lait. Le lait écrémé comme aliment lacté unique peut être utilisé à partir d'un âge de 4-5 semaines et le petit lait à partir de 2 mois à raison d'1 à 2 litres par jour. Le lait de remplacement doit contenir 15-30 % de matière grasse de bonne qualité et 20-25 % de protéines par rapport à la matière sèche. Les protéines de substitution (soja, poisson) réduisent les performances des chevreaux, particulièrement avant 3-4 semaines par rapport aux protéines lactées. Des laits de remplacement pour veaux ou agneaux peuvent être utilisés aussi pour les chevreaux. La concentration du lait reconstitué liquide peut varier de 14 à 18 % de matière sèche. Le lait acide, en particulier celui qui est acidifié bactériologiquement, peut donner de bons résultats chez le chevreau en comparaison au lait non acidifié.

L'élevage naturel avec ou sans allaitement partiel peut être recommandé pour les troupeaux orientés vers la production de viande. Dans les troupeaux laitiers, ce système est aussi utilisé si le prix de la poudre de lait de remplacement est élevé, si du lait de remplacement n'est pas disponible ou si la conduite d'élevage est de médiocre qualité. Le mode d'alimentation individuelle dans de petits récipients contenant du lait chaud à raison de 2 ou 3 repas par jour permet un contrôle individuel. L'alimentation à la tétine à volonté est une méthode bien adaptée qui peut économiser de la main-d'oeuvre mais qui augmente la consommation de lait par rapport à la distribution de lait en coupelle. Le rationnement du lait incite le jeune animal à consommer des aliments solides et évite un arrêt de croissance pendant le sevrage.

REFERENCES

AL-JAWAD A.B. and LEES J.L., 1985. Effects of ewe's colostrum and various substitutes on the serum immunoglobulin concentration, gut closure process and growth rate of lambs. Anim. Prod. 40: 123-127.

BURGKART M. and BAUER J., 1974. Artificial rearing and intensive fattening of lambs in Germany, pp. 27-41. In: Hoffmann - La Roche Ltd. public. index 1468.

CABELLO G. and LEVIEUX M., 1981. Absorption of colostral Ig G by the newborn lamb: influence of the length of gestation, the birth weight, and thyroid function. Res. Vet. Sci. 31: 190-194.

CHURCH D.C., 1976. Digestive physiology and nutrition of ruminants. Vol. I : Digestive physiology. Second edition, O. & Books, Corvallis, Oregon (USA).

DAENICKE R., 1980. Zur Anzucht von Kuh- und Bullenkälbern nach verschiedenen Verfahren. Kali-Briefe (Büntehof), 15 (6): 391-396.

EALES F.A., MURRAY L. and SMALL J., 1982. Effects of feeding ewe colostrum, cow colostrum or ewe milk replacer on plasma glucose in newborn lambs. Vet. Rec., 111: 451-453.

ECONOMIDES S., 1982. Factors affecting growth of milk fed kids and fattening of kids for meat production. Seminar of FAO Subnetwork on Goat Nutrition. Nov. 9-10, 1982, Reading (UK).

EIK L.O., 1988. Oppal av kje i uisolert hus (Rearing of kids in unisolated house). Husdyrforsøksmøtet 1988. Aktuelt frå Statens fagtjeneste for landbruket Nr. 1, 141-146.

FEHR P.M., 1971. Méthodes d'alimentation de chevrettes destinées à la production laitière (Feeding methods of female kids for dairy herd replacement). 10th Intern. Congr. of Animal Science, Jul. 20-23, 1971, Versailles (FRANCE).

FEHR P.M. and SAUVANT D., 1974. Effets séparés et cumulés du nombre de repas et de la température du lait sur les performances des chevreaux de boucherie (Separate and cumulative effects of daily meals number and milk temperature on male kid performance). Ann. Zootech., 23: 503-518.

FEHR P.M. and HERVIEU J., 1975. Effet de la distribution de 6 ou 9 kg d'aliment d'allaitement et de sa répartition dans le temps sur les performances des chevrettes (Effect of supplies of 6 or 9 kg milk replacer and mode of milk distribution on female kid performances). Journée d'étude sur l'alimentation des chevrettes, INRA-ITOVIC, Paris (FRANCE).

FOLEY J.A. and OTTERBY D.E., 1978. Availability, Storage, Treatment, Composition and Feeding Value of Surplus Colostrum; A Review. J. Dairy Sci., 61: 1033-1060.

GALL C., 1981a. Goats in Agriculture: Distribution, Importance and Development, p. 1-34. In: GALL C. (Ed.): Goat Production, pp. 1-34. Academic Press Inc., London (UK).

GALL C., 1981b. Milk production, p. 309-344. In: GALL C. (Ed.): Goat Production. Academic Press, London (UK).

GALL C., 1981c. Husbandry, p. 411-432. In: GALL C. (Ed.): Goat Production, Academic Press Inc., London (UK).

GIHAD E.A. and MORAD H.M., 1977. Development of goat's digestive tract, p. 161-163. In: Symp. on Goat Breeding in Mediterranean Countries. Oct. 3-7, 1977, Malaga, Grenada, Murcia (SPAIN).

GRØNDALEN T., GJESTANG K.E., MATRE T. and SIMENSEN E., 1984. Godt miljø - friske kalver (Good husbandry-healthy calves). Landbruksforlaget, Oslo (NORWAY).

HADJIPANAYIOTOU M., 1984. Weaning systems: Intensive fattening of Chios lambs and Damascus kids in Cyprus. World Anim. Rev., 52: 34-39.

HADJIPANAYIOTOU M., 1986. The effect of type of suckling on the pre-and post-weaning lactation performance of Damascus goats and the growth rate of the kids. J. Agric. Sci. (Camb.), 107: 377-384.

HALLIDAY R., 1978. Variation in immunoglobulin transfer from ewes to lambs. Ann. Rech. Vet. 9: 367-374.

HAVREVOLL Ø, 1984. Forsøk med bakteriologisk og kjemisk surning av mjøl- keerstatningar (Experiments with bacteriologically and chemically acidified milk replacers). Report No. 218, 66 p. Dep. Animal Nutrition, Agric. Univ. Norway, As (NORWAY).

HAVREVOLL Ø, GARMO G., HELLEBERGSHAUGEN O. and SOLHEIM J., 1985. Forsøk med bøttefôring og smokkfôring av søtt og surna mjølkefôr til kje (Experiments with bucket feeding and teat feeding of acidified or non acidified milk feeds for rearing dairy goats). Scientific reports of Agric. Univ. Norway, 64: (15), 17 p.

HAVREVOLL Ø, NEDKVITNE J.J. and GARMO T.H., 1987. Fôrslag og Fôring (Feeds and Feeding), Chap. IV, p. 70-109. In: Geitboka (The Goatbook). Landbruksforlaget, Oslo (NORWAY).

HAVREVOLL Ø., NEDKVITNE J.J. and LARSEN H.J., 1988. Råmjølk til lam (Colostrum for lambs). Husdyrforsøksmøtet 1988. Aktuelt fra Statens fagtjeneste for landbruket Nr. 1, p. 91-96.

LOUCA A., MAVROGENIS A. and LAWLOR M.J., 1975. The effect of early weaning on the lactation performance of Damascus goats and the growth rate of the kids. Anim. Prod., 20: 213-218.

MORAND-FEHR P., 1981. Growth, p. 253-283. In: GALL C. (Ed.): Goat Production. Academic Press Inc., London (UK).

MORAND-FEHR P., 1985. Seminar on Nutrition of Goats. FAO Subnetwork of Cooperative Research on Goat Production. Ann. Zootech., 34: 471-490.

MORAND-FEHR P., 1987a. Seminar on Nutrition of Goats. FAO Subnetwork of Cooperative Research on Goat Production. Ann. Zootech., 36: 319-344.

MORAND-FEHR P., 1987b. Management programs for the prevention of kid losses, p. 405-423. 4th Intern. Conf. on Goats., March 8-13, 1987, Brasilia (BRAZIL).

MORAND-FEHR P., HERVIEU J., BAS P. and SAUVANT D., 1982. Feeding of young goats, p. 90-104. 3th Intern. Conf. on Goat Production and Disease, Jan. 10-15, 1982, Tucson, Arizona (USA).

MOWLEM A., 1979. Milk replacer for kid rearing, p. 54-57. Brit. Goat Soc. Year Book (UK).

MOWLEM A., 1981. Recent advances in kid rearing. Brit. Goat Soc. Monthly J., March, p. 41-42.

MOWLEM A., 1982. Rearing dairy goat kids using milk replacer, p. 491 (Abst.). 3rd Intern. Conf. on Goat Production and Disease. Jan. 10-15, 1982, Tucson, Arizona (USA).

NITSAN Z., CARASSO Y. and NIR I., 1985. The use of starch on soybean protein in intensive rearing of veal typ kids. Ann. Zootech., 34: 487-488.

NITSAN Z., CARASSO Y., ZOREF Z. and NIR I., 1987. Effect of diet on the fatty acid profile of adipose tissue and muscle fat of kids. Ann. Zootech., 36: 339-341.

NORRMAN E., 1987. Råmjølk för kalvens hälsa (Colostrum for the health of the calf). Aktuelt från lantbruksuniversiteter 360 Husdjur, 60 p., Uppsala (SWEDEN).

OPSTVEDT J., 1968. Foringa (The Feeding), p. 74-120. In: "Geitehald" (Goat keeping), Bøndenes Forlag, Oslo (NORWAY).

ØRSKOV E.R., 1987. The feeding of ruminants. Principles and practice, 92 p. Chalcombe Publications, 13 Highwoods Drive, Marlow Bottom, Bucks SL7 3PU (UK).

ØVERÅS J., ULVUND M.J. and WALDELAND H., 1988. Anemi hos spedlam (Anemia in neonatal lambs). Norsk Veterinaertidsskrift 100: 257-264.

OWEN J.B., 1974. Artificial rearing of lambs with milk-replacers in England. In: Hoffman-La Roche Ltd. public indes. 1468, p. 3-25.

OWEN E.and DE PAIVA P., 1980. Artificial rearing of goat kids: effect of age at weaning and milk substitute restriction on performance to slaughter weight. Anim. Prod., 30: 480 (Abst.).

OWEN E. and DE PAIVA P., 1982. Artificial rearing of goat kids, p. 491 (Abst.). In: Proc. 3rd Intern. Conf. on Goat Production and Disease. Jan. 10-15, 1982, Tucson, Arizona (USA).

PENNING P.P., 1979. The artificial rearing of lambs, p. 287-296. In: The Management and Diseases of Sheep. Commonwealth Agricultural Bureaux, Slough (UK).

RADOSTITS O.M. and BELL J.M., 1970. Nutrition of the preruminant dairy calf with special reference to the digestion and absorption of nutrients: A review. Can. J. Anim. Sci., 50: 405-452.

RENNER E., 1983. Milk and Dairy Products in Human Nutrition, 450 p. Volkswirtschaftlicher Verlag, München (GERMANY).

ROBSTAD A.M. and SKRØVSETH O.I., 1979. Fôring og stell av kje (Feeding and management of kids). Reprint No. 500. Dep. Animal Nutrition, Agric. Univ. Norway, As (NORWAY).

ROY J.H.B., 1980. The Calf. Fourth Edition. Butterworths, London (UK).

SANZ SAMPELAYO M.R., MUNOZ F.J., LARA L., GIL EXTREMERA F. and BOZA J., 1987. Factors affected pre- and post-weaning growth and body composition in kid goats of the Granadina breed. Anim. Prod., 45: 233-238.

SANZ SAMPELAYO M.R., LARA L., GIL EXTREMERA F. and BOZA J., 1988a. Carcass quality of the Granadina breed goat kid under intake of a specific milk replacer. Seminar of FAO Subnetwork on Goat Nutrition and Feeding, Oct. 3-5, 1988. Potenza (ITALY).

SANZ SAMPELAYO M.R., MUNOZ F.J., GUERRERO J.E., GIL EXTREMERA F. and BOZA J., 1988b. Energy metabolism of the Granadina breed goat kid. Use of goat milk and a milk replacer. J. Anim. Physiol. Anim. Nutr., 59: 1-9.

SCHINGOETHE O.J., 1976. Whey Utilization in Animal Feeding. A Summary and Evaluation. J. Dairy Sci., 59: 556-570.

SHELDRAKE R.F. and HUSBAND A.J., 1985. Immune defences at mucosal surfaces in ruminants. Review article. J. Dairy Res., 52: 599-613.

SKJEVDAL T., 1974. Mjølkefôring av kje (Milk feeding of kids). Scientific Reports of Agric. Univ. Norway, Dep. Animal Nutrition. Vol. 53, No. 39.

STOBO I.J.F., 1981. Calf Rearing. The Feed Compounder. Dec., 1981, pp. 35-40.

TEH T.H. and ESCOBAR E.N., 1987. Evaluation of protein requirement of milk replacer for goat kids, Vol. 2, p. 1375 (Abst.). 4th Intern. Conf. on Goats, March 8-13, 1987, Brasilia (BRAZIL).

THIVEND P., TOULLEC R. and GUILLOTEAU P., 1980. Digestive adaptation in the preruminant, p. 561-585. In "Digestive Physiology and Metabolism in Ruminants", Proc. 5th Intern. Symp. on Ruminant Physiology, Sept. 3-7, 1979, Clermont-Ferrand (FRANCE).

WAYNE MODLER H., 1986. Feeding whey to ruminants. The International Whey Conference, Oct. 27-29, 1986, Chicago (USA).

Chapter 22

WEANING: A CRITICAL PERIOD FOR YOUNG KIDS

P. BAS, P. MORAND-FEHR and P. SCHMIDELY

INTRODUCTION

The techniques used to wean kids differ considerably, depending on the goat farming system used. In extensive grazing or rangeland systems, the kid may suckle its mother for several months while progressively eating increasing quantities of forage and thus be ultimately weaned between 2 and 6 months. On the other hand, on intensive dairy systems, the kid is often separated from its mother a few days after birth and fed with a milk replacer (see Chapter 21). Weaning in this case does not coincide with separation from the mother: it is defined by a progressive or abrupt replacement of milk by solid feeds.

At weaning, the kid as other ruminants must adapt to a diet rich in fibre and starch but poor in lipids and simple carbohydrates. Its metabolism is thus modified in order to synthesize glucose via gluconeogenesis, primarily from the products of ruminal digestion.

Weaning is a critical period which modifies growth rate and body composition, and which probably affects the future performance of young females during their reproductive and productive life (TROCCON and PETIT, 1989). Because of the paucity of available data on this period, we propose to analyse the characteristics of kid weaning and its effects on intake, growth, digestive and metabolic phenomena, and on body composition.

This chapter extends the information on this subject published in general articles on growth, feeding or rearing methods of kids (SKJEVDAL, 1974; FEHR, 1975; FEHR and SAUVANT, 1976; MORAND-FEHR et al., 1982; ECONOMIDES, 1986; LU et al., 1988). It will be illustrated primarily by results obtained with male Alpine kids (BAS, 1990). The kids were raised in individual pens and weaned abruptly, either early (4 weeks) or later (6 or 8 weeks) and were slaughtered 2 or 8 weeks after weaning. Techniques of artificial feeding, application of abrupt weaning, and slaughter after a short post-weaning period enable the effects of weaning on consumption, growth and body composition to be satisfactorily described. The reference group consisted of non-weaned kids reared to live weights between 22 and 25 kg, without the possibility of consuming solid feed.

EFFECTS OF WEANING ON GROWTH RATE

When milk is suppressed, it causes a decrease or more often a total arrest of growth rate, even a loss of live weight in some cases. In our experiments the abrupt suppression of milk was followed by a drop in body weight, maximal 2 to 5 days after weaning. The intensity and duration of the growth decrease were greater in the youngest and lightest kids at weaning (Figure 1). Empty body weight (EBW) (live weight minus weight of digesta), however, decreased more than live weight as a result of the increase in digestive contents resulting from the substantial increase in solid feed intake at weaning. This increase could partially mask the reduced body weight (Figure 2).

Figure 1

RELATIVE VARIATION OF LIVE WEIGHT (LW) AFTER WEANING

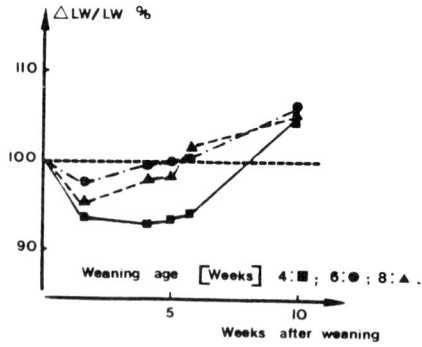

Figure 2

EVOLUTION OF KID WEIGHTS AFTER WEANING

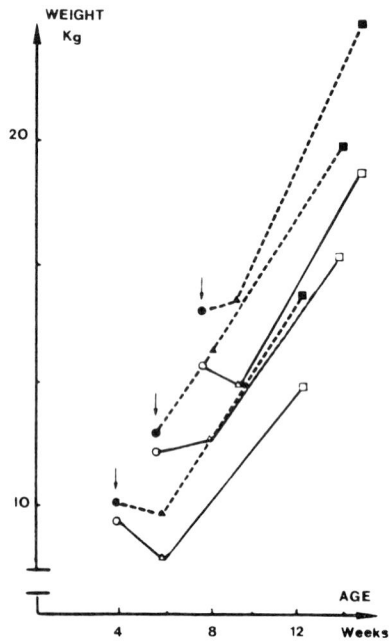

-- : live weight; ___ : empty body weight; ●,○ : unweaned kids;
▲,△ : 2 weeks after weaning; ■,□: 8 weeks after weaning

272

As in the case of lambs and calves, "weaning shock" is more severe in younger kids, as previously observed by FEHR (1975), FEHR and SAUVANT (1976) and TEH et al. (1984), and is reflected by a delay in growth. In our experiments this delay lasted 2 weeks in kids weaned early (4 weeks) in comparison to those weaned later (8 weeks). This delay is still seen at the age of about 12 weeks, as observed by TEH et al. (1984). On the other hand, kids weaned at the age of 6 weeks, tended to make up their growth delay and so reached weights comparable to those weaned later (8 weeks), this was as a result of the phenomenon of compensatory growth, which is apparently more difficult when weaning occurs at an earlier age. These results confirm those of TEH et al. (1984).

When feeding was unrestricted, weaning was responsible for only a slight difference in live weight (on average 1 kg, maximum 2 kg) around the age of 12 weeks between unweaned kids and those weaned at 2 months. However, this difference can be greater (up to 4 or 5 kg) when weaning occurs early, at 4 weeks or earlier. Between these two weaning ages, the growth delay at 12 weeks varies according to weaning conditions and depends on compensatory growth. This difference between weaned and unweaned kids was about 1 kg larger when the comparison was made on the basis of EBW.

SOLID FEED INTAKE

During the period of milk feeding, solid feed intake (hay and concentrates) was low when milk was freely available and almost nonexistent up to 4 weeks. However this subsequently increased progressively and more rapidly when milk was limited. When hay and concentrate were freely available, in our experiments hay intake increased significantly with age, while that of contentrate progressed less rapidly. At the age of 7 weeks unweaned kids ingested 440, 60 and 20 g DM of milk, hay and concentrate, respectively. These results confirm and expand those obtained by MORAND-FEHR et al. (1986) using young female Alpine kids and by LEVY and ALEXANDRE (1985) using Creole kids.

The intake of solid feed before weaning does not respond to energy or nitrogen requirements since they are satisfied primarily by milk, but depends more on behavioural or physiological factors especially from the age of 4 weeks. Intake of solid feed before weaning plays an important role in the adaptation to a change in diet, since it forces the young ruminant to salivate, chew and ruminate, and so initiates digestive processes in the rumen as seen with calves (BEN ASHER and al., 1981).

When weaning was abrupt, we observed that the total quantity of dry matter ingested during the following week by kids, raised in individual pens, was lower than that of milk during the week prior to weaning. This difference was about 30 % in kids weaned at 6 or 8 weeks and much higher (80 %) in those weaned earlier.

Between 2 and 4 weeks after weaning, the quantity of concentrate ingested increased more rapidly than that of hay (Figure 3). This was more evident in kids weaned earlier, which ingested larger quantities of concentrate per kg metabolic weight ($W^{.75}$) than those weaned later. Their intake of concentrates was also liable to decrease probably as a result of digestive disorders such as acidosis (see Chapter 12), due to the very low forage/concentrate ratio in the feed.

Metabolizable energy intake after weaning, calculated on the basis of feed composition of the diet and the theoretical energy yields in adult ruminants, was not large enough to meet energy requirements for maintenance, estimated at about 100 kcal/d/kg $W^{.75}$ (SANZ SAMPELAYO et al., 1988; BAS, 1989; see Chapter 7). This balance was more negative if the kids were younger or weighed less at weaning. As a result, kids reached their pre-weaning level of energy intake (on the basis of metabolic weight) only 8 weeks after weaning while kids weaned later, at 6 to 8 weeks, reached this level 6 weeks after weaning. In addition, it took longer to reach this level when the diet only contained concentrates.

Figure 3

CHANGES IN THE INTAKE OF SOLID FEED DRY MATTER (DMI) AROUND WEANING

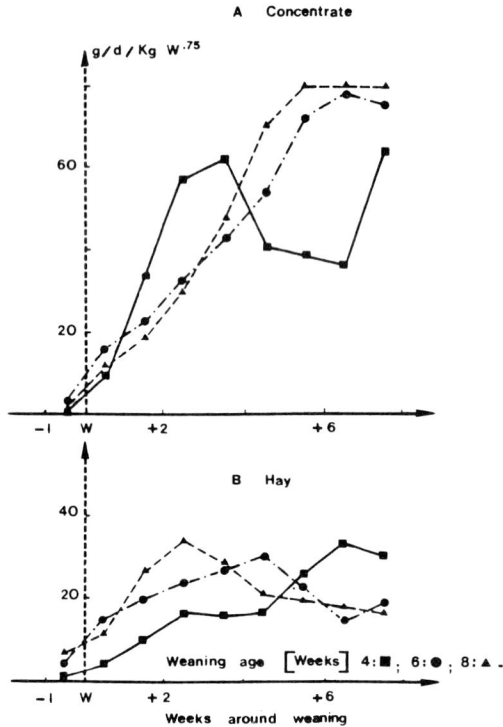

As a matter of fact, the replacement of milk at weaning was easier with concentrate than with forage, as in lambs and calves (CHARLET-LERY and ZELTER, 1955; MATHIEU and WEGAT-LITRE, 1961). Our observations nevertheless indicate that it is desirable to offer good quality forage to kids along with concentrates in order to facilitate the development of the rumen, decrease the risk of acidosis and thus reach a higher growth rate. According to our estimates the minimum forage requirement during the initial few weeks after weaning ranged between 15 and 20 g DM/d/W$^{.75}$ in Alpine goats, approximately equivalent to the recommendations for other goat breeds, expressed as lignocellulose (ADEBOWALE, 1983; Mc GREGOR, 1984).

MORAND-FEHR et al. (1982) reported that the weaning shock was less severe in females than in males, perhaps because of a higher adapting capacity to feeding changes and because of larger body fat reserves. Learning to ingest solid feeds may be facilitated by group feeding in the proximity of adults and by increasing the feeding frequency, especially that of concentrates.

274

ANATOMICAL AND PHYSIOLOGICAL MODIFICATIONS
OF THE DIGESTIVE TRACT

The most important anatomical modification at weaning involves the accelerated development of the reticulo-rumen and the omasum. In our experiments, the weight of the reticulo-rumen and the omasum on the basis of empty body weight, increased by about three (Figure 4), confirming prior observations in kids (TAMATE, 1956, 1957; CANDAU, 1972; HAMADA et al., 1976; GIHAD and MORAD, 1977).

Figure 4

CHANGES IN RETICULO-RUMEN (RR) WEIGHT RELATED TO EMPTY BODY WEIGHT (EBW)

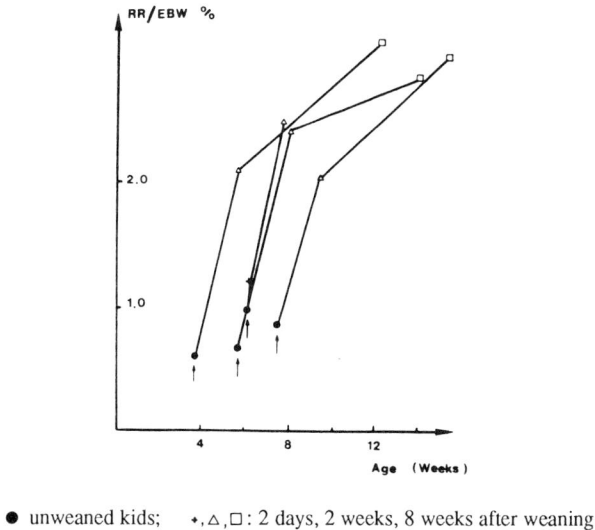

● unweaned kids;   •,△,□ : 2 days, 2 weeks, 8 weeks after weaning

Between 2 and 8 weeks after weaning, the reticulo-rumen grew faster than EBW. The weight increase of the rumen just after abrupt weaning was greater as the weight of the kid was higher at weaning leading to increased levels of solid feed intake, particularly forages, and thus to a better tolerance of weaning, as observed in lambs (MOLENAT et al., 1971) and calves (LATRILLE et al., 1983). These results in kids show that the capacity of the reticulo-rumen to increase in size depends on live weight at weaning. Also, when kid weight was between 20 and 50 % of the mean adult weight of the herd, it was observed that the reticulo-rumen developed more rapidly than the rest of the body in the absence of solid feed.

During the first 2 weeks after weaning, abomasum weight on the basis of empty body weight also increased, but to a lesser extent (40 % on average) than the reticulo-rumen and the omasum. During the same period, the relative variations in intestinal weight depended on the weight of the animal at weaning. In the conditions of our experiments, it decreased in kids weighing less than 10.2 kg live weight at weaning, equivalent to 9.3 kg body mass, but increased in kids of more than 10.2 kg. This weight loss in early weaned kids is due primarily to the small intestine whose weight decreases sharply if animals are undernourished as shown in calves and lambs (CARNEGIE et al., 1969; LEGIN, 1983).

EFFECT OF CHANGES IN BODY COMPOSITION
ON CARCASS QUALITY

After weaning, the dressing percentage relative to live weight decreased substantially (12 % on average) in comparison to unweaned kids, primarily because of the increase in digesta, but this variation in the dressing percentage was much less perceptible when the carcass value was expressed relative to EBW.

The loss in EBW observed during the first two weeks after weaning did not affect all organs in the same way. Thus, the carcass, skin and visceral organs such as heart, kidneys and liver, as well as the visceral adipose tissues lost weight, but the largest losses were those of the red organs (kidneys: - 10 %, liver: - 20 %) and abdominal adipose tissues, and to a lesser extent carcass adipose tissues. These were partially compensated by the increased weight of the empty digestive tract.

Among adipose tissues, the perirenal tissue was mobilized first and the most intensively just after weaning (Figure 5). Weight loss was detectable as early as 2 days after weaning, which was also the case for kids put on a 24 h fast (MORAND-FEHR et al., 1985). Weight loss by the various adipose tissues was accompanied by a decrease in their lipid contents. Thus, lipid reserves were reduced more than suggested by the weight changes of adipose tissues. This mobilization of lipids in fat deposits was more marked when weaning was at 4 weeks. In this case, 90 to 99 % of visceral adipose tissue lipids were mobilized, but only 20 to 65 % when weaning was at 8 weeks (BAS et al., 1985, 1986, 1987). If this mobilization was expressed relative to metabolic weight, however, its extent was similar regardless of age at weaning (20 to 30 kcal/d/kg $W^{.75}$). Carcass lipids were less mobilized, in terms of quantity and proportion, than those of the visceral adipose tissues in energy-restricted lambs (BUTLER-HOGG, 1984). In early weaned kids which mobilized almost all their abdominal fat reserves, carcass fat decreased however, more than in kids weaned later. As a result, for the same carcass weight, the carcasses of kids weaned early appeared leaner than those of kids weaned later.

Subsequently, at 6 to 7 weeks after weaning, the carcass fat content of kids weaned at 6 to 8 weeks of age was close to that of unweaned kids of the same weight. On the other hand, the carcasses of kids weaned earlier remained lean. After weaning, the levels of lipids in the visceral adipose tissues never reached that recorded just before weaning.

In the post-weaning period, the flow of nutrients was directed preferentially towards protein synthesis rather than towards that of lipids. As a result, kids appeared to be leaner as their slaughter weight was closer to weaning weight. The same results have been obtained in lambs (THOMSON et al., 1982). This phenomenon is more pronounced as the weaned kids are younger or weigh less and as the weaning shock is harsher (DREW and REID, 1975; LITTLE and SANDLAND, 1975).

CHANGES IN BLOOD METABOLITES

Blood metabolites related to energy metabolism, e.g. glucose, non-esterified fatty acids (NEFA) and beta-hydroxybutyrate (BHB) undergo substantial variations during weaning (Figure 6).

NEFA, reflecting the mobilization of fat reserves, increased considerably (more than 1mM) as early as the first day after weaning, then decreased progressively during the next 2 weeks reaching a final level, lower than that during the milk feeding period. BHB, which may be derived from the breakdown of mobilized fatty acids, changed in the same way as the NEFA, but its peak occurred later, around 5-7 days after weaning. Changes in glucose, were opposite to those of NEFA, decreasing rapidly from 1.2 to about 0.6 g/l during the day following weaning, then slowly increasing to around 0.9 to 1.0 g/l during the next 4 weeks. Packed cell volume, sensitive to water intake, varied less than the former three metabolites. It increased sharply during the first week after weaning because of the considerable drop in ingested quantities, then decreased slowly.

During the first week after weaning, changes in NEFA, BHB and glucose were explained primarily by the energy deficit. The level of metabolizable energy intake (kcal/d/kg $W^{.75}$) could be predicted by NEFA, logarithmically expressed, for the entire post-weaning period. But the prediction of energy intake during the first week after weaning was better with BHB. Subsequently, the decreased relationship between BHB and energy intake or even its total absence, was explained by the progressive substitution of endogenous BHB by exogenous BHB produced from volatile fatty acids produced in the rumen. Similarly, the post-weaning increase in glycemia reflects an increase in gluconeogenesis from propionate and lactate produced in the rumen. In addition, increase in packed cell volume after weaning shows the water deficit of the kid and could supply information on the level of dehydratation of the animal.

Figure 5

CHANGES IN FOUR ADIPOSE TISSUE WEIGHTS RELATED TO EMPTY BODY WEIGHT

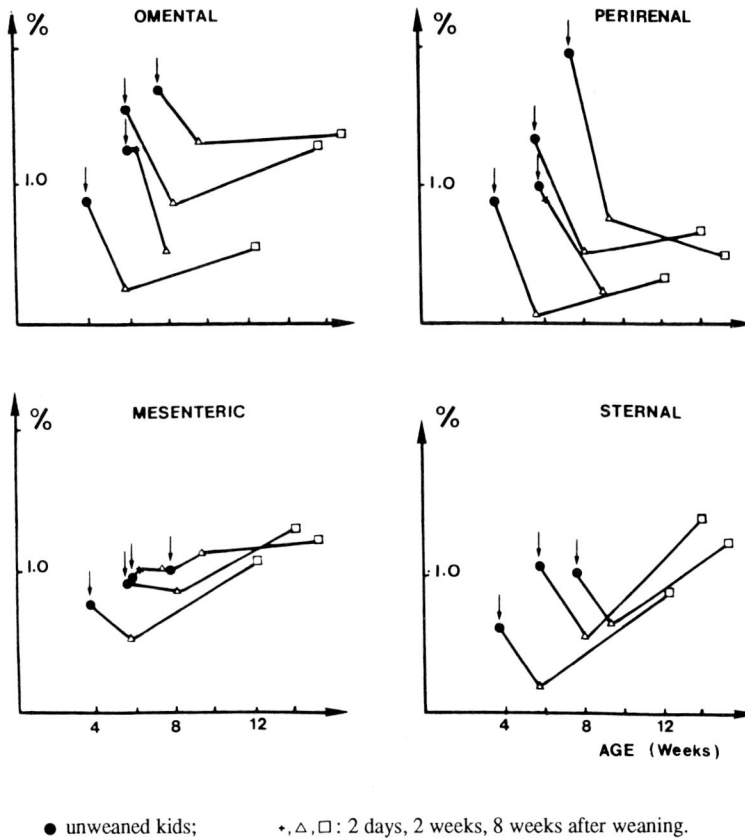

● unweaned kids;   •, △, □ : 2 days, 2 weeks, 8 weeks after weaning.

Figure 6

CHANGES IN BLOOD PARAMETERS AROUND WEANING (W)

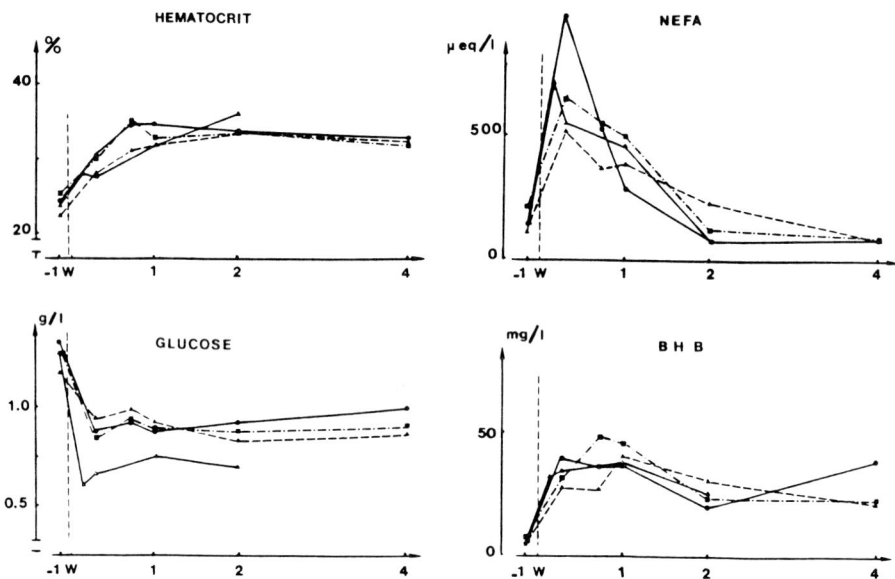

Weaning age (week): 4 (●), 6 (△) light weaning weight; 6 (▲) heavy weaning weight; 8 (■)

PRACTICAL CONDITIONS OF WEANING

From experiments carried out for the past 10 years in our laboratory (FEHR, 1975; FEHR and SAUVANT, 1976; MORAND-FEHR et al., 1982; BAS, 1990) we have established a set of practical rules for kid weaning.

The decrease in growth or even the loss of weight caused by weaning was reduced by the sufficient intake of solid feed before weaning. This was facilitated by increasing the weaning age and from 2 to 3 weeks of age offering concentrates or good quality forage . As a matter of fact, the weaning shock is highly dependent on the rapidity with which the kid increases its level of intake of solid feeds starting at the end of milk feeding. In batch feeding, the weaning shock depends on the "dominant" kid(s) which initiate(s) the others to eat solid feeds more or less rapidly. To facilitate this training, it is very effective to multiply the feeding of concentrate feeds and of hay on the day of weaning, up to 3 or 4 times per day (HERVIEU, unpublished). However, the mother can be considered to be the best teacher (LEVY and ALEXANDRE, 1985).

In general, kids can be weaned without undergoing a pronounced weaning shock if their birth weight has multiplied by 2.5 or if it has consumed 7 kg DM of milk, for females and 8.6 kg for males in the case of artificial feeding starting at the age of 4-6 days (or 1.5 kg DM/kg $W^{.75}$), and if it has consumed 10 g/d/kg $W^{.75}$ concentrates just before weaning (MORAND-FEHR et al., 1982). Furthermore, our latest observations show that kids will only tolerate weaning if their level of fattening is sufficient.

Sex, as we have seen, and breed can affect the age or weight at which kids should be weaned. TEH et al. (1984) indicated that Anglo-Nubian kids were more difficult to wean than Alpine kids, perhaps because of the presence of less body fat. The use of a target-weight for weaning, not in absolute terms but relative to the adult weight of the breed, considerably reduces breed-related disparities.

An abrupt weaning is often preferable to progressive weaning when the above conditions are followed. The major consequence of limiting milk with the aim of favoring the intake of solid feeds before weaning is to decrease the growth rate. Thus, at 8 or 12 weeks after weaning, the live weight of kids weaned progressively tends to be equal to or less than that of kids weaned abruptly.

In addition, it is recommended not to wean too early kids with a poor health status (digestive disorders, abnormal decrease in growth rate). Coccidiosis in kids often causes disorders, in particular at weaning (MOWLEM, 1981; LU et al., 1988).

Very late weaning, at 4 to 5 months, generally leads to very small improvements in performance in comparison to kids weaned earlier, but these systems are generally not satisfactory from an economic point of view, as also noted by LU et al. (1988).

Just after weaning, the diet should contain a good quality forage for the level of intake to remain above 15 to 20 g $DM/d/W^{.75}$ and thus prevent digestive disorders and also increase the quantities of energy ingested.

Weaning of nursed kids generally does not give rise to problems. In the case of insufficient milk production by the mother or mastitis, it is desirable to supplement the kid with concentrates at a very early age. The kid will thus be weaned naturally without a too large delay in growth. In the case of kids raised on pastures or on rangelands with their mothers, their health state related to parasite infestations can make the weaning period delicate. This is why it is recommended not to graze kids and their mothers on the same pastures when possible.

CONCLUSION

Kids may be weaned early (3 weeks for females, 4 weeks for males), but in this case they may have a 2-3 kg live weight growth delay at 12 weeks of age, compared to kids weaned later. In addition, this very early weaning requires excellent technical expertise by the farmer. Weaning at 6-8 weeks is generally better tolerated and avoids excessively pronounced weaning shock.

Weaning is prepared during the suckling period (see Chapter 21). The weaning period can affect subsequent growth performance (see Chapter 23) and the quality of the carcass (see Chapter 24). In particular, it is recommended not to slaughter kids soon after weaning in order to avoid excessively lean carcasses.

The modifications described here concerning growth, intake, digestion and metabolism in kids during weaning enable weaning techniques to be adapted to the special conditions of each production system or each husbandry practice.

SUMMARY

The effects of weaning on growth rate, feed intake, body composition and blood metabolites were analysed in male Alpine kids weaned between 4 and 8 weeks of age.

Weaning reduced live weight and delayed growth more in younger than in older kids. In comparison to unweaned kids, live weight differences decreased in older kids at weaning (6 or 8 weeks).

Just after weaning, kids ingested more concentrate than hay. It is suggested that kids should eat at least 15 to 20 g DM of hay/d/W.75 in order to express their growth potential. The intake of metabolizable energy was not large enough to meet maintenance requirements at 1 or 2 weeks after weaning. Expressed relative to metabolic weight, the intake of metabolizable energy reached pre-weaning levels after 6 to 8 weeks, depending on the age at weaning.

During the first two weeks after weaning, the weights of the reticulo-rumen and omasum increased two or three times. The liver and abdominal adipose tissues, on the other hand, decreased more rapidly than the empty body weight.

This adaptation to solid feed produced a drop in blood glycemia and a rise in non esterified fatty acids (NEFA), betahydroxybutyrate (BHB) and packed cell volume (PCV), which continued only for BHB and PCV.

Early weaning is thus possible in kids, but its success requires more technical expertise than late weaning. Kids must be slaughtered later after weaning in order to avoid excessively lean carcasses.

Keywords : Young goat, Weaning, Level of intake, Growth, Body composition, Blood metabolites.

RÉSUMÉ

L'incidence du sevrage sur la vitesse de croissance, l'ingestion, la digestion, le métabolisme et la composition corporelle a été analysé notamment en s'appuyant sur des résultats récents obtenus sur chevreaux alpins sevrés entre 4 et 8 semaines.

Le sevrage provoque une chute de poids et un retard de croissance d'autant plus important que les chevreaux sont plus jeunes ou plus légers au sevrage. Après le sevrage, l'écart de poids avec les chevreaux non sevrés n'a tendance à se réduire que chez les chevreaux sevrés relativement tardivement.

Juste après le sevrage, l'ingestion d'aliments concentrés est plus importante que celle du foin. Il est suggéré que la ration contienne une quantité minimale de foin, comprise entre 15 et 20 g MS/kg $P^{0.75}$ pour que s'exprime le potentiel de croissance des chevreaux après le sevrage. La quantité d'énergie métabolisable ingérée est insuffisante pour couvrir les besoins d'entretien des chevreaux pendant 1 à 2 semaines après le sevrage. Elle n'atteint le niveau observé juste avant le sevrage (exprimé par rapport au poids métabolique) qu'après un délai post-sevrage de 6 à 8 semaines selon l'âge au sevrage.

Au cours des deux premières semaines après le sevrage, le réticulo-rumen et l'omasum doublent ou triplent leur poids vide. Par contre, pendant cette période, le foie et les dépôts adipeux et plus particulièrement ceux de la cavité abdominale régressent plus vite que la masse corporelle totale.

Cette adaptation à l'alimentation solide se traduit par une baisse de la glycémie et une augmentation des acides gras non estérifiés, du béta-hydroxybutyrate (BHB) et de l'hématocrite, durable pour le BHB et l'hématocrite seulement.

Le sevrage précoce est donc possible chez le chevreau mais sa réussite nécessite plus de technicité qu'un sevrage tardif et un stade d'abattage relativement éloigné du sevrage pour que les carcasses ne soient pas trop maigres.

REFERENCES

ADEBOWALE E.A., 1983. The performance of West African dwarf goats and sheep fed silage and silage plus concentrate. World Rev. Anim. Prod., 19: 15-20.

BAS P., 1990. Influence de l'âge au sevrage sur la croissance, la composition corporelle et le métabolisme lipidique des chevreaux mâles de race alpine (Influence of weaning age on growth, body composition and lipid metabolism of Alpine male kids). Doctorat Thèse, Univ. Paris 6 (FRANCE) (in press).

BAS P., ROUZEAU A. and MORAND-FEHR P., 1985. Poids et métabolisme des réserves lipidiques au cours de la croissance chez le chevreau (Weight and metabolism of lipid deposits in growing kids). Reprod. Nutr. Dévelop., 25: 275-285.

BAS P., ROUZEAU A. and MORAND-FEHR P., 1986. Lipogenèse des tissus adipeux de chevreaux sevrés à 4, 6 ou 8 semaines (Lipogenesis in adipose tissues of kids weaned at 4, 6 or 8 weeks). Reprod. Nutr. Dévelop., 26: 649-658.

BAS P., MORAND-FEHR P., ROUZEAU A. and HERVIEU J., 1987. Evolution de la composition des tissus adipeux du chevreau mâle sevré à 4, 6 ou 8 semaines (Evolution in composition of adipose tissue in male kids weaned at 4, 6 or 8 weeks). Reprod. Nutr. Dévelop., 27: 313-314.

BEN ASHER A., NITSAN Z. and NIR J., 1981. Comparison of ruminal and post-ruminal digestion of a concentrate feed in the young calf. Reprod. Nutr. Dévelop., 21: 999-1007.

BUTLER-HOGG B.W., 1984. Growth patterns in sheep: changes in the chemical composition of the empty body and it constituent part during weight loss and compensatory growth. J. Agric. Sci., (Camb.), 103: 17-24.

CANDAU M., 1972. Stimulation physico-chimique et développement du rumen (Physical and chemical stimulation and rumen development). Thèse Doctorat, Univ. Paris 6 (FRANCE).

CARNEGIE A.B., TULLOH N.M. and SEEBECK R.M., 1969. Developmental growth and body weight loss of cattle. II - Changes in the alimentary tract. Aust. J. Agric. Res., 20: 405-415.

CHARLET-LERY G. and ZELTER S.Z., 1955. Essais d'application pratique d'une technique d'élevage d'agneaux précocement sevrés (Field trials of a breeding technic for early weaned lambs). Bull. Techn. Inf. Serv. Agric., 103: 559-564.

DREW K.R. and REID J.T., 1975. Compensatory growth in immature sheep. I - The effect of weight loss and realimentation on the whole body composition. J. Agric. Sci., (Camb.), 85: 193-204.

ECONOMIDES S., 1986. Comparative studies of sheep and goats: milk yield and composition and growth rate of lambs and kids. J. Agric. Sci., (Camb.), 106: 477-484.

FEHR P.M., 1975. L'allaitement artificiel des agneaux et des chevreaux (Artificial suckling of young goats), p. 83-105. In: SEI-CNRA (Eds.): The artificial suckling of lambs and kids. Versailles (FRANCE).

FEHR P.M. and SAUVANT D., 1976. Production de chevreaux lourds. I - Influence de l'âge et du mode de sevrage sur les performances de chevreaux abattus à 26.5 - 29 kg (Production of heavy kids. I - Influence of the age and mode of weaning on the performances of kids slaughtered at 26.5 - 29 kg). Ann. Zootech., 25: 243-257.

GIHAD E.A. and MORAD H.M., 1977. Development of the goat's digestive tract, p. 161-163. Symp. on Goat Breeding in Mediterranean Countries. Oct. 3-7, 1977, Malaga, Grenada, Murcia (SPAIN).

HAMADA T., MAEDA S. and KAMEOKA K., 1976. Factors influencing growth of rumen, liver and other organs in kids weaned from milk replacers to solid food. J. Dairy Sci., 59: 1110-1118.

LATRILLE L., PARE J.P., St-LAURENT G. and POMAR C., 1983. Heavy veal production with Holstein calves-raised by multiple suckling of milk replacers and fattened with whole corn, barley or oats. J. Anim. Sci., 63: 643-653.

LEGIN I., 1983. Effect of restricted feeding and realimentation on a compensatory growth, carcass composition and organ growth in lambs. Swedish J. Agric. Res., 13: 175-187.

LEVY F. and ALEXANDRE G., 1985. Le comportement alimentaire du cabri créole élevé en stabulation libre de la naissance au sevrage (Feeding behaviour of Creole kids kept in loose housing conditions from birth to weaning). Ann. Zootech., 24: 181-192.

LITTLE L.A. and SANDLAND R.L., 1975. Studies on the distribution of the body fat in sheep during continuous growth and following nutritional restriction and rehabilitation. Aust. J. Agric. Res., 26: 363-374.

LU C.D., POTCHOIBA M.J. and TEH T.H., 1988. Milk feeding and weaning of goat kids. A review. Small Rum. Res., 1: 105-112.

MAC GREGOR B.A., 1984. The food intake and growth of Australian feral x Angora kids when fed whole grain barley-lupins diets with three levels of roughage intake. Aust. J. Exp. Agric. Anim. Husb., 24: 77-82.

MATHIEU C.M. and WEGAT-LITRE E., 1961. Mise au point d'une méthode d'alimentation des veaux d'élevage. I - Détermination de la quantité de lait nécessaire (Method of feeding for raising calves. I - Determination of the amount of milk necessary). Ann. Zootech., 10: 161-175.

MOLENAT G., THERIEZ M. and AGUER D., 1971. L'allaitement artificiel des agneaux. Détermination de l'âge minimum au sevrage pour la production de chevreaux de boucherie (Artificial rearing of lambs. I - Determination of the minimum age at weaning for meat production). Ann. Zootech., 20: 339-352.

MORAND-FEHR P., HERVIEU J., BAS P. and SAUVANT D., 1982. Feeding of young goats, p. 90-104. 3rd Intern. Conf. on Goat Production and Disease, p. 90-104, Jan. 10-15, 1982. Tucson, Arizona (USA).

MORAND-FEHR P., BAS P., ROUZEAU A. and HERVIEU J., 1985. Development and characteristics of adipose deposits in male kids during growth from birth to weaning. Anim. Prod., 41: 343-357.

MORAND-FEHR P., HERVIEU J., FAYE A. and ROUASSI M., 1986. Adaptation comportementale à l'alimentation sèche des chevrettes pendant le sevrage (Adaptation of feeding behaviour of young goats during the weaning period). Reprod. Nutr. Dévelop., 26: 281-282.

MOWLEM A., 1981. Recent advances in kid rearing. Brit. Goat Soc., Monthly J., March, p. 41-42.

SANZ SAMPELAYO M.R., MUNOZ F.J., GUERRERO J.E., GIL EXTREMERA F. and BOZA J., 1988. Energy metabolism of the Granadina breed goat kid. Use of goat milk and a milk replacer. J. Anim. Physiol. Anim. Nutr., 59: 1-9.

SKJEVDAL T., 1974. Milk feeding of kids. Report n° 173, Vol. 52, No 39. Dept. of Anim. Nutr., Agric. Univ. Norway, As (NORWAY).

TAMATE H., 1956. The anatomical studies of the stomach of the goat. I - The post-natal development of the stomach with special reference to the weaning and prolonged suckling. Tohoku J. Agric. Res., 7: 209-223.

TAMATE H., 1957. The anatomical studies of the stomach of the goat. II - The post-natal changes in the capacities and the relative sizes of the four divisions of the stomach. Tohoku J. Agric. Res., 8: 65-71.

TEH T.M., POTCHOIBA M.J., ESCOBAR E.N. and LU C.D., 1984. Weaning methods of goats kids. J. Dairy Sci., 67: Suppl., 137-138.

THOMSON E.F., BICKEL H. and SCHURCH A., 1982. Growth performance and metabolic changes in lambs and steers after mild nutritional restriction. J. Agric. Sci. (Camb.), 98: 183-194.

TROCCON J.L. and PETIT M., 1989. Croissance des génisses de renouvellement et performances ultérieures (Growth of dairy heifers and consequences on performances). INRA, Prod. Anim., 2: 55-64.

Chapter 23

POSTWEANING FEEDING OF YOUNG GOATS

M. HADJIPANAYIOTOU, S. ECONOMIDES, P. MORAND-FEHR
S. LANDAU and O. HAVREVOLL

INTRODUCTION

Early breeding of female kids has been practised (MORAND-FEHR, 1981; MORAND-FEHR et al., 1982; MAVROGENIS and CONSTANTINOU, 1983; HAVREVOLL et al., 1988) as a simple method of intensifying production through increase of annual flock output, limitation of non productive periods and increase in the number of kid crops. Early breeding however is closely related with the level and quality of feed fed during the post-weaning period. The present paper will describe feeding of young goats from weaning to mating or slaughtering with emphasis on available feedstuffs, their feeding value and their allocation during different growth stages, and discuss how new research findings on growth promoters, cereal grain processing, protein nutrition and others might affect post-weaning performance and production efficiency.

FEEDSTUFFS

Postweaning feeding of young goats is similar to that of other young ruminants; their diet is composed of roughages and concentrates. Under semi-intensive or intensive feeding conditions growing goats rely on good quality roughages, cereal grains, protein supplements and by-products of moderately good quality.

Cereal grains

The price of barley grain has been much lower than that of corn grain in recent years. As a result, corn grain has been replaced in ruminant diets. In a study conducted in Israel (CARASSO et al., 1988), three concentrate mixtures containing different ratios of ground corn and barley grain were compared using early weaned Saanen intact male kids. The levels of corn and barley in the concentrate mixture were: 10 and 69.3 %, 69.2 and 4.7 %, 39.6 and 37.0 %, respectively. The three concentrate mixtures were isoenergetic (13.18 MJ of ME/kg DM) and isonitrogenous (17.3 % CP, DM basis). The kids on the high-corn diet consumed more feed, grew faster and had better feed conversion efficiency than those on the barley-rich diet. No difference was found in feed intake, feed efficiency and average daily gain between kids offered the medium or high-corn diet (Table 1).

Cereal grains frequently constitute a major part of concentrate mixtures used in growing goats. Therefore, efficient and economic utilization of cereal grains will maximize economic returns to the farmer. Processing of cereal grains has been practised with the purpose of ensuring better mixing of the ingredients and for improving their digestibility. However, ORSKOV (1979) reviewing the available literature concluded that any processing of cereal grains given to adult sheep and goats and early-weaned fattening lambs is likely to be of no value, and suggested that barley should be fed whole to these animals. Moreover, the feeding of whole cereal grains prevented the development of rumenitis (ORSKOV, 1973), soft fat syndrome, altered rumen fermentation pattern in lambs (ORSKOV et al., 1974) and increased milk fat content and yield in ewes and goats (ECONOMIDES et al., 1989). Studies on the effect of cereal grain processing and form of concentrate mixture on

284

the performance of growing kids are lacking. In preliminary studies conducted at the Cyprus Agricultural Research Institute (ECONOMIDES et al., unpubl.), kids on the mash diet grew slower and required considerably more feed per kg bodyweight gain compared to pelleted and WGP diets (Table 2). Feed to gain ratio was consistently lower on the pelleted diet compared to mash (trial 1) or WGP and RGP diets (trial 1) or WGP and RGP diets (trial 2, 3, 4, 5 and 6). Differences in feed/gain ratio between WGP and RGP diets were relatively small. Contrary, in similar studies with lambs offered the same diets, no difference between pelleted and WGP or RGP diets was observed supporting the findings of ORSKOV (1979). Response of female lambs and kids to processing was similar to that of males.

Table 1

PERFORMANCE OF EARLY WEANED KIDS FED CONCENTRATES
DIFFERING IN CORN TO BARLEY GRAIN RATIO
(CARASSO et al., 1988)

| | Concentrate | | |
	Low corn	High corn	Medium corn
No. of animals	8	8	6
Initial weight (kg)	11.5	11.5	11.4
Final weight (kg)	16.3 [a]	17.2 [b]	16.9 [ab]
Average daily gain (g/day)	178 [a]	211 [b]	204 [b]
Feed intake (kg)	15.2 [a]	16.9 [b]	16 [b]
Feed to gain ratio	3.29 [a]	2.97 [b]	2.9 [b]

Means in the same line with different superscripts differ significantly (P<0.05)

Table 2

PERFORMANCE OF MALE GROWING DAMASCUS KIDS OFFERED DIFFERENT FORMS OF
CONCENTRATE MIXTURES
(ECONOMIDES et al., unpubl.)

	Mash(1)	P(1)	WGP(1)	RGP(1)
Trial 1				
Weight gain (g/d)	159b(3)	294a(3)	263a(3)	-
Feed/gain ratio (2)	8.38	4.39	5.22	-
Trials 2.3 and 4 (mean)				
Weight gain (g/d)		284 a	211 b	-
Feed/gain ratio (2)	-	3.37	4.32	-
Trial 5				
Weight gain (g/d)	-	246	255	258
Feed/gain ratio (2)	-	4.10	4.56	4.67
Trial 6				
Weight gain (g/d)	-	277	254	235
Feed/gain ratio (2)	-	4.97	5.47	5.94

(1) **Mash**: all ingredients were ground and mixed; **P**: all ingredients were ground and then pelleted in 5 mm cubes; **WGP**: whole barley grain was mixed with pellets made from the other ingredients of the concentrate mixture; **RGP**: barley grain was rolled and then mixed with pellets made from the other ingredients of the concentrate mixture.

(2) Group feeding no statistics.

(3) Means in the same line with different superscripts differ significantly (P<0.05).

Dietary lipid supplementation

Low subcutaneous adipose deposits on kid carcasses reduce their commercial value on certain markets of the world (see Chapter 24). Dietary fat supplies resulted in an improvement of postweaning growth rate in Alpine kids. Offals and intermuscular adipose deposits were increased especially in weaned kids, but subcutaneous adipose tissue was not affected (BAS et al., 1987). In another study (LANDAU et al., 1988), inclusion of 4 % of long chain fatty acids to the concentrate mixture did not improve growth performance and had no effect on the partition of fat on kid carcasses.

By-product utilization

Studies conducted in Cyprus (HADJIPANAYIOTOU, 1984) showed that kids fed diets containing dried poultry litter (15 and 30 % in concentrate) perform equally well as those on the control diets. In a comparative study among growing Damascus goats, Friesian heifers and Chios sheep given constant quantities of concentrates, barley straw and barley hay to meet 70 % of their requirements, and silage (4 parts citrus pulp to 1 part poultry litter) *ad libitum*, there were significant differences in silage DM intake among species (HADJIPANAYIOTOU, unpubl.). Silage intake by growing goats was negligible (3 g DM/ kg $W^{.75}$) even after they were offered silage for 28 days. Silage intake by heifers and growing ewes at the same period reached 63 and 20 g DM/kg $W^{.75}$ daily, respectively.

FEEDING SCHEDULES

Feeding schedules used for young goats at the present time are mostly similar to those employed in sheep production (MORAND-FEHR et al., 1982). It must be emphasized however, that feeding schedules for young goats should be adapted to their nutritional characteristics which might be modified whether young goats are intented for reproduction or slaughter, and the type of animal demanded by the market.

The quality of solid feed offered during the early post-weaning period greatly depends on the age at weaning. Because of low dry matter intake in the early postweaning period, the use of solid feed of high crude protein content (240 g CP/kg DM) is recommended during the first 2 weeks after weaning for the early-weaned (5 to 6 weeks) kids. Damascus kids did not suffer any growth delay, but gained weight (males 171 g/day, females 122 g/day) during the first 2 weeks (43-56 days) after weaning when they were offered concentrate (230 g CP/kg DM) *ad libitum* along with 100 g of lucerne hay. The daily intake of concentrates and lucerne hay was 160 and 100 g, respectively. Information on the preweaning feeding and management of kids is given in Chapters 21 and 22.

Replacements

Female kids are reared to give birth as yearlings. Studies conducted in Cyprus (MAVROGENIS and CONSTANTINOU, 1983) with Damascus kids showed that lifetime performance, in terms of kid liveweight output and total milk production, is higher in goats kidding as yearlings than those kidding as 2-year old goats; the difference being about 17 % in milk production and 6 % in total kid liveweight at weaning. This is in line with French observations on Alpine kids reared under intensive conditions.

In the high-concentrate feeding system prevailing in Cyprus where early-weaned female kids are given concentrates (180 g CP/kg DM) *ad libitum* along with 100 g of good quality leguminous hay up to the age of about 3 months, and then followed by a controlled feeding, almost all kids reach mating weight (70 % of mature weight) at the age of 7 months. High feeding schedules from 2 months of age to the appearance of first oestrus resulted in lower milk yield in first lactation by dairy heifers (JOHNSON, 1986). The effect of feeding schedules on the milk yield of yearling Norwegian goats is now investigated (HAVREVOLL et al., 1988); such data are urgently needed and for other breeds in order to develop appropriate feeding systems.

In France, kids are generally mated when they reach 50 to 60 % of their mature bodyweight (MORAND-FEHR et al., 1982). Alpine kids weaned at 5 to 6 weeks of age and then offered hay *ad libitum* and variable amounts of concentrates (100-500 g/day) reached 31 kg bodyweight at the age of 30 weeks (MORAND-FEHR et al., 1982). From the age of 7-8 months onwards, Alpine young goats are mated if they weigh more than 30 kg.

Norwegian growing goats are also mated at the age of 7-8 months. HAVREVOLL et al. (1987) reported that young goats should be fed to gain at least 100 g daily from birth to first kidding in order to obtain early puberty and kidding at the age of one year. A postkidding liveweight for young goats of around 35-40 kg or 75 % of mature size is desired. Since there is a positive correlation (0.54) between postkidding liveweight and subsequent milk yield special emphasis should be given on the feeding plan of growing goats.

Fattening

The feeding management of Alpine kids destined for slaughter depends on the age at slaughtering (MORAND-FEHR et al., 1982). For slaughter weights up to 18 kg feeding entirely on milk is satisfactory (FEHR et al., 1976). Contrary, at higher slaughter weights weaning of kids and feeding on good quality diets is recommended. Lower post-than pre-weaning growth rates were reported by MORAND-FEHR et al. (1982). On the other hand, in studies with Damascus kids (ECONOMIDES, 1986) greater post-than preweaning growth rates were obtained when the early-weaned Damascus kids (50-56 days of age) were offered balanced concentrate mixtures (180 g CP/kg DM) *ad libitum* along with 100 g of lucerne hay daily.

Growth rate is a function of feed composition and level of feed intake. Earlier data on growth rates reported by DEVENDRA and BURNS (1970) were disappointingly low. Recent data however, from Norway (SKJEVDAL, 1974, cited by NAUDE and HOFMEYR, 1981), France (FEHR et al., 1976; BAS et al., 1987), South Africa (NAUDE and HOFMEYR, 1981), United Kingdom (TREACHER et al., 1987), Israel (CARASSO et al., 1988; LANDAU, 1987) and Cyprus (HADJIPANAYIOTOU, 1982; ECONOMIDES, 1986; HADJIPANAYIOTOU et al., 1988; ECONOMIDES et al., unpubl.) have shown that kids have relatively satisfactory growth potential when fed on good quality diets, and in some instances growth rates similar to lambs may be obtained. Growth rates, carcass characteristics and feed to gain ratio of Damascus kids offered concentrate *ad libitum* along with 100 g of lucerne hay/head/day and slaughtered at different bodyweights are shown in Table 3. Feed conversion (kg feed per kg bodyweight gain) improved with successive substitution of concentrate for roughage (NAUDE and HOFMEYR, 1981). Early data showed that fattening kids are less efficient than lambs. In relatively recent studies however, with Damascus (HADJIPANAYIOTOU, 1982; HADJIPANAYIOTOU et al., 1988), Boer (NAUDE and HOFMEYR, 1981), Norwegian (SKJEVDAL, 1974, cited by NAUDE and HOFMEYR, 1981), Alpine (FEHR et al., 1976), British Saanen (TREACHER et al., 1987) and Israel Saanen (CARASSO et al., 1988) kids, feed conversion efficiencies comparable to those obtained in lambs were attained. The performance of Chios lambs and Damascus kids from weaning (45 days of age) to 40 kg from males and 35 kg for females was studied by ECONOMIDES (1986). Lambs grew faster and kids required about 2 months more to reach the same liveweight as lambs. Males grew faster than females. Feed intake of lambs was higher than that of kids and feed conversion efficiency of kids was poorer than that of lambs.

ENERGY REQUIREMENTS

NRC (1981) gave an energy allowance for maintenance of 101.38 Kcal/ kg $W^{.75}$ with an additional supplement of 7.25 Kcal ME/g of gain. ZEMMELINK et al. (1985) gave a value of 9.09 Kcal ME/g gain by West African kids. Recommended energy allowance for breeding female goat kids 4 to 7 months old with a growth rate of 104 g/day was 105 Kcal NE/ kg $W^{.75}$ (MORAND-FEHR, 1981). Energy requirement for growing 12 to 34-week old goats with a growth in various breeds of Indian goats weighing 13 kg with an average weight gain of 88 g/day was 7.9 Kcal digestible energy per g of gain (SENGAR, 1980). LU et al. (1987) suggested that 9.0, 14.1 and 13.7 Kcal ME was required per g of gain in growing dairy goat kids fed diets containing 3.05, 2.77 and 2.46 Mcal ME/kg DM, respectively. Basic information about energy requirements of growing goats is given in Chapter 7.

Table 3

CONVERSION OF MILK OR SOLID FEED TO KID CARCASS
FROM BIRTH TO 40 kg LIVE WEIGHT
(ECONOMIDES, unpubl.)

			Slaughtered at			
	Birth	Weaning	25kg	30kg	35kg	40kg
No. of animals	10	5	5	5	5	5
Slaughter liveweight (kg)	4.20	15.5	25.0	30.0	35.0	40.0
Age at slaughter (days)	2	52	93	115	130	151
Cold carcass weight (kg) (IS)	1.85	8.0	11.5	14.5	17.5	20.0
Cold carcass weight (kg) (CYS)	2.40	9.7	14.0	17.4	20.7	23.8
Kg milk/kg carcass gain (IS)	-	12.8	-	-	-	-
Kg milk/kg carcass gain (CYS)	-	11.0	-	-	-	-
Kg concentrates + hay/kg carcass gain : (IS)	-	-	9.0	9.1	9.0	8.9
(CYS)	-	-	7.4	7.5	7.3	7.5
Intestinal fat (g)			185	290	430	720
Kidney fat (g)			48	130	220	300

(IS) :   International standards for carcass.
(CYS) : Cyprus standards for carcass (head, lungs, heart and liver included).

PROTEIN NUTRITION

Protein nutrition of growing goats has been outlined in chapter 9. There is considerable experimental evidence from Cyprus (LOUCA and HANCOCK, 1977; MAVROGENIS et al., 1979; HADJIPANAYIOTOU, 1982) showing that early weaned Damascus kids respond to increasing dietary protein concentration up to 180 g CP/kg DM. Dietary CP concentration for early weaned Damascus kids (52 days of age) and Chios lambs (35 to 42 days of age) to attain maximum growth rates is similar (180 g CP/kg DM) up to 90 to 100 days of age. Thereafter, the dietary CP concentration for Chios lambs can be reduced to 160 g CP/kg DM, but in kids, dietary protein should be maintained at 180 g CP/kg DM up to 140 days of age. From 90 days of age, female kids can attain high growth rates even when offered diets conducted in Cyprus (HADJIPANAYIOTOU, unpublished data) showed that early-weaned (52 days of age) Damascus kids perform better on fish meal compared to soybean meal. The effect of protein source was not significant in female kids and in male and female early-weaned Chios lambs.

IONOPHORES IN DIETS OF YOUNG GROWING GOATS

Coccidiosis in kids over six weeks of age is considered as the main cause of diarrhea, anorexia and weight loss (AUMONT et al., 1982). The role of ionophores in suppressing clinical infection of coccidia and in improving kid performance has been underlined in Chapter 16. The significance of ionophores in improving efficiency of production is attained not only through its coccidiostatic activity, but also through changes in rumen fermentation characterized by an increased propionate production and lower levels of methane loss; greater response to ionophores should be expected with high-roughage diets and under conditions favouring coccidiosis and poor nutrition.

CONCLUSION

Early mating at 7-8 months of age is widely practised. As a result, animals are offered relatively good quality feedstuffs, like cereal grains, protein supplements and forages of high digestibility and palatability. Corn grain promoted faster growth rates and better feed conversion efficiencies than barley grain. Grinding and pelleting of cereal grains is recommended for young growing goats for promoting faster growth rates and better feed conversion efficiencies. By-products can comprise part of the diet for growing goats. Addition of ionophores to the diets of growing goats improves efficiency of production significantly.

SUMMARY

Postweaning feeding is greatly affected by the age at weaning and intensity of production. Under intensive feeding systems where female kids are mated at the age of 7 months, and male kids are grown to attain maximum growth rates, good quality feedstuffs are extensively used. As a result, feeding costs are relatively high and research efforts have been undertaken aimed at improving the economics of production through the application of new nutritional concepts during the postweaning period. Addition of ionophores and pelleting of cereal grains enhanced growth rates and feed to gain ratios. Feeding schedules have been developed based on weaning age and rearing intensity. A variety of feedstuffs and by-products have been evaluated in diets of growing goats.

Keywords : Young goat, Postweaning Feeding, Growth, Fattening, Feeding schedules, By-products, Fats, Ionophores, Protein supplementation.

RÉSUMÉ

L'alimentation du chevreau après le sevrage est fortement affectée par son âge au sevrage et le niveau d'intensification de la production. Dans des systèmes d'alimentation intensive où les chevrettes sont saillies à l'âge de 7 mois, et les chevreaux mâles atteignent des vitesses de croissance maximales, des aliments de bonne qualité nutritive doivent être largement utilisés. En conséquence, les coûts alimentaires sont relativement élevés et des efforts de recherches ont été entrepris pour améliorer les résultats économiques en appliquant des concepts nutritionnels nouveaux au cours de cette période qui débute après le sevrage. L'addition d'antibiotiques ionophores et la granulation des céréales améliorent la vitesse de croissance et l'efficacité alimentaire. Les programmes d'alimentation se sont développés selon l'âge au sevrage et l'intensification de l'élevage. Divers aliments et sous-produits ont été testés dans les régimes des chevreaux en croissance.

REFERENCES

AUMONT G., YVORE P. and ESNAULT E., 1982. Coccidiosis in young goats, p. 566 (abst.). 3rd Intern. Conf. on Goat Production and Disease. Jan. 10-15, 1982, Tucson, Arizona (USA).

BAS P., MORAND-FEHR P., SCHMIDELY P. and HERVIEU J., 1987. Effect of dietary lipid supplementation on pre- and post-weaning growth and fat depostition in kids. Ann. Zootech. 36: 339 (Abst.).

CARASSO J., LANDAU S. and NITSAN Z., 1988. The effect of corn grain in concentrates for kids on their performance and ruminal activity. Seminar of FAO Subnetwork on Goat Nutrition and Feeding, October 3-5, 1988, Potenza (ITALY).

DEVENDRA C. and BURNS M., 1970. Goat Production in the Tropics. Tech. Commun. No. 19, p. 184. Commonwealth Agricultural Bureaux, Slough (UK).

ECONOMIDES S., 1986. Comparative studies of sheep and goats: milk yield and composition and growth rate of lambs and kids. J. Agric. Sci. (Camb.), 106: 477-484.

ECONOMIDES S., GEORGHIADES D., KOUMAS A. and HADJIPANAYIOTOU M., 1989. The effect of cereal processing on the lactation performance of Chios sheep and Damascus goats and the preweaning performance of their offspring. Anim. Feed Sci. Tech., 26: 93-104.

FEHR P.M., SAUVANT D., DELAGE J., DUMONT B.L. and ROY G., 1976. Effect of feeding methods and age at slaughter on growth performances and carcass characteristics of entire young goats. Livest. Prod. Sci., 3: 183-194.

HADJIPANAYIOTOU M., 1982. Protein levels for early weaned Damascus kids on high-concentrate diets. Techn. Bull. 43, 8 p. Agric. Res. Inst., Nicosia (CYPRUS).

HADJIPANAYIOTOU M., 1984. The use of poultry litter as ruminant feed in Cyprus. World Anim. Rev., 40: 32-38.

HADJIPANAYIOTOU M., PAPACHRISTOFOROU C. and ECONOMIDES S., 1988. Effects of lasalocid on growth, nutrient digestibility and rumen characteristics in Chios lambs and Damascus kids. Small Rum. Res., 1: 217-227.

HAVREVOLL O., NEDKVITNE J.J. and GARMO T., 1987. IV - Forslag og Foring (Feeds and feeding), p. 70-109. In: Geiboka (The goat book). Landbruksforlaget, Oslo (NORWAY).

HAVREVOLL O., EIK L.O. and NEDKVITNE J.J., 1988. Preliminary results from experiments with different energy levels in rearing of dairy goats. Seminar of FAO Subnetwork on Goat Nutrition and Feeding. Oct. 3-5, 1988, Potenza (ITALY).

JOHNSON D., 1986. Proper growth management important in raising of heifers. Feedstuffs, Oct. 20, 1986, p. 14-17.

LANDAU S., 1987. Ralgro implants as a growth-promoter for Saanen male kids. Ann. Zootech., 36: 342 (Abst.).

LANDAU S., RATTNER D., GUR-ARIE S. and BRAUN A., 1988. Growth performance, slaughtering data, and fatty acid composition of selected adipose tissue in Yaez and Sinai goat kids fed a concent ate compounded with or without a protected fat. Seminar of FAO Subnetwork on Goat Nutrition and Feeding, October 3-5, 1988, Potenza (ITALY).

LOUCA A. and HANCOCK J., 1977. Genotype by environment interactions for postweaning growth in the Damascus breed of goat. J. Anim. Sci. 44: 327-931.

LU C.D., SAHLU T. and FERNANDEZ J.M., 1987. Assessment of energy and protein requirements for growth and lactation in goats, Vol. 2, p. 1229-1247. 4th Intern. Conf. on Goats, March 8-13, 1987, Brasilia (BRAZIL).

MAVROGENIS A.P. and CONSTANTINOU A., 1983. Performance of Damascus goats bred as yearlings or as two-year olds. Techn. Bull. 45, p. 5. Agric. Res. Inst., Nicosia (CYPRUS).

MAVROGENIS A.P., ECONOMIDES, LOUCA A. and HANCOCK J., 1979. The effect of dietary protein levels on the performance of Damascus kids. Techn. Bull. 27, 11 p. Agric. Res. Inst., Nicosia (CYPRUS).

MORAND-FEHR P., 1981. Growth, p. 253-283. In: GALL C. (Ed.): Goat Production. Academic Press, London (UK).

MORAND-FEHR P., HERVIEU J., BAS P. and SAUVANT D., 1982. Feeding of young goats, p. 90-104. 3rd Intern. Conf. on Goat Prod. and Disease. Jan. 10-15, 1982, Tucson, Arizona (USA).

NRC, 1981. Nutrient Requirements of Goats, Nat. Acad. Sci., Washington, DC (USA).

NAUDE R.T. and HOFMEYR H.S., 1981. Meat Production, p. 285-307. In: GALL C. (Ed.): Goat Production. Academic Press, London (UK).

ORSKOV E.R., 1973. The effect of not processing barley on rumenitis in sheep. Res. Vet. Sci., 14: 110-112.

ORSKOV E.R., 1979. Recent information of processing of grain for ruminants. Livest. Prod. Sci., 6: 335-347.

ORSKOV E.R., FRASER C. and GORDON J.G., 1974. Effect of processing of cereals on rumen fermentation, digestibility, rumination time and firmness of subcutaneous fat in lambs. Brit. J. Nutr., 32: 59-69.

SENGAR O.P.S., 1980. Indian research on protein and energy requirements of goats. J. Dairy Sci., 63: 1655-1670.

SKJEVDAL T., 1974. Milk feeding of kids. Report No. 173, Vol. 53 NR 39. Dep. Anim. Nutr. Agric. Univ. (NORWAY).

SKJEVDAL T., 1982. Nutrient requirements of dairy goats based on Norwegian research, p. 105-112. 3rd Intern. Conf. on Goat Prod. and Disease. Jan. 10-15, 1982, Tucson, Arizona (USA).

TREACHER T.T., MOWLEM A., WILDE R.M. and BUTLER-HOGG B., 1987. Growth, efficiency of conversion and carcass composition of castrate male Saanen and Saanen x Angora kids on a concentrate diet. Ann. Zootech., 36: 341-342.

ZEMMELINK G., TOLKAMP B.J. and MEINDERTS J.H., 1985. Feed intake and weight gain of West African Dwart goats, p. 25-34. In: SUMBERG J.E. and CASSADAM K. (Eds.): Sheep and goats in humid West Africa. ILCA, Addis Ababa (ETHIOPIA).

Chapter 24

INFLUENCE OF FEEDING AND REARING METHODS ON THE QUALITY OF YOUNG GOAT CARCASSES

P. MORAND-FEHR, Ø. HAVREVOLL, P. BAS, P. COLOMER-ROCHER, A. FALAGAN, M.R. SANZ SAMPELAYO, D. SAUVANT and T.T. TREACHER

INTRODUCTION

Although the statistics must be treated with caution, world goat meat production is estimated to be about 2,400 million tons (FAO, 1988). This is probably an underestimate, and goat meat production holds a potential that is not to be ignored in comparison with other meats. Furthermore, in certain regions especially in arid zones, goat meat makes an indispensable contribution to animal protein supplies.

Unfortunately, because goat meat producers account for much of the consumption themselves and because the market is small and production systems varied and complex, few scientific, economic or technical studies have so far been carried out.

Several reports have described the main characteristics of goat carcass composition (FEHR et al., 1976; MAC DOWELL and BOVE, 1977; NAUDE and HOFFMEYER, 1981; GALL, 1982; COLOMER-ROCHER, 1987). However little work has been done to try to manipulate carcass quality by varying rearing techniques or feeding systems. Below we give an account of the state of the art in this field, basing our report on work undertaken over the past few years by members of the FAO sub-network of research on goat production.

THE YOUNG GOAT CARCASS: CHARACTERISTICS AND QUALITY

According to official definitions that apply mainly to sheep, the carcass is defined as the product of slaughtering, after bleeding out, drawing, skinning and separation of head and tail, the kidneys and kidney fat being included in the carcass. Unfortunately, goat carcasses are traditional products, their composition and presentation varying widely from country to country and even from region to region of the same country. In France, for example, young goat carcasses are put on the market with the pluck (head, kidney, heart, lungs, spleen, kidneys and caul). Under these conditions, it is hard to compare different studies of carcass production, especially as regards weight, carcass characteristics and dressing percentage.

However, COLOMER-ROCHER et al. (1987) have given a detailed description of a standard carcass and a cutting method that would allow researchers and goat specialists to compare their findings and assess different genotypes for meat production. These proposed standards draw on the sheep standards established by BOCCARD and DUMONT (1955). They are well suited to goat carcasses, however, and have the advantage of being simple to apply.

A carcass must be kept well in a cold store, with the least possible loss. Butchers want a carcass that enables them to market as much meat as possible; consumers buy according to their tastes, cooking traditions and eating habits. The notion of carcass quality is a complex one, generally expressed in terms of the proportion of top grade cuts, the distribution of the main fatty tissues and the proportions of muscle and fat (BOCCARD and DUMONT, 1976). These parameters must be estimated with the carcass intact, using carcass measurements and subjective scores for level of fatness and conformation. As a rule, a short, broad carcass is sought, with plump muscles and fatness according to consumer demand; sufficient surface fat is also usually required to avoid loss of meat by trimming after cold storage.

How, then, should the quality of a young goat carcass be assessed ? This depends on age at slaughter, mode of production, and consumer demand. Moreover, young goat carcasses are generally light (4 to 15 kg) and are rarely cut into more than four pieces. The notion of conformation therefore seems to be less important, in many countries, than with beef or mutton meat. A survey carried out in France's national meat market at Rungis showed that where a lightweight carcass (4-5 kg) is sought, the main quality criteria applied by professionals are sufficient fat, followed by white fat colour and pale muscle (MORAND-FEHR et al., 1980). There is no reference to the notion of conformation, which is already included in the assessment of fatness.

Goats are characteristically poor in conformation, especially in comparison to sheep, because their carcasses are long and narrow and fat deposition tends to take place later than in other ruminants. Many authors (FEHR et al., 1976; OWEN et al., 1978; GALL, 1982; GAILI et al., 1985; CASEY, 1987) have observed low fat proportions in young goats, and especially a very light covering of subcutaneous fat. In fact quite high proportions of fat can be attained in goats, but only in adults or at the end of the growing period. COLOMER-ROCHER (1986, unpubl.) notes total fat proportions of as much as 30 % on adult Saanen goat carcasses - i.e. four times as much fat as the proportions generally recorded for carcasses of about 5 kg from young Saanen goats slaughtered at five or six weeks. The proportion of muscle, the proportion of bone to a lesser extent, and the muscle: bone ratio are generally higher in goat carcasses than in sheep.

This lack of fat, especially in the subcutaneous adipose tissue, and to a lesser extent the poor conformation, are the main characteristics of young goat carcasses. Both features, however, fatness particularly, vary a great deal according to the genotype and husbandry method employed. To attain the high or low fat levels a particular market may demand, goat farmers can vary feeding techniques and husbandry methods.

THE EFFECT OF WEIGHT AND AGE AT SLAUGHTER

As a young goat gets older and heavier, conformation improves: body width increases faster than length or depth (Figure 1) (FALAGAN, 1986; FEHR et al., 1976). The carcass becomes more compact as it gets heavier (weight: length ratio). Moreover, the proportion of "Extra" category cuts (leg, saddle, ribs) increases at the expense of "First" category (shoulder), while there is little variation in the proportion of "Second" category cuts (middle neck, point of breast) (Figure 2).

The fatness of the carcass, as graded on the basis of a visual assessment, the proportion of fatty tissue in the carcass, or the weight of the caul (FEHR et al., 1976)), shows marked improvement as carcass weight increases (Figures 3, 4, 5, 6) (OWEN and MTENGA, 1980; BORGHESE et al., 1985; FALAGAN, 1986; MORAND-FEHR et al., 1986; SANZ SAMPELAYO et al., 1987; TREACHER et al., 1987), provided the animal has not suffered any nutrient deficiency. The proportion of bone decreases, the proportion of muscle may increase or decrease slightly. As a result, the muscle:bone ratio increases very considerably. All these results confirm and complement certain earlier findings, those of WILSON (1958) particularly.

Figure 1

EFFECT OF LIVE WEIGHT ON CARCASS
MEASUREMENTS
OF MURCIANA BREED KIDS

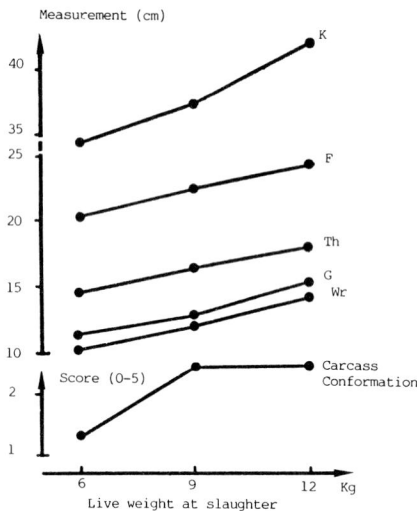

(from FALAGAN, 1986)

K: distance between the base of the tail and the base
   of the neck
G: Width at the level of trochanters
F: Distance between the perineum and the surface
   of the tarso-metatarsic articulation
Wr: Width at the level of ribs
Tr: Depth of breast

Figure 2

EFFECT OF LIVE WEIGHT ON THE
PROPORTIONS OF DIFFERENT CARCASS
CUTS IN MURCIANA BREED KIDS

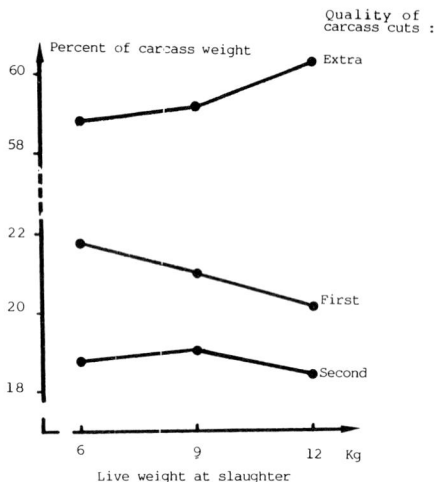

(from FALAGAN, 1986)

Extra : Long leg and ribs
First : Shoulder
Second : Flank and neck.

Under these conditions, the age and weight of young goats at slaughter can be seen to be very important factors for improving the conformation and fatness of the carcasses. These results are entirely consistent with those already recorded in other ruminants, lambs especially (BOCCARD and DUMONT, 1960, 1976), but in fact it is often the market that decides what carcass weight can be marketed and so sets the proportions of muscle, bone and fat, as shown by COLOMER-ROCHER (1987).

THE EFFECTS OF WEANING

Weaning - the shift from milk to solid feed, i.e. from the preruminant stage to the ruminant - has a major impact on carcass quality. During the adaptation phase, this feeding change often brings a slowdown, or indeed a halt, in growth, and a depletion of body reserves (see Chapter 22). Adipose deposits and body condition decline (FEHR et al., 1986; SAMPELAYO et al., 1987), as does conformation. In particular, the proportion of fat in the carcass and the fattening score decline sharply, and remain low for a variable period after weaning, depending on how quickly the young goat begins to consume sufficient energy from the solid feed (Figures 5, 7). In fact, it seems that abdominal fat is mobilized at weaning more than carcass fat, and that the post-weaning check in growth effects carcass fat slightly.

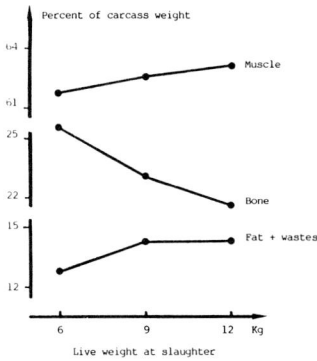

Figure 3

EFFECT OF LIVE WEIGHT ON CARCASS TISSUE COMPOSITION IN MURCIANA KIDS (from FALAGAN, 1986)

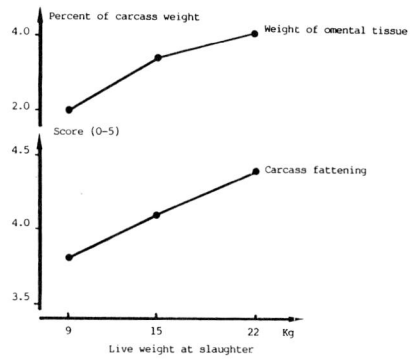

Figure 4

EFFECT OF LIVE WEIGHT ON CARCASS FATTENING IN ALPINE KIDS (from MORAND-FEHR et al., 1986)

The slaughter of weaned kids should be delayed until at least fifty days after weaning. Too short an interval from weaning to slaughter may cause a significant decline in carcass quality. It may be useful to wean early so as to prolong the period between weaning and slaughter. MAIORANA et al. (1984) observed a tendency for young goats weaned at six weeks to show a better conformation when slaughtered at fifteen weeks than did kids weaned at eight weeks.

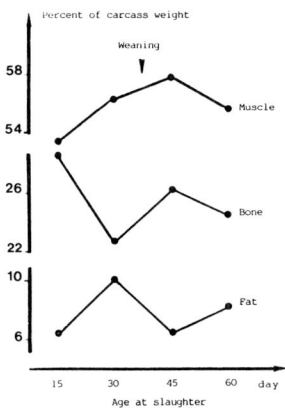

Figure 5

EFFECT OF LIVE ON CARCASS TISSUE COMPOSITION IN GRANADINA KIDS (from SANZ SAMPELAYO, 1987)

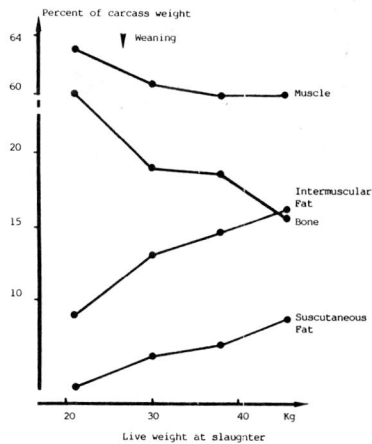

Figure 6

EFFECT OF LIVE WEIGHT ON CARCASS TISSUE COMPOSITION IN SAANEN KIDS (from TREACHER et al., 1987)

295

Figure 7

EFFECT OF WEANING ON CARCASS FATTENING IN ALPINE KIDS

(from MORAND-FEHR et al., 1986)

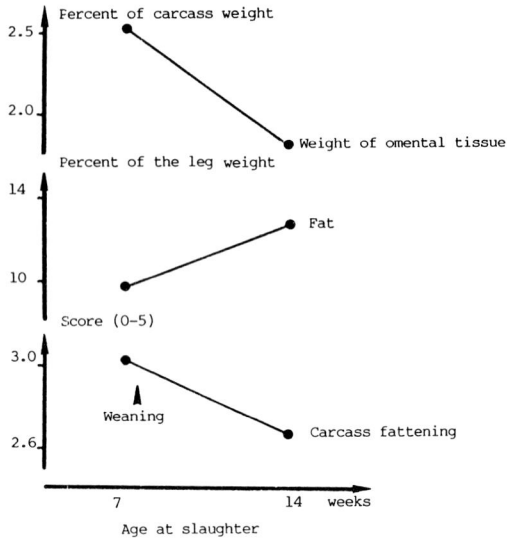

THE EFFECTS OF CASTRATION

As with other ruminants, castration of young male goats produces a clear increase in fatness, and a less marked improvement in conformation (LOUCA et al., 1977; OWEN and MTENGA, 1980; CASTELLANOS RUELAS et al., 1982; MORAND-FEHR et al., 1986) (Figures 8, 9). The proportion of adipose tissue increases, that of muscle decreases. In some cases it would seem that the subcutaneous adipose tissue develops slightly more than fat at other sites, which is particularly advantageous for cold storage of the carcass. Furthermore, the lipid content of the muscles also increases, which would tend to make the meat more succulent (MORAND-FEHR et al., 1986). Thus the characteristics of carcasses from castrated young male goats are apparently similar to those of young females, which usually have markedly higher proportions of adipose tissue and muscle lipids than do entire males (WILSON, 1960; KIRTON, 1970; BORGUESE et al., 1985).

These adipose deposits would appear to develop in inverse proportion to the secretion of androgens (KUMAR et al., 1980); this explains why late castration, at 7.5 months, is far less effective than castration at 15 days, independent of the fact that it is far more disturbing for the animal concerned (LOUCA et al., 1977). In fact, the level of fatness of the young goat increases because castration slows down growth while maintaining satisfactory feed intake levels, and because from puberty onwards castrated animals are much calmer than entire males. The differences between entire and castrated males therefore become increasingly marked as carcass weight increases (OWEN and MTENGA, 1980; TREACHER et al., 1986).

In some countries, however, castration of young male goats is difficult for reasons of custom or religion. Sometimes the slowdown in growth rate and the higher cost of feeding per kg of liveweight gain can also be obstacles to wider use of this method.

Figure 8	Figure 9
EFFECT OF CASTRATION ON CARCASS TISSUE COMPOSITION IN SAANEN MALE KIDS SLAUGHTERED AT 36.5 kg LW (from OWEN and MTENGA, 1980)	EFFECT OF CASTRATION ON CARCASS FATTENING IN ALPINE KIDS SLAUGHTERED AT 21 kg LW (from MORAND-FEHR et al., 1986)

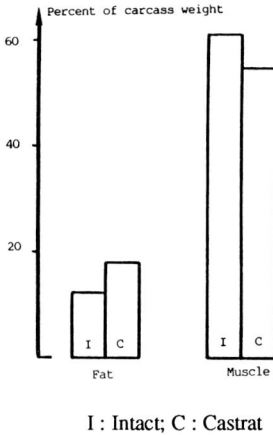

I : Intact; C : Castrat

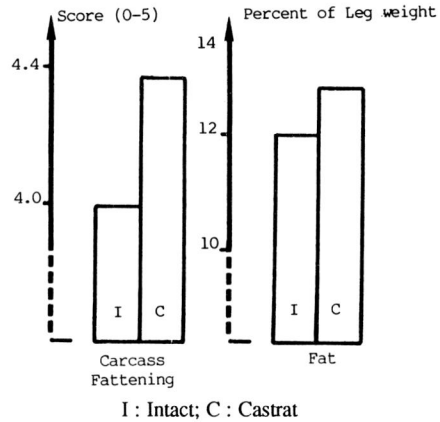

I : Intact; C : Castrat

THE EFFECTS OF DIET

During growth, young goats receive a diet based on milk prior to weaning, and on forage and concentrates after weaning. As the two phases have specific effects, we shall discuss them separately.

Milk feeding

In dairy systems, young goats left to suckle or fed on goat milk will reduce milk sales. As a result, rearing systems have developed over recent years in which unweaned goats are fed on milk replacers, even in extensive systems. The earliest experiments were conducted between1966 and 1980 (MORAND-FEHR et al., 1982), and these showed that use of a milk replacer in place of goat milk slowed down growth and reduced carcass quality, particularly fatness and fat colour when the young goats were slaughtered early, at a liveweight of 8-10 kg.

However, the results of several more recent experiments (THEODORO and SOUSA, 1987; SANZ SAMPELAYO et al., 1987; FALAGAN, 1986; MORAND-FEHR et al., 1986) differ somewhat from these earlier results. They show that, even when dry matter intake levels are similar and carcass weight the same, fatness tends to be less and muscle development greater with milk substitutes (Figures 10, 11, 12). In the case of a milk replacer diet these results can easily be explained by its lower fat content and generally lower digestibility than with goat milk in early lactation, and hence by a lower net energy supply. But when energy intake is similar in the two cases, certain fattening characteristics observed with milk replacer are very close to those recorded with goat milk. The fat of young goats fed on goat milk is generally whiter, however. Lastly, FALAGAN (1986) observed poorer conformation (length, grade) in carcasses of Murciana breed goats fed on milk replacer, even though their weight was identical to those of young goats fed on goat milk. This result has not been reported by other authors and needs corroboration.

Figure 10

EFFECT OF THE KIND OF MILK
ON CARCASS TISSUE COMPOSITION IN
SERRANA KIDS
(from TEODORA and SOUSA, 1987)

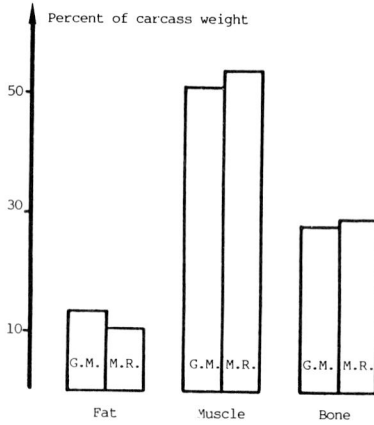

G.M. : Goat Milk; M. R. : Milk replacer

Figure 11

EFFECT OF THE KIND OF MILK
ON CARCASS TISSUE COMPOSITION IN
GRANADINA KIDS
(from SANZ SAMPELAYO et al., 1987)

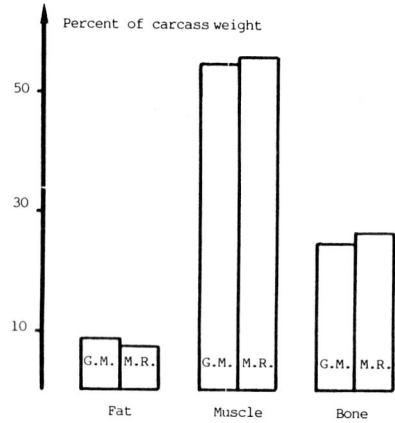

G.M. : Goat Milk; M. R. : Milk replacer

Figure 12

EFFECT OF THE KIND OF MILK
ON CARCASS FATTENING
IN ALPINE KIDS
(from MORAND-FEHR et al., 1986)

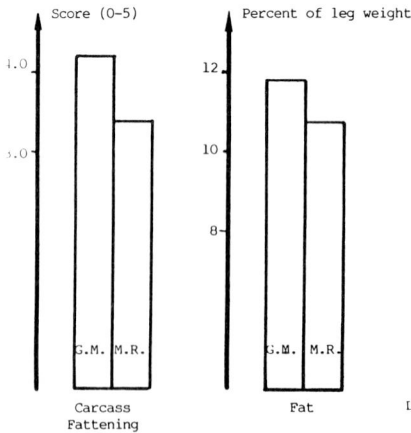

G.M. : Goat Milk; M. R. : Milk replacer

Figure 13

EFFECT OF THE LEVEL OF INTAKE
ON CARCASS TISSUE COMPOSITION IN
GRANADINA KIDS
(from SANZ SAMPELAYO et al., 1987)

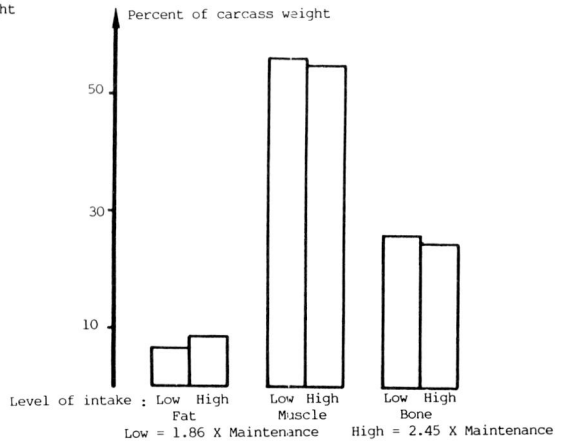

G.M. : Goat Milk; M. R. : Milk replacer

The importance of energy intake is confirmed in experiments in which young goats are fed differing quantities of the same milk (Figure 13) (SANZ SAMPELAYO et al., 1987). Fattening increases as milk intake, and hence energy intake, rise. It would therefore seem that with young goats, higher energy intake leads not only to faster growth but also to greater fattening.

An experiment with Alpine goats slaughtered at 14-16 kg compared two milk substitutes with different fat contents (Figures 14, 15) (MORAND-FEHR et al., 1986; BAS et al., 1986). An increase in carcass fatness with the high fat milk seems to occur only when there is a sufficiently marked difference in energy concentration between the two milks. SOLHEIM and HAVREVOLL (1985, unpubl.) confirm this, obtaining better carcass quality and increased fatness from young goats fed a milk replacer with 19 % fat than from those fed a defatted milk powder. However, as the milk replacers available on the market differ little in their energy levels, this seems to be limited as a way to improve fattening, particularly as young goats fed *ad libitum* can regulate their own consumption and energy intake. In economic terms, this would seem to offer little advantage in most cases.

Post-weaning feeding

The longer the weaning-to-slaughter interval, the weaker the impact of the milk diet on carcass quality and the greater the influence of the post-weaning diet.

As early as 1958, WILSON had shown with East African dwarf goats that different feed intake levels, and hence energy supply levels, did not alter conformation or the distribution of cuts after cutting when carcasses of equal weight were compared. On the other hand, they had a strong impact on fatness, affecting the proportion of adipose tissue in the carcass. The latest results have merely confirmed these early findings. Conformation differences found in the same genotype are generally due to comparing carcasses of different weights, even though in some cases the animals were slaughtered at exactly the same age.

OWEN and MTENGA (1980) observed no significant differences in the body composition of young Saanen goats receiving lucerne-based or barley-based diets. CASTELLANOS RUELAS et al. (1982) report no difference at all between the carcass conformation scores of young goats fed a maize silage diet and those of goats fed a concentrate diet. On the other hand, fatness was significantly better in young goats fed on maize silage, owing to the higher energy intake during fattening (Figure 16).

BAS et al. (1986) fed goats with pellets containing 2 or 20 % fat respectively, and confirmed that energy supply plays an essential role in determining fatness. All the parameters used to assess fatness, when applied to carcasses from young goats fed the very fat-rich diet, reached values rarely observed with young weaned goats (Figure 17).

Thus despite the fundamental change of diet at weaning, the same observations have been made as with milk-fed kids: conformation is not much affected by dietary factors, but fatness does react to energy intake levels.

As young goats have little subcutaneous fat, there is less danger than with lambs of producing a covering of soft fat that will diminish carcass quality. BAS et al. established the facts in this regard in 1982.

Additives in the ration can also alter carcass quality (see Chapter 16).

Figure 14

EFFECT OF FAT PERCENT
IN MILK REPLACER ON CARCASS
FATTENING IN ALPINE KIDS
(from MORAND-FEHR et al., 1986)

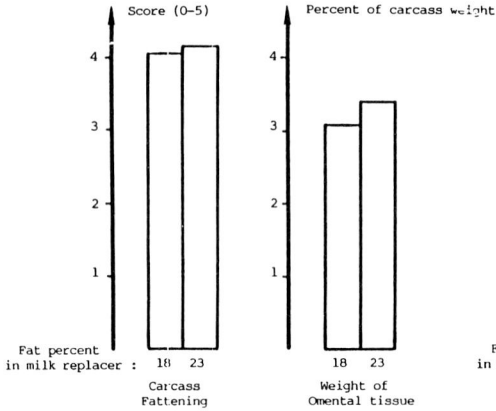

Figure 15

EFFECT OF FAT PERCENT IN MILK
REPLACER ON CARCASS FATTENING
IN ALPINE KIDS
(from BAS et al., 1986)

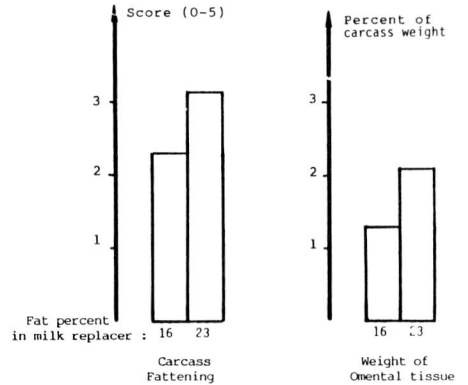

Figure 16

EFFECT OF THE TYPE OF DIET AFTER
WEANING ON CARCASS CONFORMATION
AND FATTENING IN ALPINE KIDS
(from CASTELLANOS RUELAS et al., 1982)

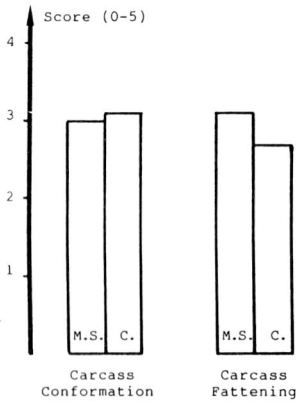

Figure 17

EFFECT OF THE FAT CONTENT IN
CONCENTRATE AFTER WEANING ON
CARCASS FATTENING IN ALPINE KIDS
(from BAS et al., 1986)

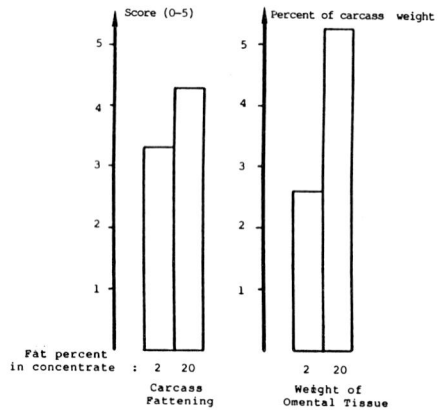

M. S. : Maize Silage; C. : Concentrate

CONCLUSION

Carcass quality is a very complex notion, depending as it does on each country's habits as regards slaughtering, marketing and cooking (COLOMER-ROCHER, 1987). It is therefore a delicate matter to draw definitive conclusions or guidelines for goat meat production. In goats, as we have seen, the notion of carcass quality is even harder to apply as conformation is less important than with other animals because the carcass is seldom cut and because a high proportion of goat meat is consumed at home by its producers. Under these conditions, fatness is usually the main criterion for judging young goat carcass quality.

From the data in this chapter it would seem that conformation can be improved mainly by prolonging fattening and slaughtering at heavier weights. Conformation can also be improved, obviously, by choosing an improved meat breed (CASEY, 1987).

Fatness is also increased where carcass weight is heavier, especially with regard to subcutaneous fat deposits, which develop late. Any dietary change that tends to increase energy consumption, either in the pre-weaning or post-weaning stage, increases the development of adipose deposits and hence fatness. Unfortunately, we cannot yet say whether diets rich in crude protein tend to restrict fatness, as insufficient data are available.

Furthermore, a very severe post-weaning check and slaughtering too soon after weaning reduce fatness. On the other hand, castration improves it, though unfortunately often at the expense of growth rate and feed conversion efficiency.

SUMMARY

Young goat carcasses are generally longer, narrower and leaner than lamb carcasses compared at identical weight, especially as regards subcutaneous fat. As carcass weight increases, conformation improves and fatness increases. Weaning tends to have a negative impact on both these factors. Castration increases fatness, but reduces the growth rate.

Dietary factors do not affect carcass conformation, but they can have quite a significant effect on fatness. Goat milk in comparison to milk replacer, a high fat content in the milk substitute and post-weaning diets based on maize silage or rich in fats, all tend to increase carcass fatness. But the animals' response seems to depend essentially on energy intake from the dietary regimens used in the experiments, whether with weaned or unweaned kids.

Keywords : Young goat, Carcass quality, Fattening, Milk feeding, Postweaning feeding.

RÉSUMÉ

Les carcasses de chevreaux sont généralement plus longues, plus étroites et moins grasses, notamment au niveau du gras sous-cutané, que les carcasses d'agneaux lorsqu'elles sont comparées au même poids. A mesure que le poids de la carcasse augmente, sa conformation et son état d'engraissement augmentent. Le sevrage tend à détériorer ces deux critères. La castration améliore l'état d'engraissement mais réduit la vitesse de croissance.

Les facteurs alimentaires n'influencent pas la conformation de la carcasse; en revanche, ils peuvent modifier assez sensiblement l'état d'engraissement. Le lait de chèvre par rapport au lait de remplacement, un taux élevé de matière grasse dans l'aliment d'allaitement, un régime à base d'ensilage de maïs ou riche en matière grasse après le sevrage tendent à augmenter l'état d'engraissement des carcasses. Mais la réponse des animaux semble dépendre essentiellement des niveaux énergétiques des régimes utilisés dans les expériences réalisées sur chevreaux non sevrés ou sevrés.

REFERENCES

BAS P., HERVIEU J., MORAND-FEHR P. and SAUVANT D., 1982. Facteurs influençant la composition des graisses chez le chevreau de boucherie : incidence sur la qualité des gras de carcasses (Factors influencing the composition of fats in young goats: effect on the quality of carcass fats). Vol. 1, p. 90-100. In: MORAND-FEHR P., BOURBOUZE A. and DE SIMIANE M. (Eds.): Nutrition and systems of goat feeding. Symposium International, Tours (FRANCE). May 12-15, 1982, INRA-ITOVIC, Paris (FRANCE).

BAS P., ROUZEAU A. and MORAND-FEHR P., 1986. Lipogenèse des tissus adipeux de chevreaux sevrés à 4, 6 ou 8 semaines (Lipogenesis in adipose tissues of young goats weaned at 4, 6 and 8 weeks). Reprod. Nutr. Dévelop., 26: 649-658.

BOCCARD R. and DUMONT B.L., 1985. Étude de la production de la viande chez les ovins. I - La coupe des carcasses. Définition d'une découpe de référence (Study of sheep meat production: Carcasse cutting and definition of a standard cutting). Ann. Zootech., 3: 241-257.

BOCCARD R. and DUMONT B.L., 1976. La qualité des carcasses des ovins (Quality of sheep carcasses). 2e Journées de la recherche ovine et caprine. INRA-ITOVIC, Paris (FRANCE).

BORGHESE A., MAIORANA M. and RUBINO R., 1985. Caratteristiche quanti-qualitative delle carcasse dei capretti allevati con diverse tecniche (Carcass characteristics in kids reared with various management methods). Il Vergaro, 12: 17-22.

CASEY N., 1987. Meat production and meat quality from Boer goats, Vol. 1, p. 211-238. 4th Intern. Conf. on Goats, March 8-13, 1987, Brasilia (BRAZIL).

CASTELLANOS RUELAS A., BAS P., SAUVANT D. and MORAND-FEHR P., 1982. Influencia de la alimentacion, del acido proponico y de la castration sobre al crecimiento del cabrito y la composicion en acidos grasos de los depositos adiposos (Effect of the kind of diets, propionic acid and castration on kid growth and composition of fatty acids in adipose deposits). Technia Pecuria, Deciembre 1982, Suppl. 9: 13-37.

COLOMER-ROCHER P., MORAND-FEHR P. and KIRTON A.M., 1987. Standard methods and procedures for goat carcass evaluation jointing and tissue separation. Livest. Prod. Sci., 17: 149-159.

COLOMER-ROCHER P., 1987. Factors influencing carcass quality, carcass components and composition. Vol. I, p. 181-194. 4th Intern. Conf. on Goats, March 8-13, 1987, Brasilia (BRAZIL).

FALAGAN A., 1986. Note concernant l'influence de l'alimentation sur la croissance et les caractéristiques bouchères de chevreaux de races murciana-granadina (Note concerning the effects of diets on the growth and meat characteristics of Murciana-Granadina breed kids). 37th EAAP Annual Meeting, Sept. 1-4, 1986, Budapest (HUNGARY).

FAO, 1988. FAO Yearbook Production; Statistics series n° 88, Vol. 42, p. 260, Rome (ITALY).

FEHR P.M., SAUVANT D. and DUMONT B.L., 1976. Croissance et qualité des carcasses de chevreaux de boucherie (Growth and carcass quality of kids). p. 166-189. 2e Journées de la recherche ovine et caprine. INRA-ITOVIC, Paris (FRANCE).

GAILI E.S. and ALI A.E., 1985. Meat from Sudan Desert Sheep and Goats. Part 1: Carcass yield, offals and distribution carcass tissues. Meat Science 13: 217-227.

GALL C.F., 1982. Carcass composition. p. 473-477. 3rd Intern. Conf. on Goat Production and Disease, Jan. 10-15, 1982, Tucson, Arizona (USA).

KIRTON A.H., 1970. Body and carcass composition and meat quality of the New-Zealand feral goat. N.Z. J. Agr. Res., 13: 167-181.

KUMAR R., KUMAR A. and SINGH H., 1980. Note of the effect of castration on meat production in goats. Indian J. Animal Sci., 50: 1160-1162.

LOUCA A., ECONOMIDES S. and HANCOCK J., 1977. Effects of castration on growth rate feed conversion efficiency and carcass quality in Damascus goats. Anim. Prod., 24: 387-391.

MAC DOWELL R.E. and BOVE L., 1977. The goat as a producer of meat. Cornell Intern. Agr. Manograph. No 56, 40 pp.

MAIORANA M., RUBINO R. and PIZZILLO M., 1984. L'allattemento artificiale nell'allevamento caprino. Ann. Ist. Sperim. Zootech. 2: 117-136.

MORAND-FEHR P., SAUVANT D., HERVIEU J. and BAS P., 1980. Quality of young goat carcasses, technical and commercial aspects. 31th EAAP Annual Meeting, Sept. 1-4, 1980, Munich (GERMANY).

MORAND-FEHR P. BAS P., SCHMIDELY P. and HERVIEU J., 1986. Facteurs influençant la qualité des carcasses et en particulier son état d'engraissement (Factors influencing carcass quality and fattening status). p. 236-252. 11e Journées de la recherche ovine et caprine. INRA-ITOVIC, Paris (FRANCE).

NAUDE R.T. and HOFMEYER H.S., 1981. Meat production. p. 285-307. In: GALL C. (Ed.): Goat Production. Academic Press, London (UK).

OWEN J.E., NORMAN B.A., PHILENROOKS C.A. and JONES N.S.D., 1978. Studies on the meat production characteristics of Botswana goats and sheep. Part III - Carcasse tissue composition and distribution. Meat Science, 2: 59-74.

OWEN J.E. and MIENGA L.A., 1980. Effect of weight, castration and diet on growth performances and carcass composition of British Saanen goats. Anim. Prod., 30: 479.

SANZ SAMPELAYO M.R., MUNOZ F.J., LARA L., GIL EXTREMERA F. and BOZA J., 1987. Factors affecting pre- and post-weaning growth and body composition in kid goats of the Granadina breed. Anim. Prod., 45: 233-238.

THEODORO M.J. and SOUSA J., 1987. Effet de différents systèmes d'allaitement dans les courbes de croissance et dans la qualité de la carcasse des chevreaux de la race Serrana (Effect of milk rearing methods on growth and carcass quality of Serrana breed kids). Vol. 1, p. 985 (Abst.). 38th EAAP Annual Meeting, Sept. 28 - Oct. 1, 1987, Lisboa (PORTUGAL).

TREACHER T.T., MOWLEN A., WILDE R.M. and BUTLER HOGG B., 1987. Growth, efficiency of conversion and carcass composition of castrate male Saanen and Saanen Angora kids on a concentrate diet. Ann. Zootech., 36: 341-342.

WILSON P.N., 1958. The effect of plane of nutrition on the growth and development of the East African dwarf goat. II - Age changes in the carcass composition of female kids. J. Agr. Sci., (Camb.), 51: 4-21.

WILSON P.N., 1960. The effect of plane of nutrition on the growth and development of the East African dwarf goat. III - The effect of plane of nutrition and sex on the carcass composition of the kid at two stages of growth 16 1b weight and 30 lb weight. J. Agr. Sci., (Camb.), 54: 105-130.

ABBREVIATIONS

AAT	Amino acid total nitrogen	FCMY	Fat Corrected Milk Yield
ADF	Acid detergent fiber	FG	Foetus daily Growth
ADG	Average daily gain	FU	Fill Unit
ADL	Acid detergent lignin	FW	Foetus Weight
ad lib.	*ad libitum*	FY	Fat Yield
ARC	Agricultural Research Council	GE	Gross Energy
BC	Body Condition	GM	Goat Milk
BHB	Beta-hydroxybutyrate	GSH-Px	Selenium dependent enzyme
BWG	Body Weight Gain		glutathion peroxidase
CCN	Cerebro-cortical Necrosis	GUW	Gravid Uterus Weight
CIMI	Concentrate Dry Matter Intake	IC	Intake Capacity
CF	Crude Fiber	INRA	Institut National de la
CP	Crude Protein		Recherche Agronomique
CTA	Technical Centre for Agri-	IU	International Unit
	cultural and Rural Cooperation	IVD	In Vitro Digestibility
DAPA	Diaminopimelic Acid	KI	Kidding Interval
DCP	Digestible Crude Protein	LS	Level of Significance
DOM	Digestible Organic Matter	LW	Live Weight
DM	Dry Matter	LWC	Live Weight Change
DMD	Dry Matter Digestibility	MCP	Microbial Crude Protein
DMI	Dry Matter Intake	ME	Metabolisable energy
E	Oestradiol	MEg	Metabolisable Energy for
EAAP	European Association for		growth
	Animal Production	MEm	Metabolisable Energy for
EBW	Empty Body Weight		maintenance
EBWG	Empty Body Weight Gain	MFL	Milk fat derived from lipo
ECNSGP	European Cooperative Network		mobilization
	on Sheep and Goat Production	MFN	Metabolic Fecal Nitrogen
ED	Energy Digestibility	MFU	Milk Fill Unit
Ed	Energy density	MR	Milk Replacer
EEC	European Economic	MRT	Mean Retention Time
	Communauty	MUL	Milk Urea Level
EFGU	Energy Fixed into the Gravid	MY	Milk Yield
	Uterus	N	Nitrogen
EUN	Endogenous Urinary Nitrogen	n	number of data
ER	Energy Retention	NDF	Neutral Detergent Fiber
FAO	Food and Agricultural	NE	Net Energy
	Organization of the	NEFA	Non Esterified Fatty Acid
	United Nations	NEm	Net Energy for maintenance
FDMI	Forage Dry Matter Intake	NIRS	Near Infrared Reflectance
FC	Fat Content		Spectroscopy
FCM	Fat Corrected Milk		

NPN	Non Protein Nitrogen		t	time
NRC	National Research Council		T4	Tetraiodothyroxine
NS	Non Significant		T3	Triiodothyroxine
OM	Organic Matter		TBA	Trembolone Acetate
OMD	Organic Matter Digestibility		TCP	Total Crude Protein
PDI	True Protein Digestible in the small intestine (French nitrogen system)		TEST	Testosterone
			UDCF	Undigested crude fiber
			UDN	Undigested fecal nitrogen
PFW	Parturial Foetus Weight		UDOM	Undigested organic matter
PUL	Blood Plasma Urea Level		UET	Unitary eating time
PY	Protein Yield		UFL	Milk feed unit (French energy system)
NEm	Net Energy for maintenance			
r	Cœfficient of correlation		UK	United Kingdom
RDOM	Rumen Degraded Organic Matter		USA	United States of America
			UV	Ultra Violet
RDP	Rumen Degraded Protein		VDMI	Voluntary Dry Matter Intake
RMY	Raw Milk Yield		VFA	Volatile Fatty Acids
RNA	Ribonucleic acid		W	Weight
RR	Reticulo-rumen		$W^{.75}$	Metabolic live weight
RSD	Residual Standard Deviation		WI	Water Intake
S	Significant		WK	Week
SD	Standard Deviation		ZER	Zeranol
SE	Standard Error			

LIST OF AUTHORS

G. ALEXANDRE
INRA-CRAAG - BP 1232
97184 Pointe à Pitre, Guadeloupe (FRANCE)

G. AUMONT
INRA-CRAAG - BP 1232
97184 Pointe à Pitre, Guadeloupe (FRANCE)

P. BAS
Station de Nutrition et Alimentation (INRA)
de l'INA-PG - 16 rue Claude Bernard
75231 Paris Cedex 05 (FRANCE)

G. BLANCHART
ENSAIA Chaire de Zootechnie
2 avenue de la Forêt de Haye
54000 Vandœuvre (FRANCE)

R. BOUCHE
INRA Laboratoire de recherches sur
le développement de l'élevage
Quartier Grosseti - BP 8
20250 Corte (FRANCE)

A. BOZZINI
FAO Regional Office for Europe
Via delle Terme di Caracalla
00100 Roma (ITALY)

A. BRANCA
Istituto Zootechnico e Caseario per la Sardegna
07040 Olmedo, Sassari (ITALY)

J. BRUN-BELLUT
ENSAIA - Chaire de Zootechnie
2 avenue de la Forêt de Haye
54000 Vandœuvre (FRANCE)

J. BOZA
Estacion Experimental del Zaidin (CSIC)
Departamento de Fisiologia Animal
Professor Albareda 1
18008 Granada (SPAIN)

Y. CHILLIARD
INRA - CRVZ - Theix
63122 Ceyrat (FRANCE)

P. COLOMER-ROCHER
Department of Animal Production
SIA-DGA - Apartado 727
Zaragoza (SPAIN)

R. DACCORD
Swiss Federal Research Station for Animal
Production - Grangeneuve
CH 1725 Posieux (SWITZERLAND)

S. ECONOMIDES
Agricultural Research Institute
Nicosia (CYPRUS)

A. FALAGAN
INIA
Crida 07 - La Alberca
Murcia (SPAIN)

V. FEDELE
Istituto Sperimentale Zootechnico
106 Viale Basento
85100 Potenza (ITALY)

S. GIGER-REVERDIN
Station de Nutrition et Alimentation (INRA)
de l'INA-PG - 16 rue Claude Bernard
75231 Paris Cedex 05 (FRANCE)

E.A. GIHAD
Faculty of Agriculture - Cairo University
Giza (EGYPT)

M. HADJIPANAYIOTOU
Agricultural Institute
Nicosia (CYPRUS)

Ø. HAVREVOLL
Agricultural University of Norway
Department of Animal Nutrition - PO BOX 25
1432 Ås NLH (NORWAY)

B. HUBERT
INRA - Laboratoire d'Ecodéveloppement
Domaine de Saint Paul
84140 Montfavet (FRANCE)

J. KESSLER
Swiss Federal Research Station for Animal
Production - Grangeneuve
CH 1725 Posieux (SWITZERLAND)

S. LANDAU
Ministry of Agriculture - Extension Service
Department "Sheep and Goats"
Rehovot - Herzl ZT
76120 Rehovot (ISRAEL)

J.E. LINDBERG
The Swedish University of Agricultural
Sciences - Department of Animal Nutrition and
Management - Kungsängens Gärl
S 75323 Uppsala (SWEDEN)

C. MASSON
Chaire de Zootechnie (INRA) de l'ENSSAA
26 boulevard Petitjean
21100 Dijon (FRANCE)

M. MEURET
INRA - Laboratoire d'Ecodéveloppement
Domaine de Saint Paul
84140 Montfavet (FRANCE)

P. MORAND-FEHR
Station de Nutrition et Alimentation (INRA) de
l'INA-PG - 16 rue Claude Bernard
75231 Paris Cedex 05 (FRANCE)

A. MOWLEM
The Goat Advisory Bureau
9 Pitts Lane - Earley
Reading Berkshire RG6 1BX (UK)

H. NARJISSE
Institut Agronomique et Vétérinaire Hassan II
BP 704
Rabat Agdal (MAROCCO)

M. NAPOLEONE
INRA - Laboratoire d'Ecodéveloppement
Domaine de Saint Paul
84140 Montfavet (FRANCE)

A. NASTIS
University of Thessaloniki
Laboratory of Range Science
Thessaloniky (GREECE)

Z. NITSAN
Agricultural Research Organisation
The Volcani Center - PO Box 6
50250 Bet Dagan (ISRAEL)

S. OWEN
University of Reading
Department of Agriculture
Reading RG6 2AT (UK)

F. POISOT
INRA-CRAAG - BP 1232
97184 Pointe à Pitre, Guadeloupe (FRANCE)

F. REMEUF
Laboratoire de Technologie
Institut National Agronomique Paris-Grignon
78850 Thiverval Grignon (FRANCE)

R. RUBINO
Istituto Sperimentale Zootechnico
106 Viale Basento
85100 Potenza (ITALY)

P. SANTUCCI
INRA Laboratoire de recherches sur le
développement de l'élevage
Quartier Grosseti - BP 8
20250 Corte (FRANCE)

M.R. SANZ SAMPELAYO
Estacion Experimental del Zaidin (CSIC)
Departamento de Fisiologia Animal
Professor Albareda 1
18008 Granada (SPAIN)

D. SAUVANT
Station de nutrition et alimentation (INRA)
de l'INA-PG - 16 rue Claude Bernard
75231 Paris Cedex 05 (FRANCE)

P. SCHMIDELY
Station de nutrition et alimentation (INRA)
de l'INA-PG - 16 rue Claude Bernard
75231 Paris Cedex 05 (FRANCE)

J.L. TISSERAND
Chaire de Zootechnie (INRA) de l'ENSSAA
26 boulevard Petitjean
21100 Dijon (FRANCE)

T.T. TREACHER
The Institute for Grassland
and Animal Production
Hurley
Maidenhead Berks SL6 5LR (UK)

New adress :
International Center for Agriculture Research in
Dry Areas
ICARDA - PO Box 5466
Aleppo (SYRIA)